ANGIOTENSIN RESEARCH PROGRESS

ANGIOTENSIN RESEARCH PROGRESS

HINA MIURA AND YUUTO SASAKI
EDITORS

Nova Biomedical Books
New York

For permission to use material from this book please contact us:
Telephone 631-231-7269; Fax 631-231-8175
Web Site: http://www.novapublishers.com

Library of Congress Cataloging-in-Publication Data

Angiotensin research progress / Hina Miura and Yuuto Sasaki,editors.
p.;cm.
Includes bibliographical references and index.
ISBN 978-1-60456-537-9(hardcover)
1.Angiotensins. I. Miura, Hina. II. Sasaki,Yuuto.
[DNLM:1. Angiotensins.@. Biomedical Research—trends. QU 68 A5879 2008]
QP572.A54A555 2008
612.4'05—dc22

2008011187

Published by Nova Science Publishers, Inc. ✣ New York

Contents

Preface

Angiotensin is an oligopeptide in the blood that causes vasoconstriction, increased blood pressure, and release of aldosterone from the adrenal cortex. It is a hormone and a powerful dipsogen. It is derived from the precursor molecule angiotensinogen, a serum globulin produced in the liver. It plays an important role in the renin-angiotensin system. This new book presents the latest research advances in the field.

Chapter I - In the adrenal cortex, specifically the zona glomerulosa, angiotensin II (AngII) stimulates the secretion of aldosterone to result in increased renal sodium retention, which can lead to hypertension. Chronically, elevated aldosterone levels can also contribute to cardiac fibrosis and congestive heart failure. The stimulation of aldosterone production in response to AngII is mediated through binding to the AngII type 1 receptor, resulting in diacylglycerol formation, which activates downstream effectors such as classical and novel protein kinase C (PKC) isoforms. The activity of this family of enzymes has been proposed to mediate sustained aldosterone secretion. Indeed, a possible involvement of PKC in mediating acute aldosterone secretion is supported by the finding that the phorbol ester, phorbol 12-myristate 13-acetate (PMA), which substitutes for diacyglycerol in activating diacylglycerol-sensitive enzymes, promotes an aldosterone secretory response. However, controversy surrounds the exact role of PKC in regulating aldosterone secretion. In addition, other phorbol ester/diacylglycerol-sensitive signaling molecules have been identified of late. One such signal transduction pathway includes the serine/threonine kinase protein kinase D (PKD), which has recently been implicated as an effector of AngII and as a regulator of aldosterone secretion. In this review, the authors discuss the evidence in the literature supporting a role for diacylglycerol, PKC and PKD in regulating aldosterone secretion.

Chapter II - Angiotensin II, a product of activation of renin-angiotensin axis, is a powerful vasoconstrictor and in addition is a mediator of many physiological and patho-physiological non-hemodynamic effects in multiple organ systems including cardiovascular, renal and neuronal tissues. Angiotensin activation has been implicated in renal injury in diabetic and non-diabetic conditions. On the other hand, nitric oxide is a product derived solely from L-arginine under the effects of nitric oxide synthase, is a vasodilator and has multiple non-hemodynamic effects in both physiological and pathological conditions in many organs. Since many effects of angiotensin are counter-regulatory to the effects of nitric oxide, a delicate balance between these two signaling pathways is necessary for the proper function of many organ systems. Therefore, an imbalance between the two systems may also underlie the pathophysiology of many cardiovascular-renal syndromes including diabetic nephropathy. The effects of angiotensin as well as angiotensin blockers on renal NO synthesis has been the subject of several investigations. In all these studies it was argued that the enhanced renal NO production is a defense mechanism against the detrimental effects of Ang II. In addition, Ang II has been shown to stimulate NO production in coronary and other microvascular endothelial cells as well as aortic endothelial cells through eNOS activity. Ang II increased the expression and activity of ca^{++} dependent NOS (NOS I and III) in the cortex with a decrease in NOS II activity in the cortex. Similar studies examining the acute in vivo effects of Ang II infusion on renal NO production and on the NOS isoforms in the cortex and medulla were performed in the authors laboratory. These studies demonstrated that angiotensin II inhibited iNOS, an effect reversed by losartan, an AT_1 receptor blocker and amplified by PD123319, an AT_2 receptor blocker. The losartan effect was associated with augmented NO synthesis and increased expression of AT_2 receptors as demonstrated by immunocytochemistry. NO derived from cNOS and iNOS did not affect the basal secretion of insulin but augmented high glucose (HG) stimulated insulin secretion in pancreatic beta cells. Furthermore, studies demonstrated the influence of Ang II on insulin secretion and that angiotensin blockade with ACE inhibitors and AT_1 receptor blockers enhanced insulin release from pancreatic β cells by augmenting tissue NO production. Ang II stimulated renal NO synthesis by augmenting the Ca^{++} dependent NOS activity and eNOS expression in the kidney, which under basal conditions surpassed the iNOS inhibition by Ang II. The authors also demonstrated in in vivo studies that despite the HG inhibition of eNOS, the intrarenal NO generation as well as eNOS activity remained stimulated in the presence of systemic Ang II. The functional interaction and reciprocal regulation of angiotensin and nitric oxide pathways in basal and pathologic states is the focus of discussion in this review, although a significant part of the discussion is centered on diabetic nephropathy.

Chapter III - The renin-angiotensin system (RAS) is one of main regulators of blood pressure and electrolyte balance during pregnancy. All the components of the RAS have been reported to be expressed in the utero-placental unit. Angiotensin (Ang) II is increasingly recognized as a growth factor, both in its own right and through interactions with other growth factors. Ang II synthesized in the placenta may serve as an autocrine/paracrine modulator of placental function. In addition to its role in regulation of blood pressure during normal pregnancy, Ang II is suggested to be involved in regulation of placentation through regulation of trophoblast invasion, angiogenesis and placental vascular development. These key effects of Ang II, in addition to its capacity to induce inflammatory response, suggest that increased vascular sensitivity to Ang II may have an important role in the increase in vascular resistance and elevated blood pressure seen pregnancy complications such as intra-uterin growth retardation (IUGR) and preeclampsia. Magnesium sulfate ($MgSO_4$) is used in pregnancy as either prophylaxis/treatment for eclamptic seizures or for preterm labor tocolysis. The authors and their collaborators have shown that $MgSO_4$ may attenuate Ang II induced placental vasoconstriction, and may affect Ang II induced placental cytokine production. This review summarizes the current understanding of the physiological aspects of Ang II in the utero-placental unit and its role in the pathogenesis of pregnancy disorders such as preeclampsia. Furthermore, the possible interactions between Ang II and $MgSO_4$ in the placental unit are also discussed.

Chapter IV - Evidence is accumulating that Angiotensin IV (Ang IV or Ang II (3-8)), a major metabolite of Ang II, exerts physiological effects on its own. Ang IV accomplishes cardiovascular, renal and, more importantly, central effects. In this context, it was shown that i.c.v. administration of Ang IV increases memory and learning in different in vivo studies. The exact mechanism by which Ang IV exerts this beneficial effect is not yet understood. Some of the effects of Ang IV occur at nanomolar concentration and are not blocked by the classical AT_1 and AT_2 receptor antagonists. This, together with the discovery of binding sites with high affinity for $[^{125}I]$Ang IV in the central, vascular and renal system, has led to the concept of a novel angiotensin receptor subtype: the AT_4 receptor. Recently, this receptor was identified as the insulin regulated aminopeptidase (IRAP), a Zn^{2+} dependent gluzincin aminopeptidase. IRAP is known for it's co-localization with the GLUT4 glucose transporter in specialized intracellular vesicles. To regulate glucose homeostasis, they translocate quickly to the cell surface in the presence of insulin. Ang IV has been shown to bind IRAP with high affinity and to inhibit its catalytic activity, albeit with lower potency. This discrepancy can be attributed to the mandatory presence of chelators in the binding studies. This implies that high affinity binding of $[^{125}I]$Ang IV only takes place to the IRAP apo-enzyme. These properties of IRAP have led to three hypotheses concerning the way Ang IV exerts its effects.

Some of IRAP's in vitro substrates are known, among them oxytocin and vasopressin. These peptides have been recognised to have a positive effect on memory and learning. By inhibiting the catalytic activity of IRAP, Ang IV is supposed to prolong the half-life of these compounds, resulting in a positive effect on memory and learning.

Secondly, IRAP could be capable of triggering intracellular events and thus function as a receptor.

Finally, increased glucose uptake in neurons has been shown to improve memory and learning. It has therefore been proposed that Ang IV interferes in the recycling of IRAP and GLUT4 in such a way that it prolongs the residence time of GLUT4 at the cell surface and thus, glucose uptake.

This review focuses on the molecular and physiological aspects of the Ang IV- IRAP interaction with special attention to recent findings and the above-mentioned hypotheses.

Chapter V - Over the last two decades, it has become increasingly obvious that angiotensin II (ANG II), the most important effector peptide of the renin-angiotensin system (RAS), may play a central role not only in the etiology of hypertension, but also in the pathophysiology of cardiovascular and renal diseases in humans. ANG II is an essential hormone that exerts pleiotropic effects in the kidneys, including sodium-retaining, growth-promoting and pro-inflammatory properties, in addition to its physiological role in maintaining body salt and fluid balance and blood pressure homeostasis. There are two major types of angiotensin II receptor: type 1 (AT_1) receptors mediate the majority of the physiological actions of ANG II, while type 2 (AT_2) receptors have opposite function to that of AT_1 receptor. Although AT_1 receptor is coupled to G protein and has no intrinsic tyrosine kinase activity, recent *in vitro* evidence suggests that AT_1 receptor activation leads to activation of not only phospholipase C (PLC), but also a variety of intracellular signal transduction cascades, as well as transactivation of epidermal growth factor (EGF) or platelet-derived growth factor (PDGF) receptor. Specifically, ANG II-mediated AT_1 receptor activation regulates sodium transport, glucose metabolism, cell growth, or various bioactive substances (vasoactive hormones, growth factors, cytokines, etc.), and activates multiple intracellular signaling cascades in renal proximal tubule cells (PTCs). In the authors' previous studies, ANG II had a biphasic effect on Na^+ transport in PTCs; picomolar concentrations of ANG II stimulated Na^+ transport, whereas micromolar concentrations of ANG II inhibited it. The authors also demonstrated that ANG II inhibited Na^+/glucose cotransport (SGLT)-mediated glucose uptake. Regulation of Na^+ transport or SGLT by ANG II is commonly mediated by protein kinase C (PKC)-mitogen activated protein kinase (MAPK)-cytosolic phospholipase A_2 ($cPLA_2$) signaling cascades through the AT_1 receptor in PTCs. These actions are thought to participate in the pathophysiology of renal disease such as hypertension or diabetic nephropathy. Therefore, this review focuses on the regulatory effect of ANG II on cellular function, its physiological and pathophysiological role in PTCs, and the major ANG II-induced cell signaling cascades. Improved understanding of the regulatory effect of ANG II on renal proximal tubule cells is necessary to develop improved therapeutic strategies.

Chapter VI - In the last decade, new biologically active components of the renin angiotensin system (RAS) were found. Angiotensin-(1-7) (ANG-(1-7)), a metabolite of angiotensin I and angiotensin II (ANG II), is considered the most pleiotropic component of RAS. ANG-(1-7) lacks the pressor, dipsogenic and aldosterone secretrory actions elicited by ANG II, and, conversely to the octapeptide, systemic administration of ANG-(1-7) induces natriuresis, diuresis and vasodilation. However, this heptapeptide increases vasopressin release, prostaglandin synthesis and facilitates peripheral noradrenergic neurotransmission in the same manner as ANG II.

In addition, activity of ANG-(1-7) is enhanced in subjects with increased activity of the RAS, acting as a counteregulatory modulator of the harmful effects induced by ANG II.

Considering the physiological actions of ANG-(1-7) and its role in the pathophysiology of cardiovascular disease, an increased interest exists in the development of therapeutic agents able to enhance ANG-(1-7) endogenous levels or activation of ANG-(1-7) receptors, such as the Mas receptor. Several reports have described that angiotensin converting enzyme (ACE) inhibitors, such as captopril, increase endogenous levels of ANG-(1-7). Considering physiological actions of ANG-(1-7), enhancement of the heptapeptide levels by ACE inhibition may participate in the beneficial effects of these therapeutic agents. In addition, non-peptidic ANG-(1-7) agonists are under development and could have a role in the treatment of cardiovascular diseases.

The aim of the present chapter is to review the biochemical and physiological actions of ANG-(1-7), the therapeutic strategies designed to enhanced ANG-(1-7) activity and the possible role of ANG-(1-7) in the treatment of cardiovascular diseases.

Chapter VII - Renal fibrosis is a common pathway of progressive renal disease and parallels the reduction in organ function that necessitates dialysis or transplantation. In the last decades, accumulating experimental and clinical evidence convincingly suggests that the renin-angiotensin system (RAS) is a key player in the progression of chronic kidney disease. Angiotensin II (ANG II), the main peptide of the RAS, was classically viewed as a vasoactive factor participating in the local and systemic hemodynamic regulation. With the progression of researches, it has emerged as a multifunctional cytokine exhibiting many non-hemodynamic effects such as growth, profibrotic, matrix metabolism and even proinflammation. This new understanding of ANG II has important clinical implications. It predicted and explained why blockade of the RAS would significantly slow the progression of fibrotic disease. However, it further suggested that higher doses and/or a combination of ANG II blockade with another agent or agents might truly halt the progressive fibrosis.

This review will focus on the non-hemodynamic aspect of ANG II and its contribution in renal fibrogenesis.

Chapter VIII – The authors have identified the existence of a local renin-angiotensin system (RAS) in the pancreas which is upregulated in animal models of type 2 diabetes mellitus (T2DM) (Lau et al., 2004; Chu et al., 2006). The authors found that inhibition of this local pancreatic RAS improves islet beta-cell function and glucose tolerance and reduces pro-inflammatory factor activity; these findings suggest a mechanism for the protective effects of RAS antagonism, including those for T2DM, in humans. Upregulation of the pancreatic islet RAS leads to impaired islet beta-cell function and glucose intolerance, and increases pro-inflammatory factor gene expression and protein production in experimental animal models of T2DM.

Vitamin D has several known physiological functions beyond calcium homeostasis. Relevant to the present discussion, insulin secretion is reduced in vitamin D insufficiency (hypovitaminosis D) and can be restored by replacement of vitamin D both experimentally and in humans without established diabetes. Furthermore vitamin D supplementation has been reported to improve glycemia in human T2DM (see discussion below). Meanwhile vitamin D deficiency can amplify insulin resistance and thus contribute to beta-cell dysfunction. The association between hypovitaminosis D and glucose intolerance suggests that hypovitaminosis D may contribute to the pathogenesis of T2DM (see review by Boucher 1998; Mathieu et al., 2005).

Despite the clear association described above, the mechanistic explanation of the effects of vitamin D on beta cells is currently limited to enhancement of availability of the ionized calcium required by some insulin-releasing proteases. More than two decades ago, Resnick and Laragh (1985) found abnormal variations in circulating levels of activated vitamin D and parathyroid hormone in hypertensive patients, as defined by renin responses. On the other hand, vitamin D was recently shown to negatively regulate the RAS via suppression of renin gene activity, presumably by binding of activated vitamin D complexes with its receptor (VDR) to vitamin D receptor response elements (VDREs) in the promoter region of the renin gene (Li et al., 2002; Li et al., 2004). Both vitamin D supplementation and exposure to vitamin D synthesis-inducing wavelength ultraviolet B radiation can reduce blood pressure in humans. Some clinical studies suggest that RAS blockade can reduce the risk of developing T2DM in hypertensive patients. Whether this risk reduction varies with vitamin D repletion status is not known, but if so, it may explain the differences among reported studies.

In view of these findings, the authors have postulated that vitamin D modulates pancreatic islet RAS expression and activity in the same way that it modulates the general RAS, and that this modulation contributes to the effects of vitamin D on islet cell function. This hypothesis is being tested by *in vitro* and *in vivo* experiments investigating the effects of activated vitamin D on islet RAS activity, on RAS-induced inhibition of islet beta-cell function, and on circulating RAS activity. Consistent with the authors' hypothesis, preliminary *in vitro* works have shown that activated vitamin D reduces the abnormally high RAS gene expression seen in mouse islets incubated in high glucose concentration conditions. A discussion of the mechanism(s) by which vitamin D-dependent RAS may modulate pancreatic islet function, and thus T2DM risk, follows.

T2DM accounts for more than 90% of diabetes cases world-wide. It imposes huge demands on healthcare resources in Western countries and is also becoming a burden in developing countries. Hypovitaminosis D is prevalent in Western countries due to lifestyle and sun avoidance for reduction of skin cancer. T2DM and hypovitaminosis D are now being reported in developing countries, especially in places where people work long hours indoors or women are veiled. Elucidation of the mechanisms involved in vitamin D modulation of islet cell function, local RASs and blood pressure, could contribute to the development of novel strategies for cost-effective prevention and management of diabetes. In this article, the authors provide a critical appraisal of progress in research examining the roles of vitamin D and pancreatic islet RAS in diabetes, with particular emphasis on putative roles of vitamin D in the regulation of pancreatic islet RAS, in relation to T2DM.

Chapter IX - Angiotensin II (Ang II) type 1 (AT_1) receptor blockers (ARBs) are highly selective for the AT_1 receptor, which is a member of the G protein-coupled receptor superfamily, and block the diverse effects by Ang II. In addition to their beneficial effects against hypertension, ARBs have also been associated with a decrease in cardiovascular morbidity and mortality. Many ARBs are available for clinical use and several studies have shown that these agents are safe and effective. Over the past decade, some reports have discussed the different degrees of the beneficial effects of ARBs. Since ARBs may not have the same effects, the benefits conferred by ARBs may also be different. These differential effects may be due to differences in the molecular characteristics of ARBs, such as insurmountability, dissociation, binding affinity and inverse agonism, in addition to their chemical structures. These findings represent an exciting new area in ARB treatment. A better understanding of the differential molecular mechanisms of each ARB could be useful for the treatment of cardiovascular diseases.

In: Angiotensin Research Progress
Editors: H. Miura and Y. Sasaki
ISBN: 978-1-60456-537-9

Chapter I

Angiotensin II Signaling in the Adrenal Cortex: A Role for Protein Kinases C and D in Acute Aldosterone Secretion?

***Brian A. Shapiro*[1] *and Wendy B. Bollag*[1,2*]**
[1]Institute of Molecular Medicine and Genetics and [2]Departments of Medicine, Cell Biology and Anatomy and Orthopaedic Surgery, Medical College of Georgia, 1120 15th Street, Augusta, Georgia 30912

Abstract

In the adrenal cortex, specifically the zona glomerulosa, angiotensin II (AngII) stimulates the secretion of aldosterone to result in increased renal sodium retention, which can lead to hypertension. Chronically, elevated aldosterone levels can also contribute to cardiac fibrosis and congestive heart failure. The stimulation of aldosterone production in response to AngII is mediated through binding to the AngII type 1 receptor, resulting in diacylglycerol formation, which activates downstream effectors such as classical and novel protein kinase C (PKC) isoforms. The activity of this family of enzymes has been proposed to mediate sustained aldosterone secretion. Indeed, a possible involvement of PKC in mediating acute aldosterone secretion is supported by the finding that the phorbol ester, phorbol 12-myristate 13-acetate (PMA), which substitutes for diacyglycerol in activating diacylglycerol-sensitive enzymes, promotes an aldosterone secretory response. However, controversy surrounds the exact role of PKC in regulating aldosterone secretion. In addition, other phorbol ester/diacylglycerol-sensitive signaling molecules have been identified of late. One such signal transduction pathway includes the serine/threonine kinase protein kinase D (PKD), which has recently been implicated as an effector of AngII and as a regulator of aldosterone secretion. In this review we will discuss the evidence in the literature supporting a role for diacylglycerol, PKC and PKD in regulating aldosterone secretion.

* TEL: (706) 721-0698; FAX: (706) 721-7915; e-mail: wbollag@mcg.edu

Keywords: Phosphatidic acid, Phospholipase D, Protein kinase C, Phorbol ester, Steroidogenesis

Introduction and Model Systems

The mineralocorticoid aldosterone is responsible for sodium homeostasis *in vivo*, and through its effects on sodium reabsorption in the kidney, aldosterone regulates blood volume and pressure. Thus, excessive aldosterone secretion can result in hypertension; indeed, recent studies suggest that 8% of moderately hypertensive patients exhibit hyperaldosteronism and this proportion likely increases to approximately 13% in patients with severe hypertension [1]. In addition, aldosterone is thought to exert direct actions on cardiomyocytes to contribute to cardiac fibrosis and congestive heart failure (reviewed in [2,3]). The randomized aldactone evaluation study (RALES) and the eplerenone post-acute myocardial infarction heart failure efficacy and survival study (EPHESUS) indicate that aldosterone receptor antagonists, added to standard treatments, are able to reduce morbidity and mortality in congestive heart failure patients (reviewed in [4]), suggesting the importance of aldosterone in human health. Therefore, knowledge about the mechanisms regulating aldosterone secretion is important for a thorough understanding of cardiovascular disease in humans.

Angiotensin II (AngII) is the primary regulator of aldosterone secretion from adrenal zona glomerulosa cells *in vivo*. Glomerulosa cells express both AngII type 1 (AT1) and type II (AT2) receptors [5], with aldosterone production in response to AngII occurring as a result of AT1 receptor activation. AngII binding to the AT1 receptor activates, through a heterotrimeric G protein, a phosphoinositide-specific phospholipase C (PLC) (reviewed in [6, 7]). This PLC hydrolyzes the membrane phospholipid, phosphatidylinositol 4,5-bisphosphate (PIP_2) to generate the second messengers, inositol 1,4,5-trisphosphate (IP_3) and diacylglycerol (DAG). IP_3 binds to receptors on the endoplasmic reticulum and releases calcium (Ca^{2+}) from this intracellular storage pool. These ions, in turn, activate Ca^{2+}/calmodulin-dependent protein kinases (CaMK) (reviewed in [6,7]) to phosphorylate multiple substrates, including elongation factor 2 (EF2) [8] and voltage-dependent T-type Ca^{2+} channels [9], to trigger various pathways. The other messenger, DAG has traditionally been thought to function through the DAG-sensitive protein kinase C (PKC) isoenzymes (that is, the classical and novel PKC isoforms, see below). However, additional DAG effector enzymes have recently been identified (reviewed in [10]). These include a Ras activator (i.e., a guanine nucleotide exchange factor), Ras guanine nucleotide release protein (Ras-GRP) as well as protein kinase D (PKD). In addition, DAG can serve as a substrate for a glomerulosa cell DAG lipase that releases arachidonic acid from arachidonic acid-containing DAG [11, 12]. This arachidonic acid is then converted by lipoxygenases to hydroxyeicosatetraenoic acids (HETEs), in particular 12-HETE, which also plays a role in stimulating aldosterone secretion [13]. Finally, AngII, via the AT1 receptor [14], PKC [15] and likely Rho (our unpublished data), also activates phospholipase D (PLD) [16], an enzyme that hydrolyzes phosphatidylcholine to generate phosphatidic acid (PA) (reviewed in [17]). PA, which is essentially phosphorylated DAG, can be metabolized by lipid phosphate phosphatases to yield DAG; thus, the combined activities of PLD and lipid phosphate phosphatases can also result in DAG generation (reviewed in [17]). Data from our laboratory indicates that PLD-mediated lipid signal generation is required for maximal AngII-induced aldosterone secretion [15, 18]. In this review we will discuss the evidence for a role of DAG and its effector enzymes in AngII-mediated aldosterone production.

Prior to discussing the evidence supporting a role for DAG in mediating AngII-induced aldosterone secretion, it is worthwhile to examine the models that are used to study this process. There are three main systems: bovine adrenal glomerulosa cells (freshly isolated and primary cultures), rat adrenocortical or glomerulosa cells (freshly isolated and capsular preparations as well as primary cultures) and a human adrenocortical carcinoma cell line developed by Rainey and colleagues, NCI H295R cells [19]. [An additional cell line, the Y1 cell line, does not respond to AngII treatment and does not synthesize aldosterone, but can be used to study the production of other steroids stimulated by adrenocorticotropic hormone (reviewed in [20])]. A few studies have also been performed in normal human adrenal glomerulosa cells [11, 21, 22], although availability limits their routine use.

Freshly isolated bovine adrenal glomerulosa cells have the advantage that they are normal and non-transformed and secrete large quantities of aldosterone acutely, with detectably elevated aldosterone production within minutes of AngII stimulation. Primary cultures have the additional advantage that the short culture period (generally two days) allows the cells to recover from the trauma of isolation. These cells have low basal aldosterone secretion such that the increase in the aldosterone secretory rate can be twenty-fold or better [23]. In addition, primary bovine adrenal glomerulosa cells can be labeled with radioisotopes and/or infected with viruses overexpressing genes of interest or short-hairpin interfering RNAs to examine signaling events involved in steroidogenesis. (Note that as primary cells, the bovine adrenal glomerulosa cells are difficult to transfect with traditional transfection methods with any degree of efficiency.) We have also obtained data demonstrating that in these primary cultures of bovine adrenal glomerulosa cells, by sixty minutes of exposure AngII modulates the expression of various genes [24], including that of steroiodogenic acute regulatory (StAR) protein, a protein that is critically important in the production of all steroid hormones (reviewed in [25]). The main disadvantage of the bovine adrenal glomerulosa cells resides in the fact that these cells express a single enzyme that mediates both aldosterone and cortisol production (CYP11B), rather than the two separate enzymes (CYP11B1 and CYP11B2 or aldosterone synthase) that mediate these two processes in humans and rodents (26). The regulation of aldosterone versus cortisol production by the common enzyme is not entirely clear but seems to involve the redox status of the cell [26, 27]. An additional caveat of the use of these cells is the inability to control the nutritional and environmental conditions, such as salt and water intakes and temperature, encountered by the cattle prior to slaughter. Thus, the bovine adrenal glomerulosa cells tend to exhibit greater variability, particularly with changes in season, than a cell line and/or cells obtained from a laboratory animal like a rat.

Rat adrenocortical cells and/or rat zona glomerulosa cells have also been used as a model to study the regulation of aldosterone secretion. Like bovine adrenal glomerulosa cells, rat cells also show acute increases in steroidogenesis in response to AngII; however, unlike bovine adrenal glomerulosa cells, rat cells express the two enzymes CYP11B1 and CYP11B2 [26], which synthesize corticosterone (the rodent equivalent of cortisol) and aldosterone, respectively. Also, unlike the cattle from which the bovine cells are derived, rats can be maintained in a completely controlled environment, with proscribed amounts of dietary salt (e.g., [28]). The main drawback of the rat system is the low cell yield, requiring large numbers of animals to be sacrificed to generate sufficient quantities of cells for study. In addition, the size of the glands makes it difficult to dissect the zona glomerulosa from the other adrenocortical layers and the capsule. For these reasons, adrenocortical cells rather than enriched glomerulosa cells are often used for studies *in vitro*; however, since the zona glomerulosa represents only a small proportion of the adrenal cortex, these adrenocortical cells are largely of fasiculata, and not glomerulosa, origin. Nevertheless, rat glomerulosa cells and/or capsular preparations (e.g., [29]) can be isolated and have been used to study the effects of AngII on steroidogenic, proliferative and hypertrophic glomerulosa cell responses.

The NCI H295R cell line is another commonly used glomerulosa cell model system. These cells were derived by selecting for adherent cells from within a human adrenocortical carcinoma cell line in the American Type Culture Collection (ATCC) (reviewed in [20]). These cells have several advantages: they are pure adenocortical cells and presumably relatively homogenous, and, as a cell line, large quantities can be generated through expansion. As a human cell line these cells are presumably extremely relevant to human physiology and pathophysiology. Also as a cell line, the H295R cells are more easily transfected than primary cell cultures. On the other hand, these cells were isolated from a malignant adrenal tumor and are therefore transformed with attendant abnormal regulation of proliferation [30]. In addition, they demonstrate characteristics of multiple layers of the adrenal cortex: thus, these cells can be induced to synthesize the various steroid hormones produced by the zonae reticularis, fasiculata and glomerulosa, as well as steroids not formed normally *in vivo* (reviewed in [30]). Finally, the cells basally express detectable but low levels of aldosterone synthase, CYP11B2, which can be induced by appropriate culture conditions (e.g., exposure to AngII or elevated extracellular potassium concentrations) [19, 31]. Because of this low-level CYP11B2 expression, these cells do not generally secrete appreciable aldosterone quantities acutely, generally requiring a minimum of one-half to one hour to produce detectable aldosterone levels [19]. Thus, for addressing questions concerning the regulation of acute aldosterone secretion, this system is not ideal, although for determining issues of tumorigenesis, adrenal zonation and control of steroidogenic enzyme expression, these cells are an excellent model system.

Evidence for a Role of Diacylglycerol-Responsive Pathways in Regulating Aldosterone Secretion

As discussed above, the binding of AngII to the AT1 receptor activates phosphoinoisitde hydrolysis to generate an IP_3/Ca^{2+} message and DAG. While it is well accepted that Ca^{2+} plays an important role in mediating aldosterone secretion, and this involvement has been reviewed extensively [6, 7], the role of DAG and DAG-responsive pathways, such as PKC (see below), is somewhat controversial. AngII has been demonstrated to trigger increases in DAG content and radiolabeled DAG levels in adrenal glomerulosa cells [23, 32], suggesting a likely involvement of this signal in some aspect of glomerulosa physiology. However, Spät and Hunyady [7] have indicated that DAG-responsive "PKC has an inhibitory, rather than stimulatory, effect." Other laboratories, including our own, have obtained data consistent with a role of a DAG/PKC pathway in stimulating aldosterone secretion, as discussed below.

One line of evidence that suggests a possible stimulatory role of DAG in steroidogenesis is the finding that DAG analogs, such as oleoylacetylglycerol (OAG) or dioctanoylglycerol (DiC8), can induce aldosterone secretion [33, 34]. These DAG analogs by themselves typically increase aldosterone production only to a small extent; however, providing an additional Ca^{2+} influx signal, using for example, a Ca^{2+} ionophore or a Ca^{2+} channel agonist, results in a synergism with the lipid signal to promote secretion [34]. The interaction between the DAG and Ca^{2+} pathways suggests the possible involvement of a classical or conventional PKC isoenzyme (discussed below). Indeed, several investigators have demonstrated AngII-induced PKC activation in glomerulosa cells via various methods [35-38]. An involvement of PKC is also suggested by the effect of tumor-promoting phorbol esters on aldosterone secretion. The major cellular receptor for the phorbol esters, which include phorbol 12-myristate 13-acetate (PMA) and phorbol 12, 13-dibutyrate (PDBu), is PKC. Phorbol esters substitute for DAG in activating PKC, and these agents have also been found to stimulate aldosterone production in adrenal glomerulosa cells by some investigators [35, 39, 40]. Again, these agents trigger a small slowly developing aldosterone secretory response on their own but act synergistically with Ca^{2+} ionophores to elicit rates comparable to those induced by AngII [40, 41]. Furthermore, the combination of phorbol ester and Ca^{2+} ionophore triggers a time-dependent phosphorylation pattern that reproduces the response to AngII, with one set of proteins phosphorylated in response to the Ca^{2+} ionophore, A23187 alone or a 1-minute AngII exposure, and a different set phosphorylated after a 30-minute treatment with AngII or combined PMA and A23187 [42]. These results suggest that AngII initiates the aldosterone secretory response through a Ca^{2+}-dependent pathway but that DAG signaling is involved in the sustained response. On the other hand, an interaction between Ca^{2+} metabolism and phorbol ester-sensitive effectors has been demonstrated by Cirillo et al. [43], who showed an ability of PMA, presumably acting through PKC, to increase Ca^{2+} influx as well as efflux, i.e. to accelerate Ca^{2+} cycling across the plasma membrane.

On the other hand, other laboratories have found no significant effects of PKC activators or inhibitors on aldosterone secretion and/or have demonstrated inhibitory effects with phorbol ester pretreatment [44, 45]. One explanation for the disparity may reside in the fact that treatment with a high concentration of phorbol ester and/or for a prolonged period of time can result in degradation and therefore down-regulation of some isoforms of PKC (reviewed in [46]). Alternatively, phorbol esters can activate enzymes in addition to PKC, such as a guanine nucleotide exchange protein for Ras, Ras guanine nucleotide-releasing protein (RasGRP), chimaerin and protein kinase D (PKD) (reviewed in [10]). Depending on the role of these enzymes in glomerulosa biology, differential expression and/or activation of these different pathways in the various glomerulosa cell models could result in disparate cellular responses to phorbol ester.

It should also be noted that multiple DAG types are likely formed upon AngII stimulation of glomerulosa cells. Thus, various fatty acid moieties can be esterified to the glycerol backbone to generate diverse DAG species. Different phospholipids generally tend to incorporate different proportions of the various fatty acids; for instance, the phosphoinositides are often enriched in polyunsaturated arachidonic acid in comparison with other fatty acids (discussed in [23]). Thus, different phospholipid sources may produce different DAG species, and the activities of phosphoinositide-specific PLC and phosphatidylcholine-specific phospholipase D (PLD) likely yield disparate DAGs. In turn, the identity of the DAG species generated by signaling mechanisms may be important for several reasons: (1) different DAGs may activate different effector enzymes, e.g., different PKC isoenzymes (discussed in [32, 47]) and/or may possess different efficacy for activating a particular effector; (2) different DAGs may be produced and activate effectors at distinct cellular membranes [32]; and/or (3) different DAG species may be metabolized disparately. As an example of this latter possibility, it is known that there exits a DAG kinase (DAGK) that preferentially phosphorylates arachidonic acid-containing DAGs [48], as well as a DAG lipase that hydrolyzes such DAG species to release arachidonic acid for further metabolism [11, 12]. Thus, the source and type of DAG may be critical in regulating a particular cellular response.

In adrenal glomerulosa cells AngII induces a biphasic increase in DAG content [32]. The initial increase is mirrored by a rise in inositol phosphates production [32] and a decrease in PIP_2 levels [40], suggesting that this early increase is the result of phosphoinositide-specific PLC-mediated hydrolysis of PIP_2. AngII also activates phosphatidylcholine-specific PLD [16], an enzyme that cleaves phosphatidylcholine to produce PA (essentially phosphorylated DAG). PA can, in turn, be dephosphorylated to yield DAG by lipid phosphate phosphatases (LPP) (reviewed in [17]). By analogy with other systems (reviewed in [49]), as well as the observed disparate sensitivities of inositol phosphate versus DAG generation to removal of extracellular calcium [32], it seems likely that the late sustained increase in DAG observed with AngII stimulation derives from phosphatidylcholine through the combined action of PLD and lipid phosphate phosphatases. Indeed, we have demonstrated that inhibition of PLD-mediated lipid signaling with 1-butanol (which diverts production from PA to slowly metabolized phosphatidylbutanol) decreases AngII-induced PA and DAG production [15], as well as AngII-stimulated aldosterone secretion [15, 18]. This result indicates the importance of PLD-mediated generation of lipid signals to steroiodogenesis, although whether the key lipid messenger is DAG or PA, or both, is not yet known.

The DAG produced upon AngII stimulation of adrenal glomerulosa cells may activate other effector enzymes and/or perform other functions as well. For example, arachidonic acid-containing DAG can be deacylated by a DAG lipase present in glomerulosa cells to release arachidonic acid [11]. Arachidonic acid, in turn, serves as a precursor for the lipoxygenase-mediated synthesis of hydroxyeicosatetraenoic acids (HETEs) [11], which have been shown to underlie aldosterone secretion in response to AngII [13]. Thus, Nadler and colleagues have shown that AngII induces the 12-lipoxygenase-mediated production of 12-HETE in human adrenal glomerulosa cells [21]. Furthermore, inhibiting the activities of either DAG lipase [11, 12] or lipoxygenase [21] reduces AngII-stimulated steroidogenesis in human glomerulosa cells and in aldosterone-producing adenomas [21]. Finally, the existence of this DAG lipase that prefers arachidonic acid-containing DAGs may also explain the more rapid disappearance of these DAG species following AngII removal in bovine adrenal glomerulosa cells [23, 50]. Alternatively, the activity of a DAGK that preferentially phosphorylates arachidonic acid-containing DAG [48] could also contribute to the lack of persistence of these DAGs. The metabolism of DAG and generation of various lipid signals in adrenal glomerulosa cells is summarized in figure 1.

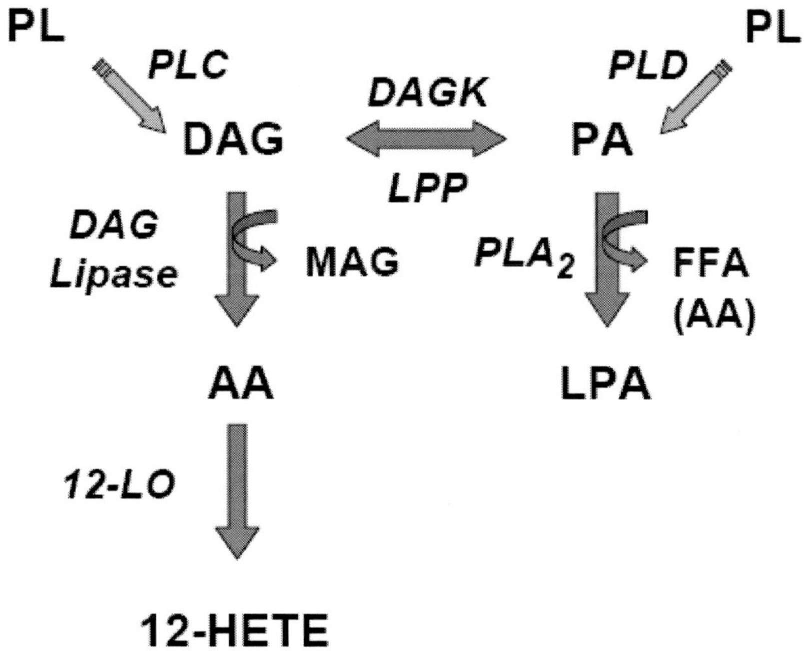

Figure 1. Diacylglycerol and phosphatidic acid metabolism in adrenal glomerulosa cells. In adrenal glomerulosa cells AngII activates phospholipase C (PLC), which hydrolyzes phospholipids (PL, in this case phosphatidylinositol 4,5-bisphosphate) to generate diacylglycerol (DAG) directly. AngII also activates phospholipase D (PLD), which hydrolyzes phospholipids (primarily phosphatidylcholine) to generate phosphatidic acid (PA). PA can be metabolized to DAG by lipid phosphate phosphatases (LPP), while DAG can be converted to PA by the action of DAG kinase (DAGK), a family of DAG-phosphorylating enzymes. DAG, in particular arachidonic acid-containing DAG, can also be deacylated by a DAG lipase to give monoacylglycerol (MAG) and arachidonic acid (AA), which can, in turn, be metabolized by 12-lipoxygenase (12-LO) to 12-hydroxyeicosatetraenoic acid (12-HETE). In addition, PA can potentially be further metabolized by phospholipase A_2 (PLA_2) (163) to yield a free fatty acid (FFA), which could be AA, and the lipid messenger, lysophosphatidic acid (LPA), which can bind to G protein-coupled LPA receptors to initiate signal transduction events.

DAG-Sensitive Effector Enzymes and Aldosterone Secretion

As discussed above, DAG can activate PKC, a family of isoenzymes with slightly different activation characteristics. Thus, the classical or conventional PKC isoforms (PKC-α, -βI, -βII and –γ) are phospholipid-dependent, DAG-activated and Ca^{2+}-sensitive; the novel isozymes (PKC-δ, -ε, -η, and –θ) are phospholipid-dependent and DAG-activated, but Ca^{2+}-insensitive. The activity of both conventional and novel PKC isoforms is stimulated by DAG or phorbol esters like PMA (reviewed in [49]). On the other hand, atypical PKCs (ζ and λ/ι) are insensitive to DAG/phorbol ester and Ca^{2+}, although they are phospholipid-dependent. There are several reports defining the PKC isoforms expressed in adrenal glomerulosa cells.

Using western analysis, Natarajan et al. [37] have demonstrated that rat adrenal glomerulosa cells express the conventional PKC, PKC-α and the novel PKC, PKC-ε but not PKC-β or –γ, and that AngII induces membrane translocation of both PKC-α and –ε. Similarly, hamster adrenal glomerulosa preparations demonstrate detectable protein levels of PKC-α and –ε as well as atypical PKC-ζ, and human adrenal glomerulosa express these isoforms and in addition PKC-λ [51]. The NCI H295R cells also express PKC-α, –ε and –ζ, but no PKC-λ, and instead demonstrate low levels of PKC-θ [51]. By immunohistochemical analysis of hamster and human adrenals, Lehoux et al. [51] also showed immunoreactivity for PKC-α, -ε and –ζ in the zona glomerulosa *in situ*. These authors showed that PMA inhibits and the conventional PKC (and PKD) inhibitor Goedecke 6976 (Gö6976) stimulates AngII-induced CYP11B2 promoter activity [51]. Overexpression by transient transfection of constitutively active PKC-α, -ε and –θ decreases, whereas the dominant negative constructs increase, the promoter activity of co-transfected CYP11B2 [51]. Interestingly, both constitutively active and dominant negative PKC-ζ increase CYP11B2 promoter activity, and while constitutively active PKC-α, -ε and –θ prevent AngII from stimulating CYP11B2 promoter activity, the dominant negative constructs have no effect [51]. These results suggest a less-than-straightforward role of the PKC isoforms in regulating CYP11B2 promoter activity.

Data in the literature indicate that AngII is able to activate PKC in adrenal glomerulosa cells. Thus, AngII has been shown to translocate PKC activity from the cytosolic (soluble) to the membrane (particulate) fraction in rat [37] and bovine [36] glomerulosa cells. (Note that since all isoforms of PKC are phospholipid-dependent, translocation to the membrane fraction is thought to represent activation of the enzyme.) AngII-induced phosphorylation (also thought to represent activation) of PKC isoforms was also reported in the NCI H295R cell line [52]. This activation may mediate steroidogenesis, as investigators have shown that selective PKC inhibitors (Ro31-8220 and a PKC substrate inhibitor peptide) can inhibit AngII-elicited aldosterone secretion from rat capsular (predominantly glomerulosa) preparations [29] and rat glomerulosa cells [37]. Similarly, other laboratories have demonstrated that PKC inhibitors can also reduce AngII-stimulated proliferation of rat glomerulosa cells [53] and c-fos expression in bovine glomerulosa cells [54]. On the other hand, Hajnoczky et al. [44] have suggested that PKC activity is inhibitory in terms of aldosterone production, based on the ability of pretreatment with PMA (to downregulate PKC), or with the relatively non-selective PKC inhibitor staurosporine [55], to enhance subsequent AngII-induced aldosterone secretion. Other investigators have also proposed that PKC is of lesser importance in regulating AngII-induced aldosterone secretion than is Ca^{2+} because of the relative ineffectiveness of non-specific PKC inhibitors and agonists such as PMA [45]. However, the fact that most of the tools used to date impact signaling enzymes in addition to PKC argues that additional research is necessary to define the role of PKC isoforms in steroidogenesis in response to AngII.

As mentioned previously, DAG can also activate Ras through Ras-GRP, a guanine nucleotide exchange factor, to activate the mitogen-activated protein kinase pathway culminating in extracellular signal-regulated kinase-1 and –2 (ERK-1/2) (reviewed in [10]). Interestingly, PA can also activate this pathway, as PLD-derived PA is able to recruit the MAP kinase kinase, Raf-1, to the membrane to interact with and be activated by Ras [56, 57] (figure 2). One or both of these mechanisms may underlie the ability of AngII to activate Ras,

Raf-1 and ERK-1/2 in adrenal glomerulosa cells, an effect that has been observed by several laboratories [52, 58-61]. The effect of AngII on ERK-1/2 activation is transient [58], likely because AngII also induces the expression of mitogen-activated protein kinase phosphatase-1 (MKP-1) [59], a feedback inhibitory mechanism. Tian et al. [58] demonstrated that activation with PMA of phorbol ester-sensitive effectors (which would include PKC, Ras-GRP and PLD, as well as PKD) is sufficient to activate both Raf-1 and ERK-1/2. The role of ERK-1/2 activity in glomerulosa biology is still debated, as some investigators have proposed that this AngII-stimulated ERK-1/2 activity underlies the proliferative or hypertrophic effect of the hormone ([60, 61] and reviewed in [62]). On the other hand, other laboratories suggest that ERK-1/2 mediates, at least in part, AngII-elicited aldosterone production [59, 61] and/or effects on the AngII-induced expression of StAR [52], which is necessary for steroidogenesis (reviewed in [25)]. In addition, ERK-1/2 has been shown to regulate the activity of cholesterol ester hydrolase, an enzyme that hydrolyzes cholesteryl esters stored in lipid droplets to release cholesterol for transport to the mitochondria and the initial step in steroidogenesis [63].

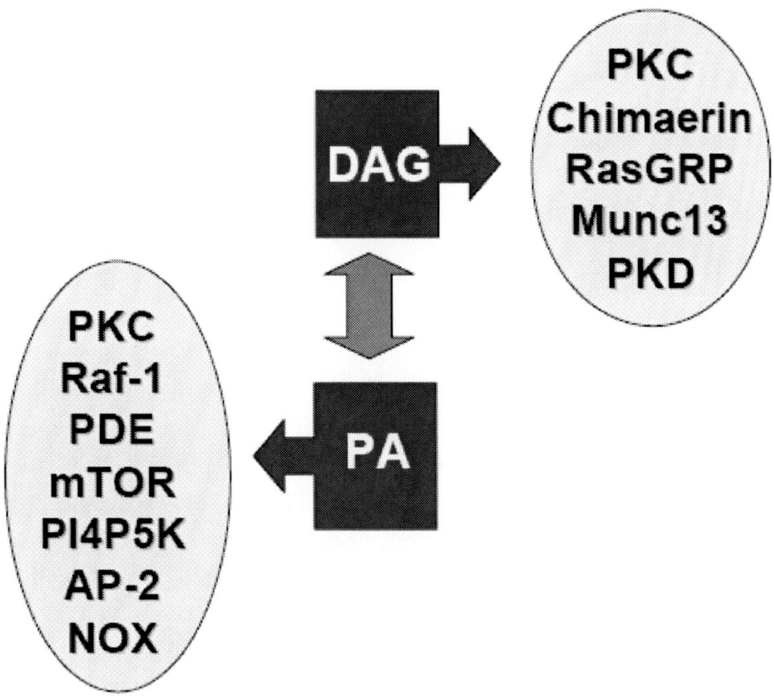

Figure 2. Targets of diacylglycerol and phosphatidic acid. Shown are some known targets/effector enzymes for the diacylglycerol (DAG) and phosphatidic acid (PA) signals. For DAG, these include protein kinase C (PKC), chimaerins, Ras guanine nucleotide releasing protein (RasGRP; a guanine nucleotide exchange protein for Ras), Munc13 and protein kinase D (PKD) (reviewed in [10]). Known PA targets comprise: PKC (certain isoforms; e.g. PKCα [164]), Raf-1, a cyclic nucleotide phosphodiesterase (PDE), mammalian target of rapamycin (mTOR), phosphatidylinositol 4-phosphate 5-kinase (PI4P5K) (reviewed in [17]), adaptor protein-2 (AP-2) and NADPH oxidase (NOX) (reviewed in [161]).

PLD can also be activated by phorbol esters/DAG through PKC. PLD consists of two isoforms of PIP_2-dependent phosphatidylcholine-hydrolyzing enzymes, PLD-1 and PLD-2, with differential expression, localization and regulation (reviewed in [17]). PLD-1 can be activated *in vitro* by small GTP-binding proteins of the Rho family, including RhoA, Rac, cdc42 and RalA ([64-66] and reviewed in [67]). PLD-1 can also be regulated by ADP ribosylation factor (ARF) and PKC (e.g., PKC-α), with synergistic activation by the small G proteins of the Rho family, ARF and/or PKC [68-71]. Note that the phosphorylating activity of PKC does not appear to be necessary for its PLD-activating action [72, 73], although PKC can phosphorylate and regulate PLD [74, 75]. On the other hand, PLD-2, when overexpressed in Sf9 insect cells and assayed *in vitro*, exhibits constitutive activity, and initially it was thought that PLD-2 might be subject to negative regulation *in vivo* [76]. Although negative regulation may occur in intact cells, particularly that mediated by actin [77], more recent results suggest that in cells PLD-2 can also be activated by ARF and PKC [78] (although there is no evidence to date that Rho family members participate in PLD-2 activation). One possible explanation for this disparity relates to the conditions for *in vitro* analysis of PLD activity: the concentration of PIP_2 included in the membrane preparations used for measurement of activity may be saturating for PLD-2. Indeed, there is evidence that PLD-2 activity might be modulated *in vivo* by changes in PIP_2 concentrations [79].

Data from our laboratory indicates that AngII induces PLD activation in both primary bovine adrenal glomerulosa [16] and NCI H295R cells [18], although in the bovine cell PLD activation is sustained [14] whereas in the H295R cells activation is transient [18]. Our studies using primary alcohol to inhibit PLD-mediated lipid signal generation support the importance of PLD to AngII-induced aldosterone secretion in both cell types [15, 18]. Research from our laboratory on the mechanism of AngII's stimulation of PLD activity indicates also that in both cell types PKC plays an important role in PLD activation, as incubation of either cell type with PMA is sufficient to induce PLD activation [15, 18]. Co-treatment with PMA and AngII indicates a common mechanism of action of the two agents [15], and PKC inhibitors can inhibit AngII-elicited PLD activation [18]. In NCI H295R cells, inhibition of phosphoinositide 3-kinase (PI3K) using the PI3K inhibitors, wortmannin or LY294002, enhances the stimulation of PLD activity by AngII, suggesting that the constitutive PI3K activity observed in these cells actually inhibits PLD's activation [18]. Also in these cells there does not appear to be a role for cAMP-dependent protein kinase or CaMK, as inhibition of the activity of these kinases has little or no effect on AngII-induced PLD activation [18]. Likewise, inhibitors of calmodulin or voltage-dependent Ca^{2+} channels also do not alter AngII-stimulated PLD activity [18]. On the other hand, our recent results suggest that Ca^{2+} influx through pathways other than voltage-dependent Ca^{2+} channels plays a role in AngII-elicited PLD activation [80] (perhaps by helping to recruit conventional PKCs to the plasma membrane to interact with and activate PLD). The PLD isoform(s) contributing to the PLD activation in response to AngII is not yet known, since both PLD-1 and PLD-2 are expressed in primary bovine adrenal glomerulosa and NCI H295R cells [18]. However, our current data indicate that inhibiting Rho action with C3 exoenzyme (a toxin that ADP ribosylates and inactivates Rho) inhibits AngII-induced PLD activation and aldosterone secretion (manuscript in preparation). This result suggests the likely involvement of PLD-1 in PLD activation in response to AngII, as well as in steroidogenesis. Importantly, AngII has been shown to activate Rho in rat [81] and bovine adrenal glomerulosa cells (manuscript in preparation), supporting the possibility of this pathway as a mechanism of AngII-induced PLD (i.e., PLD-1) activation.

Another DAG-/phorbol ester-sensitive effector enzyme is PKD, also known as PKCμ. Two groups originally described the enzyme now known as PKD1: as PKD, from Swiss 3T3 fibroblasts [82] and as PKCμ, from human YT cells [83]. In fact, PKD is now classed as a family of DAG-/phorbol ester-sensitive enzymes that includes also PKD2 and PKD3 (also known as PKCν; reviewed in [84]). Mouse PKD1 and human PKD1 are comprised of 918 and 912 amino acids, respectively, and are over 90% homologous [84]. There also exists in the NCBI database a rat 918 amino acid predicted rat PKD1 sequence 92% identical to human PKD1, and a bovine (*Bos taurus*) sequence (accession #XM_612625) that possesses 93% homology to human PKD (a predicted 96% amino acid identity), suggesting the presence of at least one bovine PKD isoform as well.

Protein Kinase D: Structure, Activation and Mediation of Cellular Responses

The PKD family members contain the following motifs: two cysteine-rich regions that bind DAG and phorbol esters [cysteine (Cys)1 and Cys2, also known as C1a and C1b], an N-terminal hydrophobic domain, an autoinhibitory pleckstrin homology (PH) domain, an activation loop, and a C-terminal catalytic domain (reviewed in [84, 85]). PKD has some similarities to the classical and novel PKC isoforms, such as sequence homology, particularly in the Cys-rich domains [82]. However, there are also profound differences. PKD lacks the pseudosubstrate domain present in conventional PKCs, and also differs from both the conventional and novel PKCs in that the Cys1 and Cys2 domains in PKD are separated by a longer stretch of amino acids [83]: 79 in PKD versus 15 in conventional PKCs and 22 amino acids in novel PKCs [83]. PKD does not efficiently phosphorylate some typical PKC substrates such as histone or pseudosubstrate-ε; because PKD is more homologous to the CaMKs in the catalytic domain, rather this enzyme phosphorylates the typical CaMK II substrate, syntide 2 [86]. In addition, like CaMKs, PKD can be persistently activated [85]: when PKD is phosphorylated by PKC on serines (Ser) 738/742 in human or Ser744/748 in mouse, PKD can then autophosphorylate Ser910 (human) or Ser916 (mouse) [87], thus rendering the enzyme persistently active [88, 89]. In addition, there is a correlation with autophosphorylation of Ser910/916 and PKD's activation status [90], such that phosphorylation at Ser910/916 is often used as a marker for PKD activation [91].

Early studies of PKD indicated that this kinase is activated by the addition of mitogenic peptides such as bombesin, vasopressin and platelet-derived growth factor (PDGF) in Swiss 3T3 cells [92], and that overexpression of PKD in fibroblasts increases DNA synthesis in response to mitogens [93]. As mentioned above, PKD activation can be initiated through PKC-mediated phosphorylation of the activation loop at Ser738/742 (744/748 in mouse) [88, 89]. This event has been observed to be mediated by the novel PKCs, PKC-δ [94, 95], -ε and -η [89, 96], as well as PKCθ [97]. In addition, this activation event is made more efficient by phosphorylation of Tyr463 by Src and Abl tyrosine kinases [98], such that maximal PKD activation likely requires both PKC- and tyrosine kinase-mediated phosphorylation. In turn,

PKD has been implicated as a signaling effector mediating a number of responses to extracellular signaling events. Thus, PKD has been implicated in such cellular processes as mitogen-activated protein kinase (MAPK) signaling [99-101] and proliferation [84, 102-105], Golgi maintenance and trafficking [106-114], cell survival in response to oxidative stress [98, 115-120], transcriptional repression [121, 122], cardiac hypertrophy [123] and steroid secretion [124, 125].

As mentioned above, AngII initiates a number of intracellular signaling pathways that mediate aldosterone secretion [126, 127]. AngII is also known to elicit PKD activation in a variety of cell types, including rabbit aortic [128] and rat [52] vascular smooth muscle cells [128], intestinal epithelial cells [91], rat cardiomyocytes [129], and NCI H295R cells [124]. AngII-elicited PKD activation appears to be mediated through the hormone's ability to activate G proteins, although this activation has been observed to be insensitive to pertussis toxin treatment (independent of G_i/G_o family protein involvement) in intestinal epithelial cells [91]. Tan et al. [95] found in rat aortic vascular smooth muscle cells that AngII-stimulated PKD phosphorylation is sensitive to treatment with losartan [antagonizes AngII type 1 (AT-1) receptor], but not PD123,319 (antagonizes the AT-2 receptor), suggesting that PKD activation occurs through the AT-1 receptor, but not the AT-2 receptor. In addition, introduction of either a dominant-negative PKC-δ construct or the PKC-δ-selective inhibitor rottlerin decreases PKD activation [95]. An inhibition of PKD activity was not observed, however, in vascular smooth muscle cells treated with dominant-negative constructs of PKC-ε and PKC-ζ [95]. Thus, at least in vascular smooth muscle cells, AngII-mediated PKD activation is dependent on PKC-δ signaling. On the other hand, in neonatal rat cardiomyocytes AngII-induced PKD activation is mediated by PKC-ε and plays a role in cardiac hypertrophy [129]. Cardiac hypertrophy, a term used to describe an enlargement of cardiac myocytes without an increase in cell number [130], is responsible for a remodeling process that can lead to heart failure. Iwata et al. [129] observed that AngII activated PKD through PKC-ε and that overexpression of constitutively active PKD in these cells resulted in increased cell size, as well as formation of Z-discs and enhanced expression of the embryonic protein, atrial natriuetic factor.

Protein Kinase D and Aldosterone Secretion

Work by the Gomez-Sanchez laboratory suggests a role for PKD in mediating aldosterone secretion [124]. These investigators found, after performing a Kinetworks protein kinase screen on AngII-treated NCI H295R cells, that PKD phosphorylation was upregulated compared with untreated control cells [124]. In contrast to the results seen in vascular smooth muscle cells by Tan et al. [95], but in agreement with Iwata and colleagues [129], Gomez-Sanchez and colleagues reported that PKC-ε mediated the AngII-induced PKD activation, rather than PKC-δ [124]. These investigators transiently transfected various PKD constructs [wild-type PKD, a reported constitutively active PKD mutant in which serines 738 and 742 are mutated to glutamate to mimic phosphorylation (PKDSer738/742Glu) and a constitutively active mutant lacking the PH domain (PKDΔPH)] into NCI H295R cells. Overexpression of these constructs resulted in increases in 11β-hydroxylase and CYP11B2 (aldosterone synthase) promoter activities, as well as enhanced cortisol and aldosterone secretion after a 24-hour incubation [124]. Similarly, Chang and colleagues [125] observed that shRNA constructs directed against PKD resulting in PKD knock-down decreased CYP11B2 expression and AngII-elicited aldosterone secretion from NCI H295R cells after 24 hours. These investigators also found that PKD is activated in tumors from patients with adenomas [125]. Together these results suggest that PKD is sensitive to AngII stimulation and is likely to be involved in regulating at least one of the mechanisms mediating chronic aldosterone synthesis.

On the other hand, our recent data suggest that the role of PKD may be complex in terms of its regulation of acute AngII-induced steroidogenesis in primary bovine adrenal glomerulosa cells. Thus, adenovirus-mediated overexpression of wild-type PKD inhibited acute AngII-elicited (but not basal) aldosterone secretion, whereas overexpression of the constitutively active $PKD^{Ser744/748Glu}$ enhanced acute AngII-stimulated aldosterone production, in comparison with an empty adenoviral vector [131]. The effect of the wild-type PKD is not thought to be due to adenovirus toxicity because: (1) the constitutively active mutant increased aldosterone secretion in response to AngII and (2) there was no difference in aldosterone secretion observed between mock-, empty vector- and PKD-infected cells stimulated with 22(R)-hydroxycholesterol, a hydrophilic cholesterol that bypasses the signaling events required for aldosterone production [131]. 22(R)-hydroxycholesterol-supported aldosterone secretion represents the overall capacity of the adrenal glomerulosa cells to synthesize the steroid hormone; therefore, the lack of effect of infection with adenovirus (either with or without PKD) indicates that this manipulation does not alter cell health and/or expression of aldosterone biosynthetic enzymes, since such changes would be reflected in effects on 22(R)-hydroxycholesterol-mediated aldosterone secretion.

At present we do not completely understand these seemingly disparate results using the wild-type and constitutively active PKD, and experiments are in progress to introduce other PKD mutants into the primary glomerulosa cells to determine their effects on aldosterone production. However, in view of the results with the H295R cells and other data in the literature, it is possible to speculate as to potential mechanisms by which PKD might regulate chronic aldosterone secretion. For instance, the NCI H295R cells express low levels of CYP112B basally [19,31]. In the experiments of Romero et al. [124] and Chang et al. [125], after transfection with the PKD constructs and an additional 24-hour incubation, AngII-induced secretion was measured after an added 24-hour time period (a total of 48 hours). Thus, if PKD can increase expression of aldosterone biosynthetic enzymes chronically, PKD overexpression could enhance chronic AngII-elicited aldosterone secretion with this long-term protocol. Indeed, there is both direct and indirect evidence that PKD can stimulate steroidogenic enzyme expression. Thus, Romero et al. [124] and Chang et al. [125] both demonstrate that PKD overexpression increases CYP11B2 promoter activity, suggesting an ability of PKD to increase expression of this enzyme. There are also data in the literature indicating that the cAMP response element binding (CREB) protein interacts with cAMP response elements to increase transcription of several steroidogenic enzymes, including CYP11A1 (the p450 side-chain cleavage complex that catalyzes the initial rate-limiting step in the biosynthesis of all steroid hormones), CYP11B2 [132] and steroidogenic acute regulatory (StAR) protein [133] [the protein that triggers cholesterol translocation from the outer to the inner mitochondrial membrane, the site of CYP11A1 localization (reviewed in [25])]. It is known that CREB levels are decreased, with upregulation of cAMP response element modulator (CREM), in the H295R cells [134].

PKD is known to phosphorylate and activate CREB to stimulate CREB-mediated transcription; thus, Johannessen and colleagues [135] identified PKD phosphorylation consensus sequences in CREB. Moreover, these authors showed that overexpression of PKD1 induces CREB phosphorylation at Ser133, resulting in activation of CRE-responsive promoters and increased CREB, activating transcription factor-1 (ATF-1) and CREM (CREB family members) target gene transcription [135]. This ability of PKD to increase CREB family-mediated transcription opens up the possibility that chronically PKD can enhance AngII-induced aldosterone production through its ability to upregulate expression of biosynthetic enzymes, for aldosterone and other steroids, controlled by CREB family members. Indeed, Romero et al. [124] observed an ability of PKD overexpression to increase not only aldosterone, but also cortisol, secretion, suggesting a more general effect of this enzyme on steroidogenesis. Nevertheless, an upregulation of steroidogenic enzymes does not appear to occur in the primary glomerulosa cells, as we observed no effect of PKD overexpression on aldosterone production in response to 22(R)-hydroxycholesterol [131].

Another mechanism by which PKD might modulate expression of steroidogenic enzymes is through effects on gene transcription regulated by histone acetylation. In many cell types PKD has been observed in the nucleus (reviewed in [136]), where it likely regulates histone deacetylase (HDAC)-mediated chromatin remodeling in response to some stimuli [137]. HDACs catalyze the removal of acetyl groups from histones, thereby increasing the histone's affinity for DNA, promoting chromatin condensation and repressing transcription of genes involved in cell growth, such as the MEF-2 transcription factor in cardiac myocytes [123] and the orphan nuclear receptor/transcription factor Nur77 in T cells [121, 137]. [Note that HDACs can also catalyze the deacetylation of other acetylated proteins in addition to the histones, including those that may be located outside of the nucleus (e.g., [138]).] Much recent work in the PKD field has focused on PKD's ability to phosphorylate class II HDACs and cause nuclear export of the enzymes, thereby allowing histones to remain acetylated and chromatin to persist in a relaxed state such that transcription of genes involved in cell proliferation can continue. For example, Vega and colleagues [123] observed in neonatal rat cardiac myocytes that PMA and constitutively active PKD constructs are able to induce export of GFP-HDAC5. Interestingly, an HDAC5 mutant in which serines at positions 259 and 498 are mutated to alanines (Ser259/498A), residues that are required for 14-3-3 protein binding, is not redistributed to the cytosol upon PMA treatment. In addition, mutation of the leucines at positions -5 to Ser259/498 [Leu254/493Gly; leucine at the -5 position is absolutely required for phosphorylation by PKD 139] renders HDAC5 completely refractory to PMA-induced nuclear export. In fact, both of these residues are directly phosphorylated by PKD [122, 123].

PKD-mediated nuclear export of HDAC also functions in T cell receptor-induced Nur77 expression and apoptosis [121]. Dequiedt and colleagues [121] determined that T cell receptor-dependent export of HDAC7, a transcriptional repressor of Nur77, is increased by PMA and decreased by Gö6976 (a conventional PKC/PKD inhibitor). HDAC7 interacts with PKD resulting in the phosphorylation of several N-terminal HDAC7 residues [121].

Additionally, these authors found that PKD potentiates Nur77 transcriptional activity. 14-3-3 binding of HDAC7 was determined to be essential for PKD-induced nuclear export, and requires PKD-mediated phosphorylation of 4 serines (amino acids 155, 181, 321 and 449) residues that were +5 of Leu and +3 of either Arg or Lys in HDAC-7. Cotransfection of GFP-HDAC7 with a constitutively active PKD mutant induces cytosolic redistribution of GFP-HDAC7 and increases Nur77 promoter activity, while expression of a dominant negative PKD reduces Nur77 promoter activity (121). Stable expression of an HDAC7 mutant (HDAC7ΔL; Leu -5 of Ser mutated to Ala as above for each of amino acids 150, 176, 316 and 444) results in impaired export from the nucleus into the cytosol. In addition, cells that express HDAC7ΔL exhibit reduced Nur77 promoter activity, as well as strong inhibition of PMA-induced apoptosis [121]. Phorbol ester-stimulated Nur77 promoter activity is also abolished by dual PKD1/PKD3 knockout in avian DT40 B-cells, while expression of an inducible PKD3 construct is able to rescue HDAC5/7 phosphorylation and nuclear export [140]. Other effectors for PKD in the nucleus include the histone chaperone SET. SET has been previously recognized to inhibit histone acetyl transferases as well as the growth suppressor and apoptosis enhancer PP2A [141]. Irie and colleagues [141] determined that in Jurkat cells PKD2 phosphorylates the SET protein, suggesting another mechanism by which PKD may be able to regulate gene transcription and steroidogenesis.

Possible Mechanisms Underlying PKD's Effects on ANGII-Induced Glomerulosa Cell Function

PKD, Proliferation and ERK-1/2 Activation

On the other hand, PKD might inhibit acute AngII-stimulated steroidogenesis through altering their commitment to proliferation, via cholesterol mobilization and/or signaling pathways. Much evidence in the literature supports the idea that PKD is a pro-proliferative signaling enzyme in many cell types (reviewed in [84]), and Foster [62] has proposed that there may be "reciprocal influences between the signaling pathways regulating proliferation and steroidogenesis" in adrenal glomerulosa cells. Therefore, PKD could potentially affect acute AngII-induced aldosterone secretion by its diversion of primary bovine adrenal glomerulosa cells to a more proliferative phenotype. On the other hand, the NCI H295R cells already exhibit dysregulated cell growth [30], in that, for example, these cells exhibit constitutive activation of pro-proliferative, pro-survival Akt [18]. Thus, the possibility exists that PKD and its downstream pathway(s) are also dysregulated or perhaps that PKD is not required for proliferation in these transformed cells. Clearly, additional studies are needed to determine the role of PKD in AngII-induced glomerulosa cell proliferation.

In terms of cholesterol content, PKD is known to potentiate and prolong the activation of ERK-1/2 in various cell types. For example, Johannes and colleagues first observed that overexpression of wild-type PKD and constitutively active PKDΔPH constructs in HEK293 cells increased p42/44ERK phosphorylation, as well as Ets-like transcription factor 1 (ELK1)-mediated serum response element transcription [100]. In fibroblasts PKD can potentiate the effects of activation of G_q on MEK/ERK/RSK activation and DNA synthesis and prolong ERK-1/2 activity [142], while siRNA knockdown of PKD reduces ERK activation and DNA synthesis in vascular endothelial growth factor (VEGF)-stimulated endothelial cells [143]. PKD antisense oligonucleotides have also been observed to inhibit basic Ca^{2+} phosphate crystal-induced p42/44 phosphorylation/activation in human fibroblasts [99]. Thus, PKD appears to function, at least in part, by enhancing activation of ERK-1/2. Since, as previously discussed, ERK-1/2 activates cholesterol ester hydrolase, the enzyme that releases cholesterol from cholesteryl esters stored in lipid droplets [63], PKD overexpression could enhance ERK-1/2-mediated activation of cholesterol ester hydrolase. This enhancement could result in an increase in the mobilization of cholesterol from its storage in lipid droplets, thereby potentially enhancing aldosterone production. Alternatively, ERK-1/2 is also known to be involved in growth regulation in several cell types, and it is possible that this ability of PKD to prolong the activation of ERK-1/2 could mediate proliferative effects of the enzyme.

Results in the literature support the idea that PKD can influence several signaling pathways, in addition to ERK-1/2, which may be important to steroidogenesis and could possibly modify aldosterone secretion. For instance, PKD inhibits the c-jun N-terminal kinase (JNK)/c-jun pathway by binding to [144] and phosphorylating c-jun on a site distinct from that phosphorylated by JNK [145]. PKD-mediated c-jun phosphorylation inhibits its transactivation activity [146]. Since c-jun is reported to bind to and activate the transcription of StAR [147], its inhibition could modify StAR expression, although it is likely that c-jun has other effects as well. Another possible pathway through which PKD could exert inhibitory effects is via interference with tyrosine kinase activation of PLC-γ. Although in adrenal glomerulosa cells AngII induces PIP_2 hydrolysis through activation of PLC-β, in other systems AngII can also activate PLC-γ (reviewed in [148]). Thus, the finding that PKD can prevent Syk's activation of PLC-γ [149] provides another potential site of PKD's modulation of AngII-elicited signaling leading to aldosterone secretion. On the other hand, in vascular endothelial cells stimulated with vascular endothelial cell growth factor (VEGF), PLC-γ activation is downstream of and requires PKD [150], again suggesting that further investigation is necessary to discover the mechanism(s) by which PKD can regulate AngII-elicited aldosterone secretion.

PKD and ROS

An exciting new aspect of PKD is its role in cell survival in response to oxidative stress. Oxidative stress can arise through many avenues, including via normal cellular respiration, inflammatory diseases and environmental conditions, as well as AngII signaling through NADPH oxidase (NOX). The "bad actors" in this system are a class of molecules known as reactive oxygen species (ROS), which are implicated in lipid peroxidation, altered protein-protein interactions and DNA damage [151]. On the other hand, it has also been recently shown that some hormones, such as AngII, can signal through the generation of such free radicals [152-155]. Indeed, redox status has been demonstrated to regulate aldosterone secretion from adrenal glomerulosa cells [148], raising the possibility that AngII's activation of NOX may modulate steroidogenesis.

Because PKD appears to play a key role in cell responses to oxidative stress, the activation of this enzyme in response to the generation of ROS has been studied extensively. It has been observed that the ROS-producing agent hydrogen peroxide (H_2O_2) induces activation of PKD [115-117, 156]. Several lines of evidence indicate that this activation is mediated through both tyrosine kinase and novel PKC signaling. For example, Toker and colleagues [98] observed that PKD activation was blocked by transfection with inactive Src or treatment with the Abl inhibitor STI-571, while coincubation of the aforementioned reagents with cells transfected with constitutively active Src or Abl mutants resulted in recovery of PKD activation. By coexpression of various PKD mutants harboring tyrosine to phenylalanine point mutations with either constitutively active Src or Abl constructs, the tyrosine residue target of Src and Abl signaling was identified by this group as tyrosine (Tyr) 463 in PKD [98].

PKD activation during oxidative stress signaling also occurs via novel PKC-mediated phosphorylation of its activation loop on Ser 738 and 742 [119]. Interestingly, Toker and colleagues [119] observed that phosphorylation of Tyr463 upregulates PKD Ser738/742 phosphorylation in response to ROS. Thus, H_2O_2 and pervanadate, as well as overexpression of constitutively active Src, constitutively active Abl, or a PKD Tyr463 to glutamate (Glu) mutant (Glu mimics phophorylation of Tyr463) increases Ser738/742 phosphorylation [119]. This phosphorylation of Ser738/742 is mediated by PKC and is required for PKD activation, as shown by the fact that selective PKC inhibitors block Ser738/742 phosphorylation and inhibit PKD activity [156]. These observations point to a model in which PKD activation requires sequential activation steps: tyrosine kinase-mediated phosphorylation of Tyr463, followed by PKC-mediated phosphorylation of Ser738/742.

The identity of the novel PKC responsible for PKD Ser738/742 phosphorylation in response to oxidative stress was determined to be PKC-δ: PKC-δ RNA interference (RNAi) reduced H_2O_2-induced Ser738/742 phosphorylation [118]. In addition, overexpression of PKC-δ and Abl results in enhanced PKD phosphorylation on Tyr463 [118]. H_2O_2 treatment also induces PKC-δ activation and the formation of a coimmunoprecipitable PKD-PKC-δ complex [118], suggesting a mechanism in which ROS activates both tyrosine kinase (Src/Abl) and PKC-δ, leading to phosphorylation on Tyr463 and Ser738/742 and inducing maximal PKD activity.

The possible effector of this PKD activation was suggested by the fact that H_2O_2 treatment also upregulates NF-κB signaling. NF-κB is a transcription factor that is retained in the cytoplasm by its association with inhibitory factors such as inhibitory κB (IκB) [154]. Serine (Ser)-specific IκB kinase (IKK)-mediated phosphorylation events induce degradation of IκB to release NF-κB from the complex thereby allowing its translocation to the nucleus. In the nucleus NF-κB functions to upregulate the transcription of products such as regulators of apoptosis (e.g., Bcl-xL and IAP) and of the cell cycle (e.g., cyclin D1 and mdm2), as well as stress response genes (e.g., manganese-dependent superoxide dismutase) [157]. The potential interaction between PKD and NF-κB was explored in experiments using coexpression of various PKC-δ or PKD constructs with a NF-κB promoter-driven luciferase reporter plasmid [118]. Overexpression of either PKC-δ or PKD increases, while treatment with PKC-δ RNAi reduces, H_2O_2–stimulated NF-κB reporter activity [118]. Overexpression of IKKβ with PKD also increases NF-κB target promoter activity [119]. Constitutively active

PKD mutants (Tyr463Glu and Ser738/742Glu) induce NF-κB activity, which can be blunted by coexpression with the kinase-dead [lysine (Lys) 612 to tryptophan (Trp)] or nonactivatable (dominant-negative) PKD [Ser738/742 to alanine (Ala)] mutants [119]. Similarly, Song and colleagues [156] observed that in H_2O_2-treated cells PKD siRNA constructs enhance intestinal epithelial cell death, as monitored by the markers of apoptosis, poly (ADP-ribose) polymerase (PARP) cleavage and activated caspase 3, again suggesting a protective effect of PKD in oxidative stress.

A biologically significant role for PKD in NF-κB signaling may be its ability to mediate mitochondria to nucleus signaling, in order to effect detoxification of the ROS produced by the mitochondria [120]. For example, treatment of HeLa cells with agents that inhibit mitochondrial respiration [diphenyleneiodonium (DPI) and rotenone, which increase oxidative stress], as well as H_2O_2, activates PKD and triggers its colocalization with mitochondria [120]. Similarly, introduction of constitutively active PKD, Src or Abl constructs increases NF-κB-mediated superoxide dismutase 2 (SOD2) reporter activity, as does treatment with H_2O_2, DPI and rotenone [120]. Also, the knock down of PKD levels using RNAi downregulates H_2O_2-, DPI- and rotenone-induced NF-κB signaling, as well as cell survival, in treated cells [120]. Thus, the proposed model of PKD signaling is that the generation of oxidative stress causes either Src to activate PKD, or Abl to activate PKC-δ, which then activates PKD and translocates the enzyme to the mitochondria [98, 119, 120]. PKD then initiates NF-κB signaling through phosphorylation of the IKK complex [119]. NF-κB signaling in the nucleus induces transcription of the SOD2 enzyme, which then moves to the mitochondria in order to detoxify ROS by scavenging free radicals, thereby leading to increased cell survival [120]. On the other hand, if ROS are required for AngII signaling to induce steroidogenesis, as it appears to be in AngII signaling for vascular smooth muscle contraction and renal cell growth (reviewed in [158]), a reduction in ROS levels by PKD might be expected to decrease aldosterone production.

Redox status is also critical in terms of regulating aldosterone biosynthesis, as the steroid synthetic enzymes require electron transfer for catalysis (reviewed in [26]). These electrons are received from a low-molecular-weight sulfur protein, adrenodoxin, which is reduced by a NADPH-dependent reductase. The importance of the redox state of the cell to steroidogenesis is highlighted by the fact that the adrenal gland is known to possess high concentrations of antioxidants, such as glutathione, α-tocopherol, selenium and ascorbic acid, and to demonstrate high expression of enzymes involved in antioxidative functions, including superoxide dismutase, glutathione peroxidase and glutathione reductase [27]. Thus, PKD's effects on ROS detoxification and cellular redox status could also potentially modulate aldosterone secretion, e.g., by promoting the reduced state of NADPH.

PKD and Golgi Function

PKD has been observed to translocate to the Golgi in a manner dependent on the functioning of the Cys1/C1a domain [109]. Thus, PKD has been observed to be activated by G protein subunits βγ ($G_{\beta\gamma}$) through interaction with the PH domain or via the C1 domain binding to DAG, which results in PKD's relocalization to the Golgi apparatus [107, 110,

111]. Jamora et al. [107] used the marine sponge compound ilimaquinone (IQ), which is known to induce vesiculation of the Golgi by activating $G_{\beta\gamma}$, to study PKD's role in regulating Golgi architecture. Incubation of cells with IQ increases PKD activity through $G_{\beta\gamma}$, thereby triggering dissolution of the Golgi [107]. On the other hand, normal rat kidney cells treated with IQ and the PKD specific inhibitor substrate mu peptide (which inhibits PKD) exhibit normal Golgi cisternal structure compared with IQ-treated cells. DAG can also trigger PKD translocation to the Golgi apparatus [111]. Also, depleting HeLa cells of DAG results in reduced PKD association with the Golgi [111]. Therefore, one mechanism by which PKD translocates to the Golgi and becomes activated has been delineated: the hydrolysis of phospholipids, presumably by a PLC or PLD, results in the generation of DAG, which then recruits PKD, through its C1a domain, to the Golgi [111], where $G_{\beta\gamma}$-binding to the autoinhibitory PKD PH domain activates the enzyme, which then induces vesiculation (i.e., vesicle budding) of the Golgi [107]. Interestingly, in HeLa cells expressing the catalytically dead PKD [Lys618 to asparagine (Asn)] mutant and incubated at 20°C (to decrease Golgi transport and accentuate the effects of PKD-Lys618Asn), tubulation of the trans-Golgi network (TGN) is observed [108]. Also, CD4, a marker of Golgi to surface transport, does not appear on the cell surface until cells are incubated at higher temperatures (25-37°C) to allow functioning of the endogenous wild-type PKD [108]. Again, this result suggests the potential involvement of PKD in Golgi trafficking. Evidence for an addition role for PKD in Golgi-mediated secretion is provided by data showing that permeabilized vesicular stomatitis virus G-protein (VSV-G)-infected normal rat kidney cells incubated with PKD inhibitors (H89, the mu peptide or the purified PH domain) exhibit inhibited VSV-G transport to the membrane [108].

PKD's βγ-mediated Golgi vesiculation was found to be mediated by a novel PKC: cotransfection of PKC-η and the β1γ2 of heterotrimeric G proteins results in activated PKC-η, which in turn triggers activation of cotransfected PKD [113]. Also, overexpression of a constitutively active PKC-η construct induces fragmention of the Golgi, which can be blocked by the coexpression of kinase-dead PKD [113]. Thus, a proposed model of PKD in the Golgi suggests that agonist-induced activation of G protein-coupled receptors releases $G_{\beta1\gamma2}$, inducing PLC-mediated hydrolysis of phospholipids to produce DAG and the activation of PKC-η, as well as the translocation of PKD to the Golgi and its activation through $G_{\beta1\gamma2}$'s interaction with the autoinhibitory PH domain of PKD. PKC-η then activates PKD, which elicits membrane fission [113] via an unknown mechanism.

One mechanism by which PKD may regulate membrane budding and Golgi trafficking may be through an ability to phosphorylate and activate a mediator of Golgi transport, phosphatidylinositol 4-kinase IIIβ (PI4KIIIβ) [114]. For example, Hausser and colleagues [114] observed that a PKD-GFP construct colocalizes with PI4KIIIβ at the Golgi complex and showed using *in vitro* kinase (IVK) assays, that PKD1 and PKD2 isoforms phosphorylate PI4KIIIβ on Ser294. Overexpression of the kinase-dead PKD reduces, while overexpression of PKD1 or the constitutively active mutant of PKD1 lacking the PH domain (PKDΔPH) enhances, PI4KIIIβ activation, resulting in increased VSV-G-GFP export [114]. Phosphatidylinositol 4-kinase is one of the enzymes required for biosynthesis of PIP_2, a lipid known to induce the recruitment of several proteins required for vesicle formation to membranes (reviewed in [159]). Thus, at least one role for PKD in Golgi vesicle trafficking may be in the regulation of the Golgi transport protein PI4KIIIβ. Alternatively, PKD3 is known to associate with vesicle-associated membrane protein 2 (VAMP2) [160] and may modulate its activity through intermediaries. Thus, another possible interaction of PKD with Golgi function may involve its regulation of VAMP2-mediated vesicle fusion. [Interestingly, PLD, another AngII-stimulated enzyme activated by DAG/phorbol ester and PKC, is also thought to play a role in vesicle trafficking (reviewed in [161]).]

On the other hand, the way in which alterations in Golgi trafficking would alter AngII-induced aldosterone secretion is at present unclear. It is possible that PKD could accelerate the trafficking of a signaling molecule involved in regulating steroidogenesis, for example, the AT1 receptor (reviewed in [162]), to the plasma membrane. Alternatively, PKD could increase the secretion of some autocrine regulator of aldosterone secretion. Obviously, whether or not the regulation of Golgi trafficking by PKD is involved in its effect on AngII-induced aldosterone secretion from adrenal glomerulosa cells requires further investigation.

Summary and Conclusions

The role of the DAG pathway and its effector enzymes, including PKC, PLD and PKD, in regulating aldosterone secretion from adrenal glomerulosa cells is supported by several lines of evidence, as discussed in this review. Nevertheless, the exact role of these signals in steroidogenesis is at present controversial, in that pharmacologic agents that substitute for DAG in activating its effector enzymes have been observed in some laboratories to stimulate aldosterone secretion and in others to inhibit this process. There are several possibilities for the discrepancy. One such explanation relates to relatively obvious differences, e.g. in the system and/or species used. Less obvious reasons for the disparity include: (1) DAG agonists might activate multiple DAG effector enzymes with opposing effects; under different conditions these enzymes may be expressed to varying levels, with the final outcome of the DAG treatment relying upon their relative amounts in the glomerulosa cell models. As an example, if PKC stimulates and PKD inhibits aldosterone production, treatment with DAG-mimicking phorbol esters, which activate both kinases (summarized in figure 3), could yield disparate results depending on whether the PKC/PKD or Ras effect predominated. Related to this idea, different DAG species, produced from different phospholipids by phosphoinositide-specific phospholipase C versus phosphatidylcholine-hydrolyzing PLD, might preferentially activate particular targets, whereas the DAG agonists may be less specific. (2) DAG not only activates effector enzymes but also serves as a precursor for additional signals. Thus, arachidonic acid-containing DAG can be hydrolyzed by a DAG lipase found in glomerulosa cells to release arachidonic acid, which is metabolized to the aldosterone stimulatory factor, 12-HETE. Most DAG substitutes cannot be so metabolized and cannot, therefore, mimic all of the effects of endogenously produced DAG. Nevertheless, it seems likely that there is a reason for the fact that AngII, upon binding to its AT1 receptor, triggers DAG production, but further investigation is necessary to define the exact mechanisms by which the DAG signaling pathway, PKC, PLD and PKD influence aldosterone secretion from adrenal glomerulosa cells.

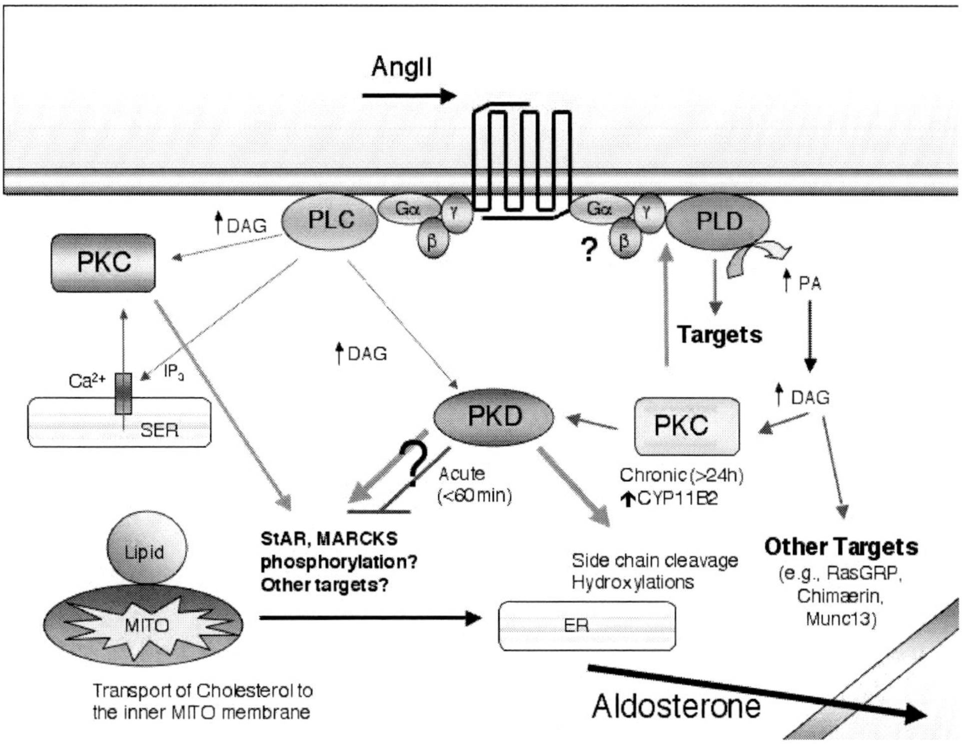

Figure 3. AngII-stimulated diacylglycerol production and protein kinase C, phospholipase D and protein kinase D activation. AngII binds to the AT-1 receptor, initiating G-protein-coupled PIP_2 cleavage by phospholipase C (PLC), to result in inositol 1,4,5-trisphosphate (IP_3) and diacyglycerol (DAG) production. IP_3, in turn, releases Ca^{2+} from a smooth endoplasmic reticulum (SER) pool to increase cytosolic Ca^{2+} concentration and activate Ca^{2+}/calmodulin-dependent protein kinases. AngII also activates PLD, presumably via a G-protein, to generate phosphatidic acid (PA). DAG, which can also be generated by the action of lipid phosphate phosphatase on PA, activates novel PKCs, whereas Ca^{2+} and DAG together activate classical PKCs. PKC then phosphorylates targets such as myristoylated alanine-rich C-kinase substrate (MARCKS) (35), which may be involved in the movement of cholesterol from lipid droplets (lipid) to the mitochondria (mito), and steroidogenic acute regulatory (StAR) protein [165], which transports cholesterol from the outer to the inner mitochondrial membrane. Both PKC types may phosphorylate and activate PKD, although research to date indicates that novel PKCs are more often involved in PKD activation. In addition, there is evidence in lymphocytes that DAG is required for PKD activation via a mechanism in addition to its ability to stimulate PKD [166]. PKD may then negatively or positively (further investigation is still required) affect acute aldosterone secretion through an as yet unknown mechanism (although possible sites of action are discussed in the text). PKD's action seems to occur upstream of side chain cleavage and hydroxylation events, as 22(R)-hydroxycholesterol-stimulated acute aldosterone secretion is unaffected by overexpression of wild-type or constitutively active PKD. On the other hand, PKD is reported to positively affect chronic AngII-stimulated aldosterone secretion, presumably by upregulating CYP11A1, CYP11B2 and/or StAR, through its ability to phosphorylate and activate cyclic AMP response element binding protein, or CREB [135], as well as perhaps through other mechanisms.

Acknowledgements

This project was supported by award #HL70046 from the National Heart, Lung and Blood Institute (NHLBI) and Grant-in-Aid #0350166N from the American Heart Association to WBB. The work was performed in partial fulfillment of the requirements for the degree of Doctor of Philosophy (BAS).

References

[1] Calhoun DA 2006 Aldosterone and cardiovascular disease: smoke and fire. *Circulation* 114:2572-2574

[2] Brilla CG, Maisch B, Zhou G, Weber KT 1995 Hormonal regulation of cardiac fibroblast function. *Eur. Heart J.* 16:45-50

[3] DiBianco R 1994 The changing syndrome of heart failure: an annotated review as we approach the 21st century. *J. Hypertens* 12:S73-S87

[4] Pitt B 2004 Effect of aldosterone blockade in patients with systolic left ventricular dysfunction: implicaitons of the RALES and EPHESUS studies. *Mol. Cell. Endocrinol.* 217:53-58

[5] Lumbers ER 1999 Angiotensin and aldosterone. *Regul. Pept.* 80:91-100

[6] Barrett PQ, Bollag WB, Isales CM, McCarthy RT, Rasmussen H 1989 Role of calcium in angiotensin II-mediated aldosterone secretion. *Endocr. Rev.* 10:1-22

[7] Spät A, Hunyady L 2004 Control of aldosterone secretion: a model for convergence in cellular signaling pathways. *Physiol. Rev.* 84:489-539

[8] Kigoshi T, Uchida K, Morimoto S 1989 Elongation factor 2 as the major substrate for Ca^{2+}/calmodulin-dependent protein kinase in rat adrenal glomerulosa cells. *J. Steroid. Biochem.* 32:381-385

[9] Yao J, Davies LA, Howard JD, Adney SK, Welsby PJ, Howell N, Carey RM, Colbran RJ, Barrett PQ 2006 Molecular basis for the modulation of native T-type Ca^{2+} channels in vivo by Ca^{2+}/calmodulin-dependent protein kinase II. *J. Clin. Invest.* 116:2403-2412

[10] Brose N, Rosenmund C 2002 Move over protein kinase C, you've got company: alternative cellular effectors of diacylglycerol and phorbol esters. *J. Cell. Sci.* 115:4399-4411

[11] Natarajan R, Stern N, Nadler J 1988 Diacylglycerol provides arachidonic acid for lipoxygenase products that mediate angiotensin II-induced aldosterone synthesis. *Biochem. Biophys. Res. Commun.* 156:717-724

[12] Natarajan R, Dunn WD, Stern N, Nadler J 1990 Key role of diacylglycerol-mediated 12-lipoxygenase product formation in angiotensin II-induced aldosterone synthesis. *Mol. Cell Endocrinol.* 72:73-80

[13] Nadler JL, Natarajan R, Stern N 1987 Specific action of the lipoxygenase pathway in mediating angiotensin II-induced aldosterone synthesis in isolated adrenal glomerulosa cells. *J. Clin. Invest.* 80:1763-1769

[14] Jung EM, Betancourt-Calle S, Mann-Blakeney R, Foushee T, Isales CM, Bollag WB 1998 Sustained phospholipase D activation in response to angiotensin II but not carbachol in bovine adrenal glomerulosa cells. *Biochem. J.* 330:445-451

[15] Bollag WB, Jung EM, Calle RA 2002 Mechanism of angiotensin II-induced phospholipase D activation in adrenal glomerulosa cells. *Mol. Cell. Endocrinol.* 192:7-16

[16] Bollag WB, Barrett PQ, Isales CM, Liscovitch M, Rasmussen H 1990 A potential role for phospholipase-D in the angiotensin-II-induced stimulation of aldosterone secretion from bovine adrenal glomerulosa cells. *Endocrinology* 127:1436-1443

[17] Bollag WB, Zheng X. 2005 The role of phospholipase D in keratinocyte biology. In: Trends in Protein Research. JW Robinson, ed. Hauppauge, New York: *NovaScience Publishers*

[18] Zheng X, Bollag WB 2003 AngII induces transient phospholipase D activity in the H295R glomerulosa cell model. *Mol. Cell Endocrinol.* 206:113-122

[19] Bird IM, Hanley NA, Word RA, Mathis JM, McCarthy JL, Mason JI, Rainey WE 1993 Human NCI-H295 adrenocortical carcinoma cells: A model for angiotensin-II-responsive aldosterone secretion. *Endocrinology* 133:1555-1561

[20] Rainey WE, Saner K, Schimmer BP 2004 Adrenocortical cell lines. *Mol. Cell Endocrinol.* 228:23-38

[21] Natarajan R, Stern N, Hsueh W, Do Y, Nadler J 1988 Role of the lipoxygenase pathway in angiotensin II-mediated aldosterone biosynthesis in human adrenal glomerulosa cells. *J. Clin. Endocrinol. Metab.* 67:584-591

[22] Gu JL, Natarajan R, Ben-Ezra J, Valente G, Scott S, Yoshimoto T, Yamamoto S, Rossi JJ, Nadler JL 1994 Evidence that a leukocyte type of 12-lipoxygenase is expressed and regulated by angiotensin II in human adrenal glomerulosa cells. *Endocrinology* 134:70-77

[23] Bollag WB, Barrett PQ, Isales CM, Rasmussen H 1991 Angiotensin-II-induced changes in diacylglycerol levels and their potential role in modulating the steroidogenic response. *Endocrinology* 128:231-241

[24] Nogueira E, Vargas C, Otis M, Gallo-Payet N, Bollag WB, Rainey WE 2007 Profiling of angiotensin II rapid response genes in human, bovine, and rat adrenocortical cells. *J. Mol. Endocrinol.* 39:365-374

[25] Stocco DM 2001 Tracking the role of a star in the sky of the new millenium. *Mol. Endocrinol.* 15:1245-1254

[26] Lisurek M, Bernhardt R 2004 Modulation of aldosterone and cortisol synthesis on the molecular level. *Mol. Cell Endocrinol.* 215:149-159

[27] Crivello JF, Hornsby PJ, Gill GN 1982 Metyrapone and antioxidants are required to maintain aldosterone synthesis by cultured bovine adrenocortical zona glomerulosa cells. *Endocrinology* 111:469-479

[28] Williams GH, Hollenberg NK 1991 Functional derangements in the regulation of aldosterone secretion in hypertension. *Hypertension* 18[suppl. III]:143-149

[29] Kapas S, Purbrick A, Hinson JP 1995 Role of tyrosine kinase and protein kinase C in the steroidogenic actions of angiotensin II, α-melanocyte-stimulating hormone and corticotropin in the rat adrenal cortex. *Biochem. J.* 305:433-438

[30] Rainey WE, Bird IM, Mason JI 1994 The NCI-H295 cell line: a pluripotent model for human adrenocortical studies. *Mol. Cell Endocrinol.* 100:45-50

[31] Bird IM, Mathis JM, Mason JI, Rainey WE 1995 Ca^{2+}-regulated expression of steroid hydroxylases in H295R human adrenocortical cells. *Endocrinology* 136:5677-5684

[32] Hunyady L, Baukal AJ, Bor M, Ely JA, Catt KJ 1990 Regulation of 1,2-diacylglycerol production by angiotensin-II in bovine adrenal glomerulosa cells. *Endocrinology* 126:1001-1008

[33] Lobo MV, Mendoza RR, Marusic ET 1990 sn-1,2-dicotanoylglycerol mimics the effects of angiotensin II on aldosterone production and potassium permeability in isolated bovine glomerulosa cells. *J. Steroid. Biochem.* 35:29-33

[34] Kojima I, Kojima K, Rasmussen H 1985 Role of calcium fluxes in the sustained phase of angiotensin II-mediated aldosterone secretion from adrenal glomerulosa cells. J. Biol. Chem. 260:9177-9184

[35] Betancourt-Calle S, Bollag WB, Jung EM, Calle RA, Rasmussen H 1999 Effects of angiotensin II and ACTH on MARCKS phosphorylation in glomerulosa cells. *Mol. Cell. Endocrinol.* 154:1-9

[36] Lang U, Valloton MB 1987 Angiotensin II but not potassium induces subcellular redistribution of protein kinase C in bovine adrenal glomerulosa cells. *J. Biol. Chem.* 262:8047-8050

[37] Natarajan R, Lanting L, Xu L, Nadler J 1994 Role of specific isoforms of protein kinase C in angiotensin II and lipoxygenase action in rat adrenal glomerulosa cells. *Mol. Cell. Endocrinol.* 101:59-66

[38] Natarajan R, Gonzales N, Hornsby PJ, Nadler J 1992 Mechanism of angiotensin II-induced proliferation in bovine adrenocortical cells. *Endocrinology* 131:1174-1180

[39] Vinson GP, Laird SM, Whitehouse BJ, Hinson JP 1989 Specific effects of agonists of the calcium messenger system on secretion of 'late-pathway' steroid products by intact tissue and dispersed cells of the rat adrenal zona glomerulosa. *J. Mol. Endocrinol.* 2:157-165

[40] Kojima I, Kojima K, Kreutter D, Rasmussen H 1984 The temporal integration of the aldosterone secretory response to angiotensin occurs via two intracellular pathways. *J. Biol. Chem.* 259:14448-14457

[41] Kojima I, Lippes H, Kojima K, Rasmussen H 1983 Aldosterone secretion: effect of phorbol ester and A23187. *Biochem. Biophys. Res. Commun.* 116:555-562

[42] Barrett PQ, Kojima I, Kojima K, Zawalich K, Isales CM, Rasmussen H 1986 Temporal patterns of protein phosphorylation after angiotensin II, A23187 and/or 12-O-tetradecanoylphorbol 13-acetate in adrenal glomerulosa cells. *Biochem. J.* 238:893-903

[43] Cirillo M, Canessa M, Quinn S, Conlin PR 1998 Protein kinase C activation stimulates calcium transport in adrenal zona glomerulosa cells. *Biochem. Biophys. Res. Comm.* 245:466-471

[44] Hajnoczky G, Varnai P, Buday L, Farago A, Spat A 1992 The role of protein kinase-C in control of aldosterone production by rat adrenal glomerulosa cells: activation of protein kinase-C by stimulation with potassium. *Endocrinology* 130:2230-2236

[45] Ganguly A, Chiou S, Fineberg NS, Davis JS 1992 Greater importance of Ca(2+)-calmodulin in maintenance of ang II- and K(+)-mediated aldosterone secretion: lesser role of protein kinase C. *Biochem. Biophys. Res. Commun.* 182:254-261

[46] Griner EM, Kazanietz MG 2007 Protein kinase C and other diacylglycerol effectors in cancer. *Nat. Rev. Cancer* 7:281-294

[47] Bollag WB, Bollag RJ 2001 1,25-Dihydroxyvitamin D_3, phospholipase D and protein kinase C in keratinocyte differentiation. *Mol. Cell. Endocrinol.* 177:173-182

[48] MacDonald ML, Mack KF, Williams BW, King WC, Glomset JA 1988 A membrane-bound diacylglycerol kinase that selectively phosphorylates arachidonoyl-diacylglycerol. *J. Biol. Chem.* 263:1584-1592

[49] Nishizuka Y 1995 Protein kinase C and lipid signaling for sustained cellular responses. *FASEB J.* 9:484-496

[50] Bollag WB, Kent P, White S, Malinova M, Isales CM, Calle RA 2007 Characterization and phospholipase D mediation of the angiotensin II priming response in adrenal glomerulosa cells. *Endocrinology* 148:585-593

[51] LeHoux JG, Dupuis G, Lefebvre A 2001 Control of CYP11B2 gene expression through differential regulation of its promoter by atypical and conventional protein kinase C isoforms. *J. Biol. Chem.* 276:8021-8028

[52] Li J, Feltzer RE, Dawson KL, Hudson EA, Clark BJ 2003 Janus kinase 2 and calcium are required for angiotensin II-dependent activation of steroidogenic acute regulatory protein transcription in H295R human adrenocortical cells. *J. Biol. Chem.* 278:52355-52362

[53] Mazzochi G, Malendowicz L, Gottardo G, Rebuffat P, Nussdorfer GG 1997 Angiotensin-II stimulates DNA synthesis in rat adrenal zona glomerulosa cells: receptor subtypes involved and possible signal transduction mechanism. *Endocr. Res.* 23:191-203

[54] Clark AJ, Balla T, Jones MR, Catt KJ 1992 Stimulation of early gene expression by angiotensin II in bovine adrenal glomerulosa cells: roles of calcium and protein kinase C. *Mol. Endocrinol.* 6:1889-1898

[55] Toullec D, Pianetti P, Coste H, Bellevergue P, Grand-Perret T, Ajakane M, Baudet V, Boissin P, Boursier E, Loriolle F, Duhamel L, Charon D, Kirilovsky J 1991 The bisindolylmaleimide GF 109203X is a potent and selective inhibitor of protein kinase C. *J. Biol. Chem.* 266:15771-15781

[56] Rizzo MA, Shome K, Vasudevan C, Stolz DB, Sung T-C, Frohman MA, Watkins SC, Romero G 1999 Phospholipase D and its product, phosphatidic acid, mediate agonist-dependent Raf-1 translocation to the plasma membrane and the activation of the mitogen-activated protein kinase pathway. *J. Biol. Chem.* 274:1131-1139

[57] Rizzo MA, Shome K, Watkins SC, Romero G 2000 The recruitment of raf-1 to membranes is mediated by direct interaction with phosphatidic acid and is independent of association with ras. *J. Biol. Chem.* 275:23911-23918

[58] Tian Y, Smith RD, Balla T, Catt KJ 1998 Angiotensin II activates mitogen-activated protein kinase via protein kinase C and Ras/Raf-1 kinase in bovine adrenal glomerulosa cells. *Endocrinology* 139:1801-1809

[59] Casal AJ, Ryser S, Capponi AM, Wang-Buholzer CF 2007 Angiotensin II-induced mitogen-activated protein kinase phosphatase-1 expression in bovine adrenal glomerulosa cells: implications in mineralocorticoid biosynthesis. *Endocrinology* 148:5573-5581

[60] McNeill H, Whitworth E, Vinson GP, Hinson JP 2005 Distribution of extracellular signal-regulated protein kinases 1 and 2 in the rat adrenal and their activation by angiotensin II. 2005 187:149-157
[61] Otis M, Gallo-Payet N 2007 Role of MAPKs in angiotensin II-induced steroidogenesis in rat glomerulosa cells. *Mol. Cell Endocrinol.* 265-266:126-130
[62] Foster RH 2004 Reciprocal influences between the signalling pathways regulating proliferation and steroidogenesis in adrenal glomerulosa cells. *J. Mol. Endocrinol.* 32:893-902
[63] Cherradi N, Pardo B, Greenberg AS, Kraemer FB, Capponi AM 2003 Angiotensin II activates cholesterol ester hydrolase in bovine adrenal glomerulosa cells through phosphorylation mediated by p42/p44 mitogen-activated protein kinase. *Endocrinology* 144:4905-4915
[64] Luo JQ, Liu X, Hammond SM, Colley WC, Feiq LA, Frohman MA, Morris AJ, Foster DA 1997 RalA interacts directly with the Arf-responsive, PIP_2-dependent phospholipase D1. *Biochem. Biophys. Res. Comm.* 235:854-859
[65] Hammond SM, Jenco JM, Nakashima S, Cadwallader K, Gu Q-m, Cook S, Nozawa Y, Prestwich GD, Frohman MA, Morris AJ 1997 Characterization of two alternately spliced forms of phospholipase D1. *J. Biol. Chem.* 272:3860-3868
[66] Kim JH, Lee SD, Han JM, Lee TG, Kim Y, Park JB, Lambeth JD, Suh PG, Ryu SH 1998 Activation of phospholipase D1 by direct interaction with ADP-ribosylation factor 1 and RalA. *FEBS* Lett 430:231-235
[67] Frohman MA, Morris AJ 1996 Phospholipid signalling: Rho is only ARF the story. *Curr. Biol.* 6:945-947
[68] Hammond SM, Altshuller YM, Sung T-C, Rudge SA, Rose K, Engebrecht J, Morris AJ, Frohman MA 1995 Human ADP-ribosylation factor-activated phosphatidylcholine-specific phospholipase D defines a new and highly conserved gene family. *J. Biol. Chem.* 270:29640-29643
[69] Colley WC, Altshuller YM, Sue-Ling CK, Copeland NG, Gilbert DJ, Jenkins NA, Bollag RJ, Bollag WB, Frohman MA 1997 Cloning and expression analysis of murine phospholipase D1. *Biochem. J.* 326:745-753
[70] Du G, Altshuller YM, Kim Y, Han JM, Ryu SH, Morris AJ, Frohman MA 2000 Dual requirement for rho and protein kinase C in direct activation of phospholipase D1 through G-protein-coupled receptor signaling. *Mol. Biol. Cell.* 11:4359-4368
[71] Zhang Y, Altshuller YM, Hammond SM, Morris AJ, Frohman MA 1999 Loss of receptor regulation by a phospholipase D1 mutant unresponsive to protein kinase C. *EMBO J.* 18:6339-6348
[72] Singer WD, Brown HA, Jiang X, Sternweis PC 1996 Regulation of phospholipase D by protein kinase C is synergistic with ADP-ribosylation factor and independent of protein kinase activity. *J. Biol. Chem.* 271:4504-4510
[73] Conricode KM, Brewer KA, Exton JH 1992 Activation of phospholipase D by protein kinase C: evidence for a phosphorylation-independent mechanism. *J. Biol. Chem.* 267:7199-7202
[74] Han JM, Kim JH, Lee BD, Lee SD, Kim Y, Jung YW, Lee S, Cho W, Ohba M, Kuroki T, Suh PG, Ryu SH 2002 Phosphorylation-dependent regulation of phospholipase D2

by protein kinase C delta in rat pheochromocytoma PC12. *J. Biol. Chem.* 277:8290-8297

[75] Kim Y, Han JM, Han BR, Lee KA, Kim JH, Lee BD, Jang IH, Suh PG, Ryu SH 2000 Phospholipase D1 is phosphorylated and activated by protein kinase C in caveolin-enriched microdomains within the plasma membrane. *J. Biol. Chem.* 275:13621-13627

[76] Colley W, Sung T-C, Roll R, Jenco J, Hammond SM, Altshuller YM, Bar-Sagi D, Morris AJ, Frohman MA 1997 Phospholipase D2, a PLD1-related isoform with novel regulatory properties and discrete subcellular localization that provokes cytoskeletal reorganization. *Curr. Biol.* 7:191-201

[77] Lee S, Park JB, Kim JH, Kim Y, Kim JH, Shin KJ, Lee JS, Ha SH, Suh PG, Ryu SH 2001 Actin directly interacts with phospholipase D, inhibiting its activity. *J. Biol. Chem.* 276:28252-28260

[78] Shome K, Rizzo MA, Vasudevan C, Andresen B, Romero G 2000 The activation of phospholipase D by endothelin-1, angiotensin II, and platelet-derived growth factor in vascular smooth muscle A10 cells is mediated by small G proteins of the ADP-ribosylation factor family. *Endocrinology* 141:2200-2208

[79] Divecha N, Roefs M, Halstead JR, D'Andrea S, Fernandez-Borqa M, Oomen L, Saqib KM, Wakelam MJ, D'Santos C 2000 Interaction of the type Ialpha PIPkinase with phospholipase D: a role for the local generation of phosphatidylinositol 4,5-bisphosphate in the regulation of PLD2 activity. *EMBO J.* 19:5440-5449

[80] Bollag WB, Kent P, Isales CM, Parker PM, Wilson MV Manuscript submitted The role of calcium influx in phospholipase D activation in bovine adrenal glomerulosa cells. J. Lipid Res.

[81] Otis M, Gallo-Payet N 2006 Differential involvement of cytoskeleton and rho-guanosine 5'-triphosphatases in growth-promoting effects of angiotensin II in rat adrenal glomerulosa cells. *Endocrinology* 147:5460-5469

[82] Valverde AM, Sinnett-Smith J, Van Lint J, Rozengurt E 1994 Molecular cloning and characterization of protein kinase D: a target for diacylglycerol and phorbol esters with a distinctive catalytic domain. *Proc. Natl. Acad. Sci. U S A* 91:8572-8576

[83] Johannes FJ, Prestle J, Eis S, Oberhagemann P, Pfizenmaier K 1994 PKCμ is a novel, atypical member of the protein kinase C family. *J. Biol. Chem.* 269:6140-6148

[84] Bollag WB, Dodd ME, Shapiro BA 2004 Protein kinase D and keratinocyte proliferation. *Drug News Persp.* 17:117-126

[85] Waldron RT, Iglesias T, Rozengurt E 1999 Phosphorylation-dependent protein kinase D activation. *Electrophoresis* 20:382-390

[86] Van Lint JV, Sinnett-Smith J, Rozengurt E 1995 Expression and characterization of PKD, a phorbol ester and diacylglycerol-stimulated serine protein kinase. *J. Biol. Chem.* 270:1455-1461

[87] Matthews SA, Rozengurt E, Cantrell D 1999 Characterization of serine 916 as an in vivo autophosphorylation site for protein kinase D/Protein kinase Cmu. *J. Biol. Chem.* 274:26543-26549

[88] Waldron RT, Rey O, Iglesias T, Tugal T, Cantrell D, Rozengurt E 2001 Activation loop Ser744 and Ser748 in protein kinase D are transphosphorylated in vivo. *J. Biol. Chem.* 276:32606-32615

[89] Waldron RT, Rozengurt E 2003 Protein kinase C phosphorylates protein kinase D activation loop Ser744 and Ser748 and releases autoinhibition by the pleckstrin homology domain. *J. Biol. Chem.* 278:154-163

[90] Matthews S, Iglesias T, Cantrell D, Rozengurt E 1999 Dynamic re-distribution of protein kinase D (PKD) as revealed by a GFP-PKD fusion protein: dissociation from PKD activation. *FEBS Lett.* 457:515-521

[91] Chiu T, Rozengurt E 2001 PKD in intestinal epithelial cells: rapid activation by phorbol esters, LPA, and angiotensin through PKC. *Am. J. Physiol. Cell. Physiol.* 280:C929-942

[92] Zugaza JL, Waldron RT, Sinnett-Smith J, Rozengurt E 1997 Bombesin, vasopressin, endothelin, bradykinin, and platelet-derived growth factor rapidly activate protein kinase D through a protein kinase C-dependent signal transduction pathway. *J. Biol. Chem.* 272:23952-23960

[93] Zhukova E, Sinnett-Smith J, Rozengurt E 2001 Protein kinase D potentiates DNA synthesis and cell proliferation induced by bombesin, vasopressin, or phorbol esters in Swiss 3T3 cells. *J. Biol. Chem.* 276:40298-40305

[94] Tan M, Xu X, Ohba M, Ogawa W, Cui MZ 2003 Thrombin rapidly induces protein kinase D phosphorylation, and protein kinase C delta mediates the activation. *J. Biol. Chem.* 278:2824-2828

[95] Tan M, Xu X, Ohba M, Cui MZ 2004 Angiotensin II-induced protein kinase D activation is regulated by protein kinase Cdelta and mediated via the angiotensin II type 1 receptor in vascular smooth muscle cells. Arterioscler Thromb Vasc Biol 24:2271-2276

[96] Rozengurt E, Sinnett-Smith J, Van Lint J, Valverde AM 1995 Protein kinase D (PKD): a novel target for diacylglycerol and phorbol esters. *Mutat. Res.* 333:153-160

[97] Yuan J, Bae D, Cantrell D, Nel AE, Rozengurt E 2002 Protein kinase D is a downstream target of protein kinase Ctheta. *Biochem. Biophys. Res. Commun.* 291:444-452

[98] Storz P, Doppler H, Johannes FJ, Toker A 2003 Tyrosine phosphorylation of protein kinase D in the pleckstrin homology domain leads to activation. *J. Biol. Chem.* 278:17969-17976

[99] Reuben PM, Sun Y, Cheung HS 2004 Basic calcium phosphate crystals activate p44/42 MAPK signal transduction pathway via protein kinase Cmicro in human fibroblasts. *J. Biol. Chem.* 279:35719-35725

[100] Hausser A, Storz P, Hubner S, Braendlin I, Martinez-Moya M, Link G, Johannes FJ 2001 Protein kinase C mu selectively activates the mitogen-activated protein kinase (MAPK) p42 pathway. *FEBS Lett.* 492:39-44

[101] Brandlin I, Hubner S, Eiseler T, Martinez-Moya M, Horschinek A, Hausser A, Link G, Rupp S, Storz P, Pfizenmaier K, Johannes FJ 2002 Protein kinase C (PKC)eta-mediated PKC mu activation modulates ERK and JNK signal pathways. *J. Biol. Chem.* 277:6490-6496

[102] Shapiro BA, Ray S, Jung EM, Allred WT, Bollag WB 2002 Putative conventional protein kinase C inhibitor Gödecke 6976 [12-(2-cyanoethyl)-6,7,12,13-tetrahydro-13-methyl-5-oxo-5H-indolo(2,3-a)pyrrolo(3,4-c)-carbazole] stimulates transglutaminase

activity in primary mouse epidermal keratinocytes. *J. Pharmacol. Exp. Ther.* 302:352-358

[103] Rennecke J, Rehberger PA, Fürstenberger G, Johannes F-J, Stöhr M, Marks F, Richter KH 1999 Protein kinase-Cμ expression correlates with enhanced keratinocyte proliferation in normal and neoplastic mouse epidermis and in cell culture. *Int. J. Cancer* 80:98-103

[104] Ristich VL, Bowman PH, Dodd ME, Bollag WB 2006 Potein kinase D distribution in normal human epidermis, basal cell carcinoma and psoriasis. *Br. J. Dermatol.* 154:586-593

[105] Dodd ME, Ristich VL, Ray S, Lober RM, Bollag WB 2005 Regulation of protein kinase D during differentiation and proliferation of primary mouse keratinocytes. *J. Invest. Dermatol.* 125:294-306

[106] Diaz Anel AM 2007 Phospholipase C beta3 is a key component in the Gbetagamma/PKCeta/PKD-mediated regulation of trans-Golgi network to plasma membrane transport. *Biochem. J.* 10:10

[107] Jamora C, Yamanouye N, Van Lint J, Laudenslager J, Vandenheede JR, Faulkner DJ, Malhotra V 1999 Gbetagamma-mediated regulation of Golgi organization is through the direct activation of protein kinase D. *Cell* 98:59-68

[108] Liljedahl M, Maeda Y, Colanzi A, Ayala I, Van Lint J, Malhotra V 2001 Protein kinase D regulates the fission of cell surface destined transport carriers from the trans-Golgi network. *Cell* 104:409-420

[109] Rey O, Rozengurt E 2001 Protein kinase D interacts with Golgi via its cysteine-rich domain. *Biochem. Biophys. Res. Commun.* 287:21-26

[110] Maeda Y, Beznoussenko GV, Van Lint J, Mironov AA, Malhotra V 2001 Recruitment of protein kinase D to the trans-Golgi network via the first cysteine-rich domain. *Embo J.* 20:5982-5990

[111] Baron CL, Malhotra V 2002 Role of diacylglycerol in PKD recruitment to the TGN and protein transport to the plasma membrane. *Science* 295:325-328

[112] Yeaman C, Ayala MI, Wright JR, Bard F, Bossard C, Ang A, Maeda Y, Seufferlein T, Mellman I, Nelson WJ, Malhotra V 2004 Protein kinase D regulates basolateral membrane protein exit from trans-Golgi network. *Nat. Cell. Biol.* 6:106-112

[113] Diaz Anel AM, Malhotra V 2005 PKCeta is required for beta1gamma2/beta3gamma2- and PKD-mediated transport to the cell surface and the organization of the Golgi apparatus. *J. Cell. Biol.* 169:83-91

[114] Hausser A, Storz P, Martens S, Link G, Toker A, Pfizenmaier K 2005 Protein kinase D regulates vesicular transport by phosphorylating and activating phosphatidylinositol-4 kinase IIIbeta at the Golgi complex. *Nat. Cell. Biol.* 7:880-886

[115] Waldron RT, Rozengurt E 2000 Oxidative stress induces protein kinase D activation in intact cells. Involvement of Src and dependence on protein kinase C. *J. Biol. Chem.* 275:17114-17121

[116] Waldron RT, Rey O, Zhukova E, Rozengurt E 2004 Oxidative stress induces protein kinase C-mediated activation loop phosphorylation and nuclear redistribution of protein kinase D. *J. Biol. Chem.* 279:27482-27493

[117] Storz P, Toker A 2003 Protein kinase D mediates a stress-induced NF-kappaB activation and survival pathway. *Embo J.* 22:109-120

[118] Storz P, Doppler H, Toker A 2004 Protein kinase Cdelta selectively regulates protein kinase D-dependent activation of NF-kappaB in oxidative stress signaling. *Mol. Cell. Biol.* 24:2614-2626

[119] Storz P, Doppler H, Toker A 2004 Activation loop phosphorylation controls protein kinase D-dependent activation of nuclear factor kappaB. *Mol. Pharmacol.* 66:870-879

[120] Storz P, Doppler H, Toker A 2005 Protein kinase D mediates mitochondrion-to-nucleus signaling and detoxification from mitochondrial reactive oxygen species. *Mol. Cell. Biol.* 25:8520-8530

[121] Dequiedt F, Van Lint J, Lecomte E, Van Duppen V, Seufferlein T, Vandenheede JR, Wattiez R, Kettmann R 2005 Phosphorylation of histone deacetylase 7 by protein kinase D mediates T cell receptor-induced Nur77 expression and apoptosis. *J. Exp. Med.* 201:793-804

[122] Huynh QK, McKinsey TA 2006 Protein kinase D directly phosphorylates histone deacetylase 5 via a random sequential kinetic mechanism. *Arch. Biochem. Biophys.* 450:141-148

[123] Vega RB, Harrison BC, Meadows E, Roberts CR, Papst PJ, Olson EN, McKinsey TA 2004 Protein kinases C and D mediate agonist-dependent cardiac hypertrophy through nuclear export of histone deacetylase 5. *Mol. Cell. Biol.* 24:8374-8385

[124] Romero DG, Welsh BL, Gomez-Sanchez EP, Yanes LL, Rilli S, Gomez-Sanchez CE 2006 Angiotensin II-mediated protein kinase D activation stimulates aldosterone and cortisol secretion in H295R human adrenocortical cells. *Endocrinology* 147:6046-6055

[125] Chang HW, Chu TS, Huang HY, Chueh SC, Wu VC, Chen YM, Hsieh BS, Wu KD 2007 Down-regulation of D2 dopamine receptor and increased protein kinase Cmu phosphorylation in aldosterone-producing adenoma play roles in aldosterone overproduction. *J. Clin. Endocrinol. Metab.* 92:1863-1870

[126] Williams GH 2005 Aldosterone biosynthesis, regulation, and classical mechanism of action. *Heart Fail Rev.* 10:7-13

[127] Capponi AM 2004 The control by angiotensin II of cholesterol supply for aldosterone biosynthesis. *Mol. Cell Endocrinol.* 217:113-118

[128] Abedi H, Rozengurt E, Zachary I 1998 Rapid activation of the novel serine/threonine protein kinase, protein kinase D by phorbol esters, angiotensin II and PDGF-BB in vascular smooth muscle cells. *FEBS Lett.* 427:209-212

[129] Iwata M, Maturana A, Hoshijima M, Tatematsu K, Okajima T, Vandenheede JR, Van Lint J, Tanizawa K, Kuroda S 2005 PKCepsilon-PKD1 signaling complex at Z-discs plays a pivotal role in the cardiac hypertrophy induced by G-protein coupling receptor agonists. *Biochem. Biophys. Res. Commun.* 327:1105-1113

[130] Gupta S, Das B, Sen S 2007 Cardiac hypertrophy: mechanisms and therapeutic opportunities. *Antioxid Redox Signal* 9:623-652

[131] Shapiro BA, Parker PM, Kent P, Wilson MV, Arun S, Bollag WB Manuscript in preparation Protein kinase D reguates acute angiotensin II-induced aldosterone secretion from bovine adrenal glomerulosa cells.

[132] Bassett MH, Zhang Y, White PC, Rainey WE 2000 Regulation of human CYP11B2 and CYP11B1: comparing the role of the common CRE/Ad1 element. *Endocr. Res.* 26:941-951

[133] Clem BF, Hudson EA, Clark BJ 2005 Cyclic adenosine 3',5'-monophosphate (cAMP) enhances cAMP-responsive element binding (CREB) protein phosphorylation and phospho-CREB interaction with the mouse steroidogenic acute regulatory protein gene promoter. *Endocrinology* 146:1348-1356

[134] Groussin L, Massias JF, Bertagna X, Bertherat J 2000 Loss of expression of the ubiquitous transcription factor cAMP response element-binding protein (CREB) and compensatory overexpression of the activator CREMtau in the human adrenocortical cancer cell line H295R. *J. Clin. Endocrinol. Metab.* 85:345-354

[135] Johannessen M, Delghandi MP, Rykx A, Dragset M, Vandenheede JR, Van Lint J, Moens U 2007 Protein kinase D induces transcription through direct phosphorylation of the cAMP-response element-binding protein. *J. Biol. Chem.* 282:14777-14787

[136] Rozengurt E 2007 Mitogenic signaling pathways induced by G protein-coupled receptors. *J. Cell. Physiol.* 213:589-602

[137] Kuo MH, Allis CD 1998 Roles of histone acetyltransferases and deacetylases in gene regulation. *Bioessays* 20:615-626

[138] Murphy PJ, Morishima Y, Kovacs JJ, Yao TP, Pratt WB 2005 Regulation of the dynamics of hsp90 action on the glucocorticoid receptor by acetylation/deacetylation of the chaperone. *J. Biol. Chem.* 280:33792-22799

[139] Nishikawa K, Toker A, Johannes FJ, Songyang Z, Cantley LC 1997 Determination of the specific substrate sequence motifs of protein kinase C isozymes. *J. Biol. Chem.* 272:952-960

[140] Matthews SA, Liu P, Spitaler M, Olson EN, McKinsey TA, Cantrell DA, Scharenberg AM 2006 Essential role for protein kinase D family kinases in the regulation of class II histone deacetylases in B lymphocytes. *Mol. Cell Biol.* 26:1569-1577

[141] Irie A, Harada K, Tsukamoto H, Kim JR, Araki N, Nishimura Y 2006 Protein kinase D2 contributes to either IL-2 promoter regulation or induction of cell death upon TCR stimulation depending on its activity in Jurkat cells. *Int. Immunol.* 18:1737-1747

[142] Sinnett-Smith J, Zhukova E, Hsieh N, Jiang X, Rozengurt E 2004 Protein kinase D potentiates DNA synthesis induced by Gq-coupled receptors by increasing the duration of ERK signaling in swiss 3T3 cells. *J. Biol. Chem.* 279:16883-16893

[143] Wong C, Jin ZG 2005 Protein kinase C-dependent protein kinase D activation modulates ERK signal pathway and endothelial cell proliferation by vascular endothelial growth factor. *J. Biol. Chem.* 280:33262-33269

[144] Hurd C, Waldron RT, Rozengurt E 2002 Protein kinase D complexes with C-Jun N-terminal kinase via activation loop phosphorylation and phosphorylates the C-Jun N-terminus. *Oncogene* 21:2154-2160

[145] Waldron RT, Whitelegge JP, Faull KF, Rozengurt E 2007 Identification of a novel phosphorylation site in c-jun directly targeted in vitro by protein kinase D. *Biochem. Biophys. Res. Commun.* 356:361-367

[146] Hurd C, Rozengurt E 2001 Protein kinase D is sufficient to suppress EGF-induced c-Jun Ser 63 phosphorylation. *Biochem. Biophys. Res. Comm.* 282:404-408

[147] Manna PR, Stocco DM 2007 Crosstalk of CREB and Fos/Jun on a single cis-element: transcriptional repression of the steroidogenic acute regulatory protein gene. *J. Mol. Endocrinol.* 39:261-277

[148] Marrero MB, Paxton WG, Schieffer B, Ling BN, Bernstein KE 1996 Angiotensin II signaling events mediated by tyrosine phosphorylation. *Cell Signal* 8:21-26

[149] Sidorenko SP, Law CL, Klaus SJ, Chandran KA, Takata M, Kurosaki T, Clark EA 1996 Protein kinase C mu (PKC mu) associates with the B cell antigen receptor complex and regulates lymphocyte signaling. *Immunity* 5:353-363

[150] Qin L, Zeng H, Zhao D 2006 Requirement of protein kinase D tyrosine phosphorylation for VEGF-A165-induced angiogenesis through its interaciton and regulation of phospholipase Cγ phosphorylation. *J. Biol. Chem.* 281:32550-32558

[151] Li N, Karin M 1999 Is NF-kappaB the sensor of oxidative stress? *FASEB J.* 13:1137-1143

[152] Touyz RM 2004 Reactive oxygen species and angiotensin II signaling in vascular cells -- implications in cardiovascular disease. *Braz. J. Med. Biol. Res.* 37:1263-1273

[153] Touyz RM, Tabet F, Schiffrin EL 2003 Redox-dependent signalling by angiotensin II and vascular remodelling in hypertension. *Clin. Exp. Pharmacol. Physiol.* 30:860-866

[154] Brasier AR, Jamaluddin M, Han Y, Patterson C, Runge MS 2000 Angiotensin II induces gene transcription through cell-type-dependent effects on the nuclear factor-kappaB (NF-kappaB) transcription factor. *Mol. Cell. Biochem.* 212:155-169

[155] Wolf G 2005 Role of reactive oxygen species in angiotensin II-mediated renal growth, differentiation, and apoptosis. *Antioxid Redox Signal* 7:1337-1345

[156] Song J, Li J, Lulla A, Evers BM, Chung DH 2006 Protein kinase D protects against oxidative stress-induced intestinal epithelial cell injury via Rho/ROK/PKC-delta pathway activation. *Am. J. Physiol. Cell Physiol.* 290:C1469-1476

[157] Tergaonkar V 2006 NFkappaB pathway: a good signaling paradigm and therapeutic target. *Int. J. Biochem. Cell Biol.* 38:1647-1653

[158] Sachse A, Wolf G 2007 Angiotensin II-induced reactive oxygen species and the kidney. *J. Am. Soc. Nephrol.* 18:2439-2446

[159] De Matteis M, Godi A, Corda D 2002 Phosphoinositides and the golgi complex. *Curr. Opin. Cell Biol.* 14:434-447

[160] Lu G, Chen J, Espinoza LA, Garfield S, Toshiyuki S, Akiko H, Huppler A, Wang QJ 2007 Protein kinase D 3 is localized in vesicular structures and interacts with vesicle-associated membrane protein 2. *Cell Signal* 19:867-879

[161] Liscovitch M, Czarney M, Fiucci G, Tang X 2000 Phospholipase D: molecular and cell biology of a novel gene family. *Biochem. J.* 345:401-415

[162] Touyz RM, Berry C 2002 Recent advances in angiotensin II signaling. *Braz. J. Med. Biol. Res.* 35:1001-1015

[163] Kojima I, Kojima K, Rasmussen H 1985 Possible role of phospholipase A_2 action and arachidonic acid in angiotensin II-mediated aldosterone secretion. *Endocrinology* 117:1057-1066

[164] Yokozeki T, Homma K, Kuroda S, Kikkawa U, Ohno S, Takahashi M, Imahori K, Kanaho Y 1998 Phosphatidic acid-dependent phosphorylation of a 29-kDa protein by protein kinase Ca in bovine brain cytosol. *J. Neurochem.* 71:410-417

[165] Betancourt-Calle S, Calle RA, Isales CM, White S, Rasmussen H, Bollag WB 2001 Differential effects of agonists of aldosterone secretion on StAR phosphorylation. *Mol. Cell Endocrinol.* 173:87-94

[166] Wood CD, Marklund U, Cantrell DA 2004 Dual phospholipase C/diacylglycerol requirement for protein kinase D1 activation in lymphocytes. *J. Biol. Chem.* 280:6245-6251

In: Angiotensin Research Progress
Editors: H. Miura and Y. Sasaki
ISBN: 978-1-60456-537-9

Chapter II

Angiotensin-Nitric Oxide Interactions

Wyatt McMahon, Oscar Iznaola and Sharma Prabhakar [1*]
[1]Division of Nephrology and Hypertension
Texas Tech University Health Sciences Center
3601 4th Street, Suite 4C-178
Lubbock, TX 79430

Abstract

Angiotensin II, a product of activation of renin-angiotensin axis, is a powerful vasoconstrictor and in addition is a mediator of many physiological and pathophysiological non-hemodynamic effects in multiple organ systems including cardiovascular, renal and neuronal tissues. Angiotensin activation has been implicated in renal injury in diabetic and non-diabetic conditions. On the other hand, nitric oxide is a product derived solely from L-arginine under the effects of nitric oxide synthase, is a vasodilator and has multiple non-hemodynamic effects in both physiological and pathological conditions in many organs. Since many effects of angiotensin are counter-regulatory to the effects of nitric oxide, a delicate balance between these two signaling pathways is necessary for the proper function of many organ systems. Therefore, an imbalance between the two systems may also underlie the pathophysiology of many cardiovascular-renal syndromes including diabetic nephropathy. The effects of angiotensin as well as angiotensin blockers on renal NO synthesis has been the subject of several investigations. In all these studies it was argued that the enhanced renal NO production is a defense mechanism against the detrimental effects of Ang II. In addition, Ang II has been shown to stimulate NO production in coronary and other microvascular endothelial cells as well as aortic endothelial cells through eNOS activity. Ang II increased the expression and activity of ca^{++} dependent NOS (NOS I and III) in the cortex with a decrease in NOS II activity in the cortex. Similar studies examining the acute in vivo effects of Ang II infusion on renal NO production and on the NOS isoforms in the cortex and medulla were performed in our laboratory. These studies demonstrated that angiotensin II inhibited iNOS, an effect reversed by losartan, an AT_1 receptor

[*] Tel : (806)743-3155; Fax: (806)743-3148; e-mail:sharma.prabhakar@ttuhsc.edu

blocker and amplified by PD123319, an AT_2 receptor blocker. The losartan effect was associated with augmented NO synthesis and increased expression of AT_2 receptors as demonstrated by immunocytochemistry. NO derived from cNOS and iNOS did not affect the basal secretion of insulin but augmented high glucose (HG) stimulated insulin secretion in pancreatic beta cells. Furthermore, studies demonstrated the influence of Ang II on insulin secretion and that angiotensin blockade with ACE inhibitors and AT_1 receptor blockers enhanced insulin release from pancreatic β cells by augmenting tissue NO production. Ang II stimulated renal NO synthesis by augmenting the Ca^{++} dependent NOS activity and eNOS expression in the kidney, which under basal conditions surpassed the iNOS inhibition by Ang II. We also demonstrated in in vivo studies that despite the HG inhibition of eNOS, the intrarenal NO generation as well as eNOS activity remained stimulated in the presence of systemic Ang II. The functional interaction and reciprocal regulation of angiotensin and nitric oxide pathways in basal and pathologic states is the focus of discussion in this review, although a significant part of the discussion is centered on diabetic nephropathy.

Physiological Considerations

Renin Angiotensin Aldosterone - Historical Aspects

In 1889 Robert Tigerstedt and his student Per Bergman published a landmark paper: "Niere und Kreistlauf" (Kidney and circulation). They documented long-lasting pressor effects from renal cortex extracts, and named the substance renin. For approximately 40 years little interest was given to renin until the early 1930's. By 1938 the idea that renin was not a constrictor, but rather acts as an enzyme to produce a pressor substance, was proposed. Goormaghtigh speculated in 1937 that the juxtaglomerular cells of the afferent arterioles were the source of renin and that these cells play a role in linking tubular and vascular functions of the kidney. In 1940 what is now known as angiotensin received the names of hypertensin or angiotonin, and for the next 18 years both terms were used. In 1958, it was agreed to name the pressor substance *angiotensin*; by extension the substrate from which angiotensin is released was called *angiotensinogen.*

In 1956, it was discovered the angiotensin converting enzyme (ACE) and it was published the amino acid sequence of the octapeptide Ang II. Gross and his colleagues began to investigate the physiological relationships between adrenal hormones and the RAS in Germany. In 1958, it was found that changes in Na+ balance caused parallel changes in the RAS and aldosterone secretion and proposing that Ang II was an important regulator of aldosterone secretion. In 1961 it was demonstrated that Ang II administration stimulated aldosterone secretion in dogs. These observations were the starting point for investigations of the physiology of the renin–angiotensin–aldosterone axis and the concept that blood pressure control by Angiotensin II is closely interrelated with its renal sodium-retaining actions. [1,2,3]

Physiology of RAAS

The renin-angiotensin-aldosterone system (RAAS) is a coordinated hormonal cascade that governs cardiovascular, renal, and adrenal functions by regulating fluid and electrolyte balance as well as arterial pressure. Traditionally it has been thought that the cascade starts with renin release, primarily by juxtaglomerular cells but also by other tissues. Renin is a step-limiting enzyme whose synthesis is influenced by the hydrostatic pressure sensed at the glomerular afferent arterioles, angiotensin II levels, and the quantity of sodium delivered to the macula densa. Plasma K+ concentration and levels of both atrial natriuretic peptide and endothelin also regulate renin synthesis.

Renin acts to cleave angiotensinogen synthezised by the liver, forming angiotensin I. Angiotensin-converting enzyme (ACE) then converts angiotensin I to angiotensin II. Recent findings in the last two decades have changed this traditional concept, and expanded our understanding of the RAAS system involving the discoveries of Angiotensin II receptors, other subtypes of angiotensins, angiotensin converting enzyme II (ACE2), and most recently the prorenin and renin receptors. (4,5)

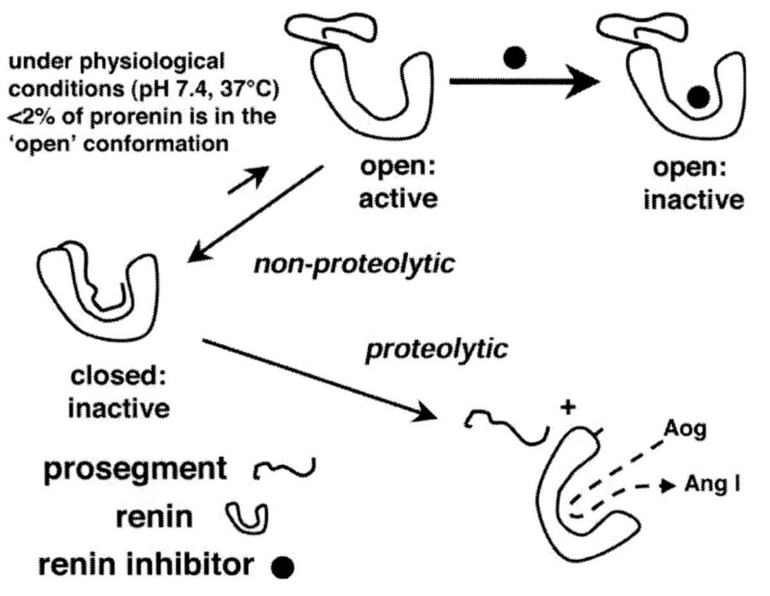

Figure 1. Prorenin contains a propetide that covers the enzymatic cleft and obstructs the access of angiotensinogen to the active site of renin. Inactive prorenin can be activated under physiologic conditions by proteolytic and non-proteolytic pathways making the 43-amino acid N-terminal propeptide to uncover the enzymatic cleft and convert angiotensinogen to angiotensin I. The renin inhibitor Aliskiren competes with angiotensinogen at this site and inhibits the formation of angiotensin I. [5]

Prorenin

By several observations in the early 70's it was postulated the existence of prorenin as an activable "big rennin" (5 kDa higher than that of renin). After the cloning of renin in 1984, prorenin was definitely proved to be the precursor of renin. Prorenin contains a 43-amino acid N-terminal propeptide that inhibit the enzymatic activity of prorenin. This propeptide covers the enzymatic cleft and obstructs access of angiotensinogen to the active site of renin. It is now known that Prorenin can be activated by proteolytic or non-proteolitic ways (see Figure 1).

Proteolytic activation in vitro can be observed by the removal of the propeptide with kallikrein, whereas in vivo enzymatic activation is not clear, however various renal processing enzymes have been proposed including proconvertase I and cathepsin B.

Prorenin does not proteolytically self-activate like pepsinogen, and its blood circulating level is 10 times higher than that of mature renin. Some investigators have recently proposed that prorenin is a useful marker of diabetic microvascular complications in which the prorenin/renin ratio is increased out of proportion (see Figure 2).

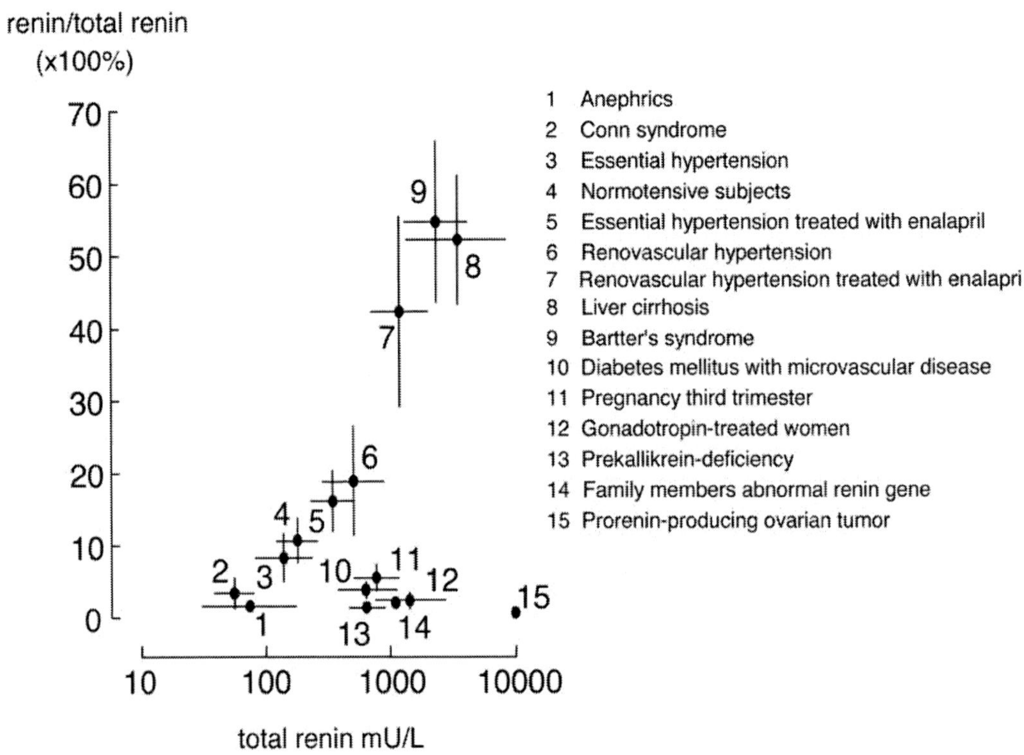

Figure 2. Relationship between the renin/total renin ratio (ordinate, mean_SD) and total renin (renin_prorenin) (abscissa, geometric mean, 95% CI) in various clinical conditions. Data from Deinum and Schalekamp. (5)

Non proteolytic activation of prorenin can be demonstrated in vitro by acidic PH or low temperatures. Recently advanced research in pro-renin has shown in vivo that when a prorenin-binding protein interacts with the "handle region" of the prorenin prosegment, the prorenin molecule undergoes a conformational change to an enzymatically active state. [6,7]

(Pro) Renin receptors

(Pro)renin receptors have been localized to glomerular mesangium and vascular smooth muscle cells within the subendothelium of glomerular and coronary arteries. The binding of renin to its receptor with a single transmembrane domain has multiple and far-reaching consequences. Receptor binding induces a 4-fold increase in the catalytic conversion of angiotensinogen to Ang I, suggesting that the cell surface is an important site of Angiotensin generation. [8] Currently, 2 (pro)renin receptors have been identified, and the existence of a third receptor has been proposed. [9,10,11] The cloned renin receptor was found to bind prorenin equally well (ie, renin's active site is not involved in the binding process), and in contrast to the described receptors, cell surface-bound renin and prorenin were neither internalized nor degraded. Importantly, binding of renin to this receptor induced a 4-fold increase of the catalytic efficiency of angiotensinogen conversion to Ang I, and receptor-bound prorenin became fully enzymatically active in a nonproteolytic manner. These data support angiotensin generation on the cell surface, allowing Ang II to bind immediately to AT_1 receptors after its synthesis, without leaking into the extracellular space. Furthermore, in the presence of the AT_1 receptor antagonist losartan, (pro)renin binding to the (pro)renin receptor resulted in rapid activation of the MAP kinases ERK1 (p44)/ERK2 (p42), thereby demonstrating for the first time Ang II-independent effects of renin and prorenin. Immunohistochemistry and in situ hybridization studies have localized the receptor in vascular smooth muscle cells in human heart and kidney, in glomerular mesangial cells and in distal and collecting tubular cells in the kidney.

Based on experiments with a series of antibodies directed against various parts of the prosegment, Suzuki et al. [6], recently proposed that human prorenin has so-called gate and handle regions for its nonproteolytic activation. According to this concept, the handle region (I11PFLKR15P) interacts with a putative receptor, which then leads to dissociation of the gate region T7PFKR10P from the renin molecule. Because this gate region is crucial for refolding and the maintenance of the inactive state, dissociation allows prorenin to display enzymatic activity.

In a subsequent in vivo study, these investigators applied a decoy peptide corresponding to the handle region to block nonproteolytic prorenin activation. This peptide reduced the renal content of Ang I and II and fully prevented the development of diabetic nephropathy in streptozotocin-induced diabetic rats. Interestingly, there were no effects on the plasma levels of Ang I and II, nor did the decoy peptide affect the tissue levels of Ang I and II in control rats. [8]

Thus, these data are the first to confirm that endogenous prorenin contributes to tissue Ang I and II generation in diabetic animals via a mechanism involving binding of its handle region to a receptor. It is tempting to speculate that this receptor is the above-mentioned (pro)renin receptor, but this remains to be proven. An explanation should also be provided for the lack of prorenin-dependent (renal) Ang I generation in nondiabetic animals.

The mannose-6-phosphate receptor also binds renin and prorenin with high affinity in human endothelial cells. This receptor is identical to the insulin like growth factor II (IGFII) receptor, and as such it contains binding domains for both IGFII and phosphomannosylated (M6P-containing) proteins like renin and prorenin. It does not bind nonglycosylated (pro)renin. After binding, both renin and prorenin are rapidly (within minutes) internalized, and internalized prorenin is proteolytically cleaved to renin. (Pro)renin binding to M6P/IGFII receptors do not result in extracellular or intracellular angiotensin generation and intracellular renin was foud to be degraded slowly. Thus, M6P/IGFII receptors most likely serve as clearance receptors for both renin and prorenin, thereby determining the extracellular levels of (pro)renin. Alternatively, because binding of M6Pcontaining proteins to M6P/IGFII receptors results in activation of second messenger pathways in a G-protein– dependent manner, it is possible that renin and prorenin act as agonists for this receptor. [12, 13]. In addition, several "(pro)renin-binding proteins"[(P)RnBP] have been investigated, either in membranes prepared from rat tissues or in intracellular compartments. Of these (P)RnBPs, only the intracellular RnBP has been cloned and characterized. [14] Mice lacking RnBP display normal blood pressure and plasma renin activity. Therefore, it is unlikely that this intracellular RnBP is a determinant of renin activity and/or metabolism in vivo. [15]

Enzymatic stimuli for angiotensin II formation

The most widely known enzyme that is capable of AngII formation is ACE, but it is not the only one. Chymase is another AngII generating enzyme, which mediates more than 80% of AngII formation in the heart and more than 60% in the vessels. ACE inhibitors do not reduce chymase activity. Upregulation of chymase, mainly in the tubules, is observed in renal biopsies of patients with diabetic nephropathy. These findings indicate that under pathologic conditions, an upregulation of chymase occurs and increased local AngII generation cannot be attenuated by ACE inhibitors. Mechanical stress of podocytes stimulates local AngII synthesis by non-ACE pathways that presumably involve chymase, yet the exact role of chymase-mediated AngII formation for renal disease in humans is unclear and ACE inhibitors clearly slow progression of disease. [16, 17, 18]

ACE-2 and other types of angiotensin

The way we think about the RAS in health and disease changed with the discovery, of ACE2, an angiotensin-converting enzyme (ACE)–related carboxypeptidase which has approximately 40% homology with ACE. [19] Previously, some of the so-called "breakdown products" of angiotensin I and II had unknown functions, and many in the field felt that these had little or no physiologic importance. ACE2, which preferentially removes carboxyterminal basic or hydrophobic acids, has as a major function, the formation of the inactive angiotensin 1-9 from Ang I and the formation of angiotensin 1-7 that has vasodilatory and antiproliferative effects. In the six years since its discovery, it would appear that an overarching function of ACE2 is to counterbalance the effects of ACE, which forms Ang II (see Figure 3).

ACE2 is not responsive to ACE inhibitors, although it is not known what stimulating ACE2 or inhibiting will do clinically. The role of ACE2 in hypertension, diabetes, renal disease, and a number of other conditions is currently in the early phases of exploration. ACE2 is primarily localized to podocyte foot processes and the parietal cells of Bowman's capsule, whereas ACE is present in endothelial cells. ACE2 may regulate the levels of Ang II within the glomerulus by degrading it to Ang 1-7, a suggestion supported by the known actions of the enzyme. Ang 1-7 has recently been shown to inhibit Ang II–stimulated phosphorylation of MAP kinases in the proximal renal tubular cells, which is protective. Ang 1-7 also increases NO and prostacycline release and potentiates the effects of bradykinin. [21]

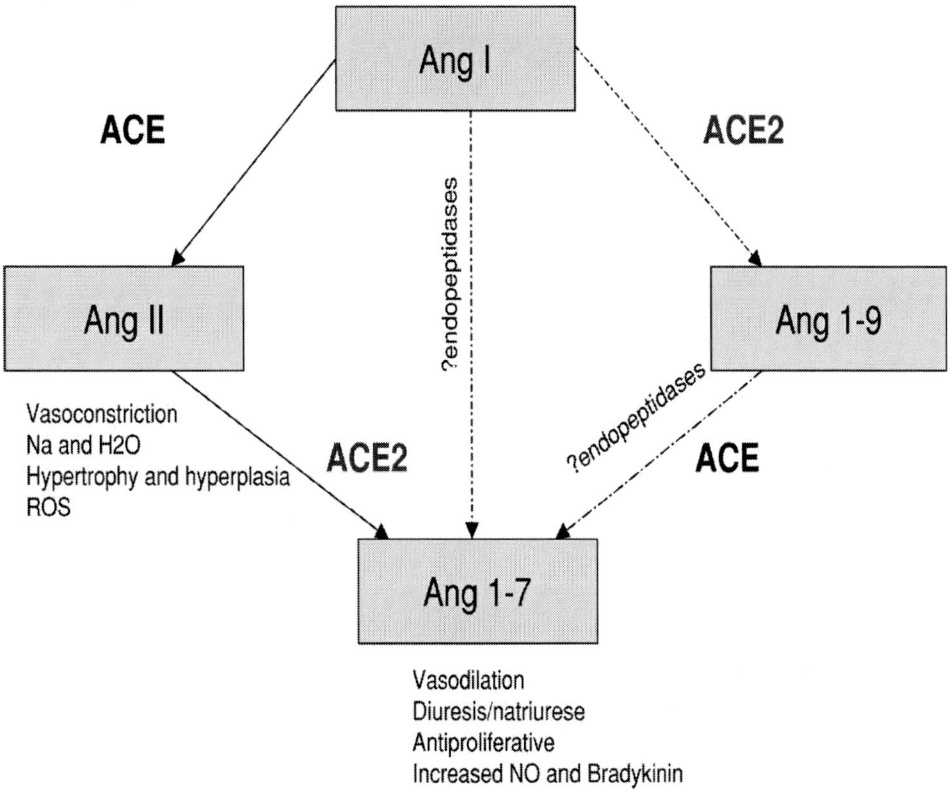

Figure 3. This diagram indicates the interplay of angiotensin-converting enzyme (ACE) and ACE2. As can be seen, ACE2 produces angiotensin (Ang) 1-7, which counterbalances many effects of Ang II. [19]

Pathways of angiotensin II degradation

AngII is metabolized by peptidases, such as aminopeptidase A, into AngIII and further into Ang IV. Ang III interacts with Ang II type 1 (AT_1) and AT_2 receptors, albeit with a lower affinity, and exhibits principally similar effects as AngII with 40% less pressor activity of Angiotensin II, but 100% of the aldosterone-producing activity. AngIV binds to a specific receptor called AT_4, which is widely expressed in the kidney, including endothelial cells and proximal as well as convoluted tubules (see Figure 4). [22]

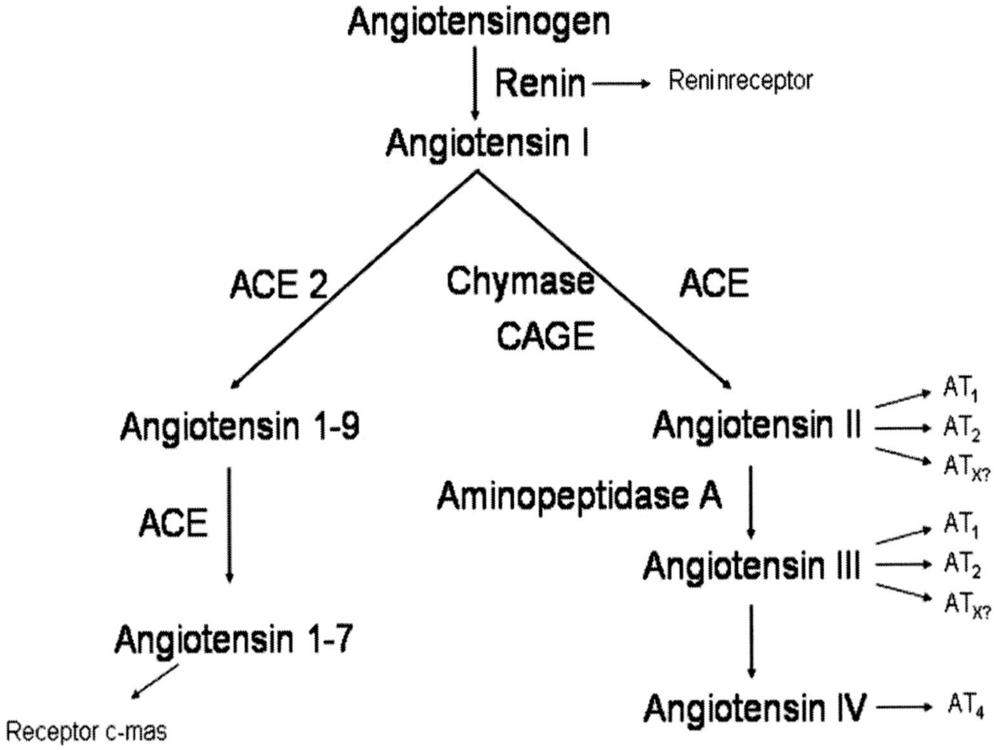

Figure 4. Overview of the renin-angiotensin-aldosterone system (RAAS). The system has become increasingly complex with alternative ways of angiotensin II (AngII) formation besides angiotensin-converting enzyme (ACE; (Chymase, chymostatinsensitive AngII-generating enzyme [CAGE]), a second form of ACE (ACE2), and novel peptides such as AngIV, angiotensin 1-9, and angiotensin 1-7. Clinically important could be that AngII can bind to AngII type 2 (AT2) receptors and AngIV to AT4 receptors that are not antagonized by sartanes, inducing proinflammatory and profibrotic effects (e.g., induction of chemokines, stimulation of plasminogen activator inhibitor-1). [16])

Angiotensin II Receptors

The multiple effects of AngII are mediated by different receptors. The two major AngII receptors, AT_1 and AT_2, are differentially expressed within in the kidney. Both share approximately 30% homology on the protein level. Also their genes reside on different chromosomes; the gene for AT1 is located on chromosome 3 and that for AT_2 is found on the X chromosome. As a result, they are functionally distinct and employ different signal transduction pathways. [22, 23]. AT_1 receptors are distributed in the vasculature, kidney, adrenal gland, heart, liver, and brain. These receptors are coupled to G proteins and mediate different, mainly second-messenger signal transduction pathways, such as activation of phospholipases, inhibition of adenylate cyclase, stimulation of tyrosine phosphorylation, extracellular signal–regulated kinases 1 and 2, the phosphatidylinositol 3-kinase–dependent kinase Akt, and the mammalian target of rapamycin/S6 kinase pathway.

AT_2 receptors in healthy adults, is present in the adrenal medulla, uterus, ovary, vascular endothelium, and distinct brain areas. The AT_2 receptor is also widely distributed in the fetus. AT_2 receptors are involved in an increase in intracellular protein phosphatase activity. The number of AT_1 and AT_2 receptors is developmentally regulated, and during maturation of the kidney, AT_1 receptor expression becomes more abundant. AT_2 receptors are not suppressed by AngII, but, interestingly, they are upregulated in injured tissue and during inflammation. In particular, AT_2 receptors are re-expressed in the kidney during renal injury and remodeling nephrons.

Almost all AngII-induced physiologic and pathophysiologic functions, such as vasoconstriction, aldosterone release, stimulation of tubular transport, proinflammatory effects, and profibrogenic and growth stimulatory actions, are mediated by AT_1 receptors. The role of AT_2 receptors seems less clear. Activation of AT2 receptors leads to a decrease in BP through release of nitric oxide, inhibits growth and induces differentiation, and also is involved in mediation of apoptosis (see Figure 5). [4, 17, 22, 23]

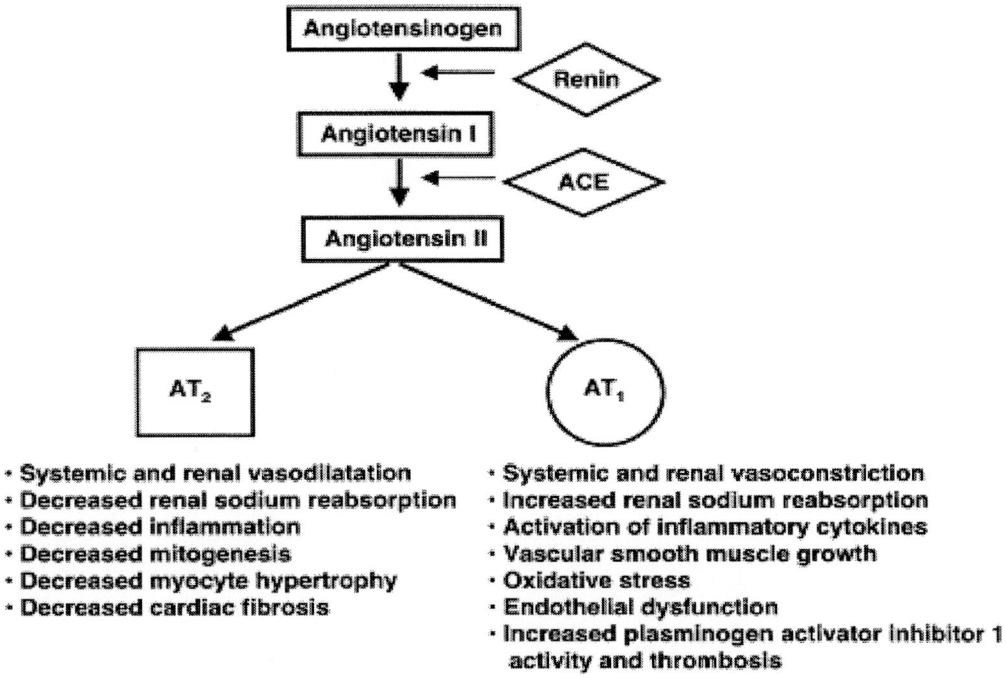

Figure 5. Properties of the angiotensin II receptor subtypes. ACE / angiotensin-converting enzyme; AT_1 /angiotensin II type 1 receptor; AT_2 / angiotensin II type 2 receptor. [4]

Tissue RAAS

During the past two decades, local RAAS have been described to operate independent from their systemic counterpart. A local RAAS including all its components could have been shown in the proximal tubular cells of the kidney. Proximal tubular cells actively produce AngII and also secrete angiotensinogen into the urine. Intraluminal angiotensinogen may be converted in the distal tubules to AngII, and recent observations suggest that it leads to induction of sodium channels independent of aldosterone. More recently, the discoveries of (pro)renin receptors and its interactions with prorenin/renin have clarified the understanding of the activation of the tissue renin-angiotensin system. Under pathologic circumstances hyperglycemia and proteinuria could stimulate local AngII synthesis, mainly by oxygen species as signal transducers. Renal injury activates the local RAAS directly and indirectly. For example, a reduction in calcitriol stimulates renin transcription accompanied by local increase of AngII, demonstrating an indirect activation.

Of clinical relevance is the observation that complete systemic inhibition of the AngII formation by an ACE inhibitor is not accompanied by a significantly reduced intrarenal AngII production. Intrarenal AngII is found regionally compartmentalized. Intact AngII is found in endosomes that are derived from receptor-mediated endocytosis. This might be an important mechanism, because observations in certain cells demonstrated that AngII can be translocated into the nucleus, where it directly regulates the gene transcription. [16, 24, 25, 26, 27]

L-Arginine -Nitric Oxide Pathway

The discovery that mammalian cells generate nitric oxide, a gas previously considered to be merely an atmospheric pollutant, has provided important information about many biologic processes. Nitric oxide is synthesized from the amino acid l-arginine by a family of enzymes, the nitric oxide synthases, through the l-arginine-nitric oxide pathway. The synthesis of nitric oxide by vascular endothelium is responsible for the vasodilator tone that is essential for the regulation of blood pressure (see Figure 6).

Nitric oxide synthase (NOS), expressed in three isoforms. The NOS isoforms were named upon the cell types in which they were first isolated: neuronal NOS (nNOS or NOS1); inducible or macrophage NOS (iNOS, NOS2); and endothelial NOS (eNOS, NOS3). These three enzymes, which are cytochrome p450-like proteins, produce NO and water. [29

]NO is a paracrine mediator, produced and released by individual cells, penetrates the biological membranes of neighboring cells, modulating a number of signaling cascades. Since it has an extremely short half-life, it exerts its effects locally and transiently.

Expression of Nitric Oxide Synthases in the Kidney

nNOS is principally expressed in the macula densa and the inner medullary collecting duct. iNOS has been localized to several tubule segments, like thick medullary ascending limb, the distal convoluted tubule and proximal tubule; it is also expressed in the glomerulus, and the interlobular and arcuate arteries. eNOS is expressed in the endothelium of glomerulus, afferent and efferent arterioles, and intrarenal arteries. The greatest enzymatic activity for NO production in the kidney is found in the inner medullary collecting duct (IMCD). IMCD NOS activity is three to six fold higher than that observed in the glomeruli (see Figure 7). [30]

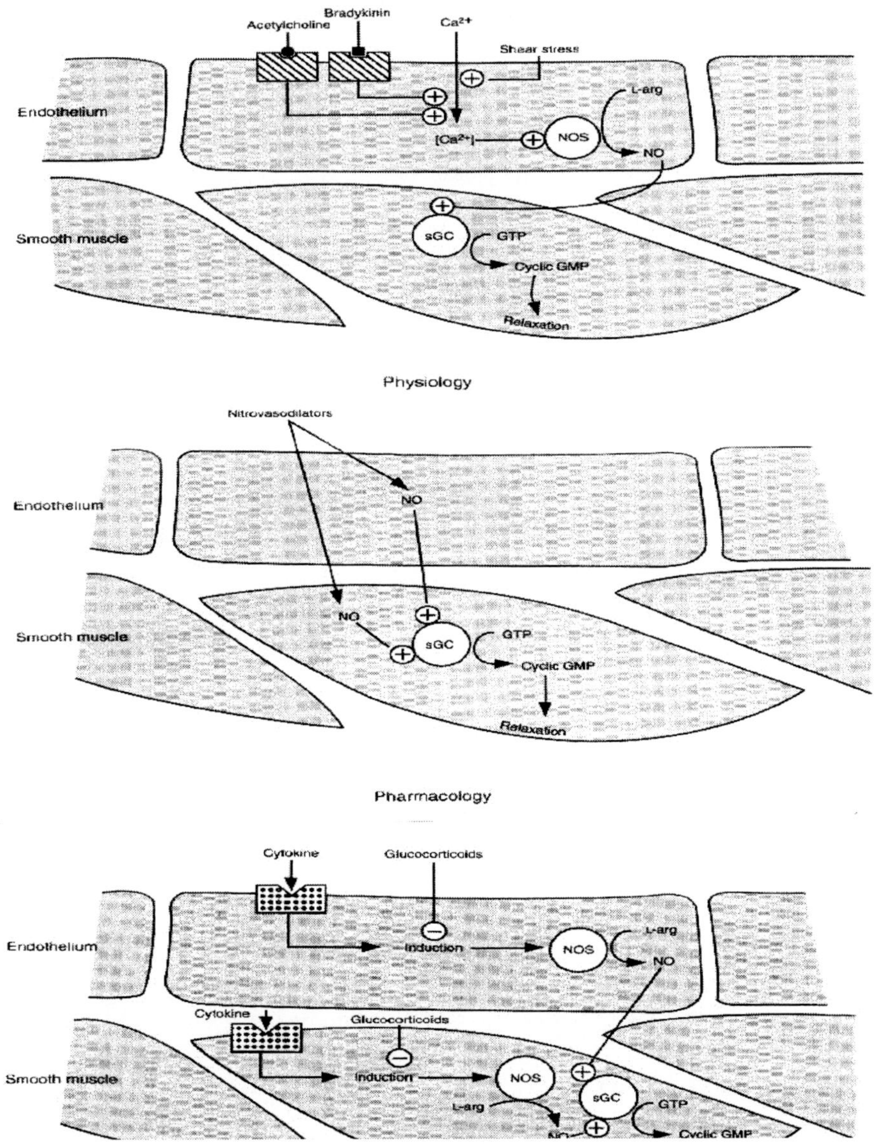

Figure 6. In Panel A, shear stress or receptor activation of vascular endothelium by bradykinin or acetylcholine results in an influx of calcium. The consequent increase in intracellular calcium stimulates the constitutive nitric oxide synthase (NOS). The nitric oxide (NO) formed from l-arginine (l-arg) by this enzyme diffuses to nearby smooth-muscle cells, in which it stimulates the soluble guanylate cyclase (sGC), resulting in enhanced synthesis of cyclic GMP from guanosine triphosphate (GTP). This increase in cyclic GMP in the smooth-muscle cells leads to their relaxation. In Panel B, nitrovasodilators such as sodium nitroprusside and nitroglycerin release nitric oxide, spontaneously or through an enzymatic reaction. The liberated nitric oxide stimulates the soluble guanylate cyclase in the vascular smooth-muscle cell, resulting in relaxation. In Panel C, the interaction of cytokines with their receptors on endothelial and smooth-muscle cells results in the induction of calcium-independent nitric oxide synthase. This induction is inhibited by glucocorticoids. Once induced, the nitric oxide synthase produces nitric oxide continuously, resulting in sustained activation of soluble guanylate cyclase and leading to prolonged smooth-muscle relaxation, reduced responsiveness to vasoconstrictors, and possibly tissue damage. The soluble guanylate cyclase in the endothelial cell is also stimulated by nitric oxide (not shown), but the importance of this is not clear at present. Plus signs indicate stimulation, and minus signs inhibition. [28]

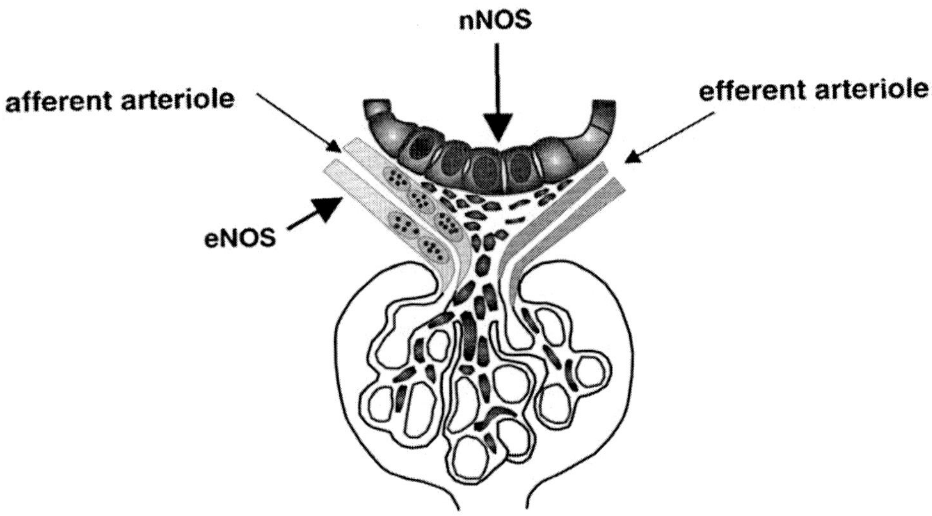

Figure 7. Renal juxtaglomerular granular cells are encircled by endothelial, mesangial, and tubular macula densa cells, which have a high capacity for nitric oxide (NO) formation. eNOS, NO synthase type III; nNOS, NO synthase type I. [32]

nNOS is principally expressed in the macula densa and the inner medullary collecting duct. iNOS has been localized to several tubule segments, like thick medullary ascending limb, the distal convoluted tubule and proximal tubule; it is also expressed in the glomerulus, and the interlobular and arcuate arteries.

eNOS is expressed in the endothelium of glomerulus, afferent and efferent arterioles, and intrarenal arteries. The greatest enzymatic activity for NO production in the kidney is found in the inner medullary collecting duct (IMCD). IMCD NOS activity is three to six fold higher than that observed in the glomeruli. [31, 32]

It is well known that NO plays an important role in the regulation of renal blood flow, glomerular hemodynamics, and tubular function. The close association between MD cells and vascular components of the juxtaglomerular apparatus (JGA) form the anatomic basis for the physiological response observed during flow rate changes in the thick ascending limb (TAL) of the loop of Henle, which implies changes of NaCl load at the MD. MD cells are thought to sense these changes and to mediate a tubulovascular information transfer that serves to control glomerular filtration rate via the tubuloglomerular feedback response (TGF) and to adjust renin gene expression and renin secretion.

The mediator of this dual effect of the MD signal is not clear at present. Among numerous locally acting paracrine mediators, NO is currently considered an important candidate. NO at the JGA was recently reported to play an important role by affecting glomerular capillary pressure and by reducing the TGF sensitivity. In particular, MD-derived NO is thought to mediate a vasodilating component of the TGF response, shown to be active under high NaCl concentrations at the MD, thus implying salt-dependent regulation of NO synthesis at the MD. Similar to the adjustment of the TGF, regulation of the renin-angiotensin system (RAS) is a critical step in the maintenance of whole body volume homeostasis.

Renin, the rate-limiting enzyme of the RAS, is synthesized and stored in the granular cells of the preglomerular afferent arteriole. There is accumulating evidence that NO is a potent modulator of renin secretion from the kidney. The precise effect of NO on renin synthesis and secretion, however, is still a matter of controversy; on one hand, there is data suggesting an inhibitory effect of NO on renin synthesis and secretion, and on the other, data from isolated perfused rat kidneys, from isolated renal juxtaglomerular cells, and from the conscious dog suggest that NO is a potent stimulator of renin. [33, 34]

Angiotensin - Nitric Oxide Interactions in Physiological and Pathological States

The ANG II/NO Axis: Master Regulators of Vascular Tone

As described above, ANG II is a potent vasoconstrictor, and NO is an important vasodilator. These counterbalancing forces have proven extremely important in regulating vascular tone, especially in the kidney. However, more recent discoveries have indicated that these two vasoactive components regulate each other, thereby forming a regulatory loop between the two. Thus, as ANG II increases, NO production also increases and thereby inhibits ANG II levels and activities. The initial observations that ANG II activates NO production came from studies in neuroblastoma cells, where ANG II treatment of these cells *in vitro* resulted in an increase in cGMP levels, an index for NO production [35,36]. However, these observations spread from a limited number of *in vitro* experiments to the detailed analysis of the receptors involved and the signaling pathways activated. Together, the picture that is painted shows that ANG II and NO are part of a powerful regulatory axis that is essential for maintaining local blood pressure and plays a major role in many pathologic states.

Opposing Receptor Activities - AT1 VS AT2

The ANG II peptide is recognized predominantly by at least G-protein coupled receptors, AT1 and AT2 [37, 38]. Both receptors have been cloned and are G-protein coupled receptors; however, despite being of the same class and binding the same peptide, they share only 34% identity with each other, suggesting divergent functions. Traditionally, it is believed that most of the vasoconstrictive (and thereby physiological) activities of Ang II are performed via the AT1 receptor. The AT2 receptor, in contrast, was originally believed to only be expressed during fetal development [39], and the physiological role of this receptor in the adult was therefore considered to be minimal. However, not only has AT2 been found in numerous adult tissues [37, 38, 40], but the function of AT2 has come to be regarded as an important counterbalancing force to the AT1-mediated functions of ANG II. Early evidence of this stemmed from the observation that stimulation of AT2 blunted an ANG II-mediated vasoconstriction in uterine arteries, where AT2 receptors have been shown to be expressed at high levels [41]. At the time, these studies were in line with previous observations indicating that AT1 and AT2 counterbalance each other in other vascular functions, such as proliferation and vasoreactivity. This, however, was the first demonstration of counterbalancing effects in terms of vasoconstriction. Still, the mechanisms of this interaction were far from clear.

More recent evidence using transgenic mice gives powerful support for the opposing effects of the two receptors on blood pressure and vasoconstriction. Unlike humans, rodents have two different AT1 receptors, AT1A and AT1B, which are encoded by different genes on different chromosomes [42, 43]. AT1A null mice show a mild phenotype [44], while AT1B null mice show little to know observable phenotype [45]. However, mating between the two strains generated the AT1 DKO mice, which lack both versions of the AT1 receptor. These mice have growth defects, abnormal kidney morphology, and marked hypotension [45]. The authors noted that these mice closely mimic the phenotype of angiotensinogen and ACE knockout mice [46,47] in terms of hypotension and abnormal kidney morphology. The implication is that as originally suspected, much of the vasoconstrictive activities of the renin-angiotensin pathway are mediated by the AT1 receptor. Important evidence for the opposing functions of the AT1 and AT2 receptors came from treatment of the AT1 DKO mice with ACE inhibitors. This led to a counter-intuitive increase in blood pressure. The authors suggested that this increase was because of diminished signaling via the AT2 receptor, and it was later shown that inhibition of the AT2 receptor with PD 123319 caused a significant (albeit not complete rescue) increase in blood pressure in the AT1 DKO mice. The model that has therefore emerged from over twenty years of research now states that the AT1 receptor activity is offset by the function of the AT2 receptor.

NO Production via the AT2 Receptor

As described above, the current model for the Ang II receptors states that the AT1 receptor acts as a vasoconstrictor, but that the AT2 receptor acts as a vasodilator. The question that arises, therefore, is what mediates this vasodilation activity associated with the AT2 receptor? Clues began to accumulate in 1996, when Siragy and Carey showed an increase in cyclic guanosine 3′, 5′ monophosphate (cGMP) in renal interstitial fluid (RIF) following Ang II perfusion in conscious rats [48], an index for NO production. Additionally, this accumulation was blocked by PD123319, an inhibitor to the AT2 receptor, indicating that cGMP accumulation in RIF was mediated by the AT2 receptor. These studies were followed by a demonstration that L-NAME, a NO inhibitor, prevented the cGMP accumulation, indicating that the increase in cGMP RIF was indeed caused by NO produced in response to activation of the AT2 receptor [49]. It was also shown that NO was produced via the AT2 receptor in rat aortas as well [50]. Together, these data gave the first evidence that the AT2 receptor was responsible for activating NO production to counterbalance AT1 activation.

Mice lacking the AT2 receptor were generated by multiple groups [51, 52, 53]. Surprisingly, not all strains of AT2-deficient mice showed the same phenotype; namely, the mice generated by Siragy and Ichiki were both hypertensive, while those generated by the Hein group were normotensive. These data suggest the possibility that the systemic role of the AT2 receptor could interact with the genetic background, further complicating matters. One phenotype that was common to all three groups was Ang II insensitivity; injection of mice with Ang II had little difference on mean arterial pressure when compared with wild type controls. The diminished response to ANG II was eliminated by treatment with NO and bradykinin inhibitors in the Siragy mice, suggesting that the counteractive activity on vasoconstriction mediated through AT2 is largely attributable to NO production. Together, these data indicated that ANG II activates a signaling pathway through the AT2 receptor that leads to a balancing effect mediated by NO and bradykinin, thus forming a negative regulatory loop between NO and ANG II.

The mechanisms for the AT2-receptor mediated nitric oxide production has been actively researched in recent years. In the kidney (as well as other tissues with endothelial cells) the predominant mode of NO production is made by the endothelial nitric oxide synthase (eNOS or NOS-III). The eNOS protein is approximately 150,000 Daltons in size and contains a heme group which is necessary for the conversion of L-arginine to NO [54]. The activity of eNOS is regulated by numerous mechanisms, and these are well-reviewed [55]. Specifically, eNOS activity is regulated by Ca++, protein-protein interactions, change in subcellular localization, increase in protein expression, and phosphorylation. The activation of the AT2 receptor (as well as the AT1 receptor) uses many of these mechanisms to upregulate the activity of eNOS in both the short and long terms.

Much has been done to understand the role AT2 has on the transcriptional regulation of eNOS. In myocardial cells, Ritter et al. [56] showed that eNOS is transcriptionally upregulated following AT2 receptor activation via an NF-AT-dependent transcriptional upregulation. In pulmonary endothelial cells, ANG II mediates an AT2-dependent mechanism of eNOS upregulation by activating Src kinase, phosphorylation of sequence homology to collagen (Shc), and an association between Shc and Grb2, all of which are dependent on the pertussis-toxin dependent AT2 receptor [57], indicating that a complex signaling pathway downstream of the AT2 receptor leads to an upregulation of eNOS protein.

eNOS also is regulated by multiple post-translational modifications, one of which is phosphorylation. The eNOS protein is phosophorylated at multiple sites, most of which serve as activating phosphorylations [58]. As an example, activation of eNOS by phosphorylation at serine 1177 (S1177 in the human sequence) has probably received the most attention. It was shown by Dimmler and Fulton that Akt is capable of activating eNOS by phosphorylating S1177 [59, 60]. To this end, Hiyoshi et al. recently showed that in thoracic aortas of mice subjected to two kidneys one clip hypertension (which elevates blood pressure), there was an elevated level of phosphorylated eNOS (at S1177), phsophorylated Akt (suggesting activated Akt), as well as higher levels of total eNOS and Akt proteins when compared with mice with sham surgeries [61]. These were reversible by PD123319, suggesting that these increases were due to signaling via the AT2 receptor. Additionally, Yayama et al. showed that in mice subjected to aortic banding (which increases Ang II levels) eNOS became phosphorylated at both S1177 and S633 and this was inhibited by treatment with the AT2 receptor inhibitor PD123319 [62]. However, this appears to be mediated by PKA in aortas, rather than the PI3-k/Akt pathway, since wortmannin (an inhibitory of the PI3-k/Akt pathway) had no effect on these phosphorylation events, but treatment with protein kinase A inhibitors (H89 and KT5720) prevented these phosphorylations [62]. These are a small sampling of the various studies performed to understand the mechanism of NO production via the AT2 receptor. However, it is important to note that the AT2 receptor operates via numerous pathways to upregulate NO production, and thereby blunt the effect of AT1 receptor activation (see Figure 8).

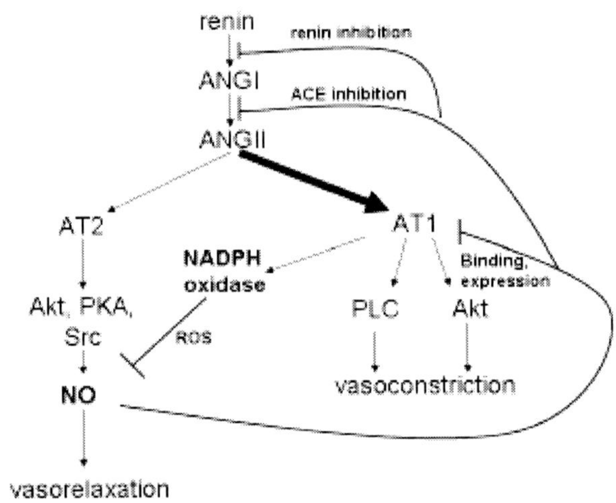

Figure 8. Diagram of the Ang II/NO axis. Arrowheads at the end of a line indicate stimulation, while perpendicular lines indicate inhibition. The mechanism of interaction is indicated next to each line.

NO Production via Shear Stress

As discussed, ANG II is a potent vasoconstrictor that is responsible for elevations in blood pressure. As discussed in the previous sections, one of the results of elevated ANG II is increased activation of the eNOS protein and elevated levels of NO production to counteract the ANG II-mediated vasoconstriction. However, another consequence of elevated ANG II is increased shear stress on endothelial cells, and this also causes an increase in the production of NO. Many of the details of the molecular mechanism of this activation are unclear, but it appears that shear stress causes an increase in NO in two different phases: an early, Ca++-dependent phase and a longer term phosphorylation-dependent phase. eNOS activation as a result of shear stress extends beyond the scope of this book chapter, but it was recently reviewed by [63].

NO Production and the AT1 Receptor

As described above, current evidence points to the AT2 receptor as playing the most important role in Ang II-mediated NO release. This creates a convenient model in which Ang II binds both AT1 and AT2 receptors and the former activates vasoconstrictive signaling, while the latter activates vasodilation. However, some previous reports have suggested that in carotid arteries and endothelial cells, ANG II can activate cGMP production (an index of NO production via the AT1 receptor blocker losartan [64, 65, 66]. Later, these results were repeated when Thorup et al., showed that ANG II perfusion of tertiary renal arteries resulted in increased NO production that was blocked by losartan [67]. Isolated glomerular afferent arterioles also release NO following ANG II treatment, and this effect is inhibited by treatment with AT1 blocker, but not AT2 blocker [68], suggesting that in glomerular afferent arterioles, AT1 is the predominant receptor responsible for NO stimulation. These studies indicate that ANG II - via both of its predominant receptors - is capable of increasing NO production in numerous tissues.

Another means by which the AT1 receptor increases NO production is through the upregulation of eNOS mRNA and protein expression. Hennington et al. showed that short term treatment of rats with Ang II caused an increase in the eNOS mRNA expression, and long term doses caused an increase in eNOS protein expression in rat kidney [69].

Further evidence for AT1-mediated activation of NO production stems from the presence of an inhibitory protein-protein interaction eNOS has with the AT1 receptor. A recently identified and interesting aspect of eNOS regulation indicates that these two proteins can interact *in vitro* [70], and this association has an inhibitory effect on eNOS activity. However, the implications of this observation remain unclear; thus far, no investigators have shown any *in vivo* role for this interaction. Despite this lack of evidence, it is exciting to hypothesize that this interaction is an important regulatory step in eNOS regulation via the AT1 receptor. One can visualize the AT1 receptor working as a scaffold sequestering eNOS in its inactive state, until binding of Ang II, upon which numerous activating phosphorylations occur, thereby mediating the activation of NO synthesis in response to Ang II. However, experimental data to support this concept have not yet been published.

ANG II Inhibition of NO

Despite the large volume of information indicating that ANG II can enhance NO production and thereby offset some of its own vasoconstrictive effects, ANG II can also work to inhibit NO production through the production of superoxide, albeit indirectly. Ang II increases the levels of reactive oxygen species [64] via the upregulation of NADPH oxidase [71] (Griendling et al. 1994), which includes oxygen free radicals, hydrogen peroxide and superoxide. The increase in the amount of NADPH oxidase results in an increase in the amount of reactive oxygen species, which act as a double-edged sword against NO production [72] (Beckman and Koppenol 1996). First of all, superoxide will bind tightly to NO, producing peroxynitrate, which has no vasodilatory activities and is quite toxic. Additionally, superoxide can bind to the heme group that is present in all NO synthases, and thereby inactivate the synthase activity. Therefore, through its upregulation of the oxidative phosphorylation pathway, ANG II can act to produce superoxide anions that can inhibit NO that is already present and also prevent the generation of more NO. It should be noted, however, that direct inhibition of NO activity by any angiotensin receptor has yet to be reported, and all ANG II-based inhibition of NO activity has been indirect.

NO Effects on ANG II

The above-mentioned data make it clear that ANG II activates NO to offset its own vasoconstrictive activities. However, NO release counterbalances this effect by down-regulating the ANG II pathway at multiple stages (discussed below). In principle, this has two effects. First of all, an ANG II-induced activation of NO further counterbalances the effects of ANG II activity by inhibiting the pathway. Second, the decrease in ANG II results in a decrease in the amount of NO that is produced, and therefore decreases the counterbalancing effects of NO. Therefore, a key component of the ANG II/NO regulatory axis is the inhibitory effects that NO has on the ANG II system.

NO Inhibits Renin Production

One level at which NO inhibits Ang II production is by inhibiting the production of renin, the precursor to ANG II [73, 74]. Recent studies have indicated that NO production decreases the production of renin in the kidney. Sodium-depleted rats (thereby expressing higher levels of renin mRNA) given L-NAME showed higher levels of renin mRNA in the kidney. Furthermore, treatment of sodium depleted rats with PD123119 (the AT2 receptor blocker) also resulted in an increase in renin production [75, 76]. The existence of this pathway indicates that NO is capable of inhibiting renin production, which consequently reduces the amount of ANG II present in the local milieu.

NO Inhibits ACE

Another means by which NO inhibits Ang II is at the level of the angiotensin converting enzyme (ACE). As described above, ACE is the enzyme that is responsible for the conversion of angiotensin I to Ang II, which is the active peptide. Takemoto et al. showed that *in vivo* treatment with L-NAME, a NOS inhibitor, increased ACE activity in heart and aorta. Further, Michel et al. showed that treatment of rats with L-NAME greatly increased aortic ACE activity [80]. In this study no affect was seen on kidneys. These studies were followed by Persson and Andersson showing that in porcine iliac arteries the NO donor SNAP can significantly decrease the activity of ACE [77]. Also, both aqueous NO and the NO donors SIN-1 and diethylamine/NO were capable of inhibiting ACE from plasma as well as purified from lung [78]. Thus, it appears that NO can inhibit ACE activity both *in vitro* and *in vivo*, indicating the existence of yet another pathway by which NO can inhibit Ang II activity.

NO Blocks Prevents AT1-Mediated Signaling

As described above, most of the vasoconstrictive activity ascribed to ANG II is carried out through activation of the AT1 receptor. It is interesting to note, therefore, that NO expression can lead to a downregulation in the abundance of the AT1 receptor. Ichiki et al, noticed that treatment with the NO donor S-nitroso acetyl DL-penicillamine (SNAP) caused a decrease in the abundance of the AT1 receptor mRNA in cultured vascular smooth muscle cells [79]. This decrease was independent of transcription, as it occurred in the presence of actinomycin D. More recent investigations into the effect of NO on Ang II signaling indicated that rats treated with L-NAME showed an increased expression of AT1 receptor after one month [80]; however, these investigators also showed that short term treatment of renal mesangial cells with nitric oxide donors caused a marked decrease in Ang II binding, AT1 transcript abundance, and calcium signaling in response to Ang II treatment. This suggested that the AT1 receptor downregulation in the rats was an indirect effect of NO loss. Together, these data indicate the possibility that AT1 receptor mRNA may be downregulated by NO.

An interesting recent report also indicates that there might be another mechanism for NO-mediated inhibition of the renin-angiotensin system. Treatment of AT1-expressing HEK-293 cells with the NO donor sodium nitroprusside (SNP) diminished the affinity of the AT1 receptor for Ang II; furthermore, this report indicated that mutation of cysteine 289 to serine abolished the SNP sensitivity [81]. Thus, excess NO may prevent Ang II binding to the AT1 receptor, further blocking the Ang II pathway.

ANG II and NO Interactions in Diabetic Nephropathy

Together, the above data indicate that not only does Ang II upregulates NO production to offset its own the vasoconstrictive activities, but NO also inhibits Ang II activity, in an apparent effort to further downregulate Ang II. The interaction between these pathways suggests an intricate physiological mechanism to decrease local blood pressure, and thereby prevent damage to vascular tissues. The importance of this interaction is underscored when it is observed that pathogenic states, like diabetes, can cause a gross disruption of the Ang II/NO axis, and exacerbate the damage done in the diabetic state.

Diabetes is characterized by high levels of glucose in the blood, either because of lack of pancreatic function (insulin-dependent diabetes mellitus, IDDM) or because of insulin resistance (non-insulin-dependent diabetes mellitus, NIDDM). While complications from the elevated blood sugar may not be immediate, lack of appropriate glycemic control can have numerous pathophysiological consequences, primarily on the vascular system. One very important complication associated with diabetes is diabetic nephropathy (DN) – the progressive worsening of proteinuria and loss of glomerular filtration rate, ultimately leading to end stage renal disease. Approximately 40% of diabetes patients eventually develop DN. However, approximately 90% of DN patients have NIDDM; since NIDDM is continuing to increase and reach epidemic proportions, understanding DN in NIDDM is becoming a high priority.

DN is associated with a loss of balance in the Ang II/NO axis, with pathogenic consequences. Patients with diabetes often show high tissue levels of ANG II (generally via renal Ang II production and also hypertension in addition to high glucose. Further, the high glucose environment also upregulates AT1 receptor expression, leading to increased sensitivity to ANG II activity. Another characteristic of DN is an elevated level of renal NO production (measured by urinary nitrate concentration) during the early stages, followed by a progressive loss of NO production during later stages. The importance of this change in NO is not yet clear; however, obese mice that develop diabetes do not develop nephropathy, but obese mice lacking the gene for eNOS develop pronounced nephropathy, suggesting that NO production plays an important role in the development of nephropathy in diabetics.

Another abnormality in DN that affects the Ang II/NO axis is an increase in oxidative stress. While this stress derives from many sources, as discussed above, ANG II increases the local ROS content through an upregulation of NADPH oxidase subunits, contributing to the overall renal oxidative stress. Accordingly, it has been shown that in a rodent model for IDDM, the renal expression of p47phox is increased, and this increase is eliminated by treatment with ACE inhibitors and candesartan, the AT1 receptor blocker [82]. The increased expression of this protein can lead to an increase in ROS and, as discussed above, inactivate NOS and NO itself, ultimately preventing the ability for NO to contribute to increased local blood pressure and further ROS. Another contributor to oxidative damage in diabetes is, ironically, from ROS derived from the eNOS protein. High glucose content has been shown to "uncouple" NO production, specifically from eNOS [83,84]; instead of generating more NO, eNOS begins generating superoxide anion. As mentioned above, superoxide can inactivate the endogenous NO generating peroxynitrite, resulting a higher level of local vasoconstriction. It is no wonder that the role of oxidative stress in DN has been the subject of numerous investigators.

Physicians often prescribe ACE inhibitors and AT1 blockers to diabetics to prevent diabetic nephropathy with great benefit – and how such an intervention interplays with ANGII -NO pathways is a subject of significant research interest. However, these treatments generally only delay nephropathy, and do not necessarily prevent it. The ANG II/NO axis clearly seems to play a major role in the progression of DN, and future therapies will likely work to modulate these pathways. Hopefully, further research into the mechanisms of this axis will lead to superior preventative therapies in the future.

Research Work from Our Laboratory

Effects of Acute Angiotensin II Infusion on Nitric Oxide Synthesis and NOS Isoforms in the Rat Kidney

Previous studies from our laboratory in renal cell cultures suggested that angiotensin II (Ang II) inhibited nitric oxide (NO) derived from inducible nitric oxide synthase (iNOS) and stimulated endothelial nitric oxide synthase (eNOS) . However, the effects of Ang II on systemic and renal NO production in intact animals and various NOS isoforms in the kidney are controversial and currently not well understood. Therefore we examined the acute effects of systemic administration of ang II on renal NO synthesis and NOS isoforms in rats. Adult Sprague-Dawley rats (20 weeks) were anesthetized and either ang II infusion (10^{-8}M) or 5 mM glucose(control) was given for 4 hours after which the rats were sacrificed and kidneys harvested. Nitrites and nitrates (NO_x) were measured in serum and urine samples collected before and after the infusion as an index of NO production. Renal cortices were seperated from medullae and homegenates examined for Ca^{++} dependent NOS (cNOS) and iNOS activity.

Group	Plasma NOx (M/kG)	Urine NOx (M/kG)	cNOS activity-cortex	cNOS activity-medulla	iNOS activity-cortex	iNOS activity-medulla
Contro l	1.7±0.3	3.5±0.6	256±34	413±53	124±27	156±55
Ang II	3.60±.8	9.6±4.3*	524±68*	533±112	48±14*	131±22

N=6 in each group, P<0.05, *P<0.01vs. control. NOS activity expressed in cpm/mg.protein/min

The expression of iNOS and eNOS in homogenates was examined by Western blot analysis using anti-iNOS and anti-eNOS antibodies. The cNOS activity was determined by conversion of (^{3}H)L-arginine to (^{3}H)L-citrulline in the presence and absence of EGTA and iNOS activity by (^{3}H) L-citrulline generation in the presence of EGTA alone or EGTA and N^G-nitro-L-arginine methyl ester (L-NAME).

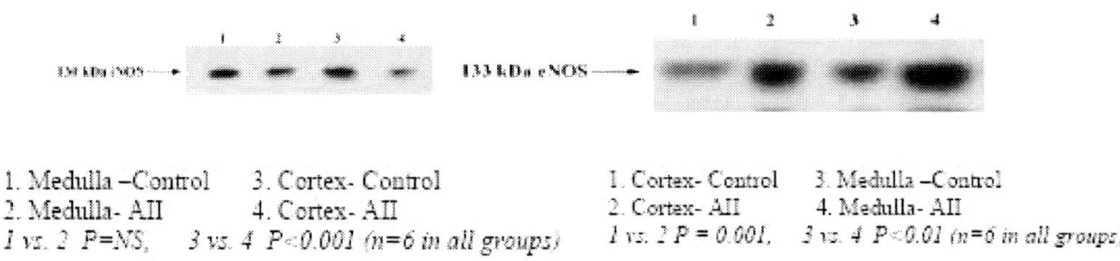

Figure 9. Effects of acute AII infusion on iNOS (right) and eNOS (left) expression in the rat kidney

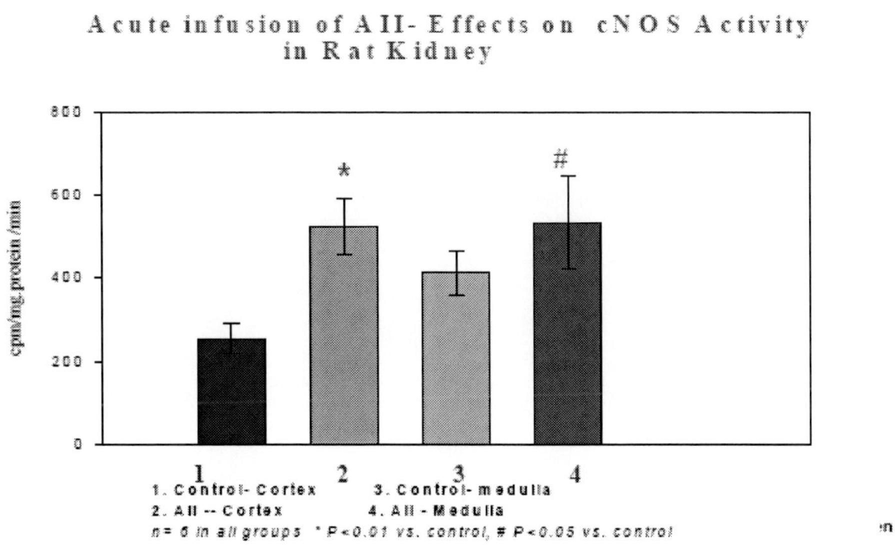

Figure 10. Angiotensin effects on renal NO generation.

Our data showed that Ang II inhibited iNOS expression and activity in the cortex with no effect in the medulla. In contrast, eNOS activity and expression was increased particularly in the cortex. The significant increase in urinary NO_x with a modest increment in plasma NO_x implied enhanced renal NO synthesis. To summarize the data, Ang II stimulated renal NO synthesis by augmenting the Ca^{++} dependent NOS activity and eNOS expression in the kidney, which under basal conditions superceded the iNOS inhibition by Ang II. [85]

Angiotensin II Stimulated Intrarenal eNOS Expression and Nitric Oxide Generation in Anesthetized Rats Even in the Presence of Hyperglycemia

Previously we have demonstrated that high ambient glucose (HG) inhibited inducible NO synthesis in cultured mesangial cells (Prabhakar SS, Am J Physiol, Renal 281, F0179-F0188, 2001) while others demonstrated HG induced eNOS inhibition as well. In vivo and in vitro studies showed that Angiotensin II (Ang II) enhanced eNOS expression and activity in the kidney. But the net effects of Ang II in the presence of HG on eNOS are presently unknown. Therefore we examined the eNOS expression in the renal cortex and medulla of rats infused with Ang II and HG. Adult male Sprague Dawley rats (20 wks old) were anesthetized and infused through femoral veins for 4 hours with either 5 mM glucose (Gr I-control), Ang II (10^{-8}M) (Gr II), 20 mM glucose (Gr III) and a combination of 20 mM glucose and Ang II (Gr IV). The rats were then sacrificed, the kidneys harvested and medulla separated form cortex. Systemic and renal NO generation during the study was assayed by measuring the stable NO metabolites, nitrites and nitrates (NO_x) in the serum and urine before and after the study, using a chemi-luminescence technique. The eNOS expression in the kidney was evaluated by western blot technique using anti-eNOS antibody. Ca^{++} dependent NOS (cNOS) activity in the kidney homogenates was measured by (^{3}H) L-citrulline from (^{3}H) L-arginine in the presence and absence of EGTA.

Group	I	II	III	IV
Urinary NOx(M/kG)	2.1±0.6	13.6±3.8*	-6.7±4.4*	9.3±1.7
Plasma NOx(M/kG)	1.8±0.2	3.2±0.7*	-0.6±0.3*	2.8±0.9*
cNOS activity- cortex	243±25	514±67*	198±23	515±43*
cNOS activity-medulla	38566	44679	16651*	543103*

cNOS activity measured as cpm/mg.protein/min, n= 6 in each group, *P<0.01 vs. Gr I

The eNOS expression in the Gr III and IV was greater than in Gr I (higher in cortex than medulla) while Gr II was significantly decreased. These data suggest that despite the high glucose inhibition of eNOS, the intra-renal NO generation as well as eNOS activity continues to be enhanced with systemic Ang II exposure. We concluded that Ang II is a powerful stimulant of eNOS activity in the renal tissues even in the presence of high glucose [86].

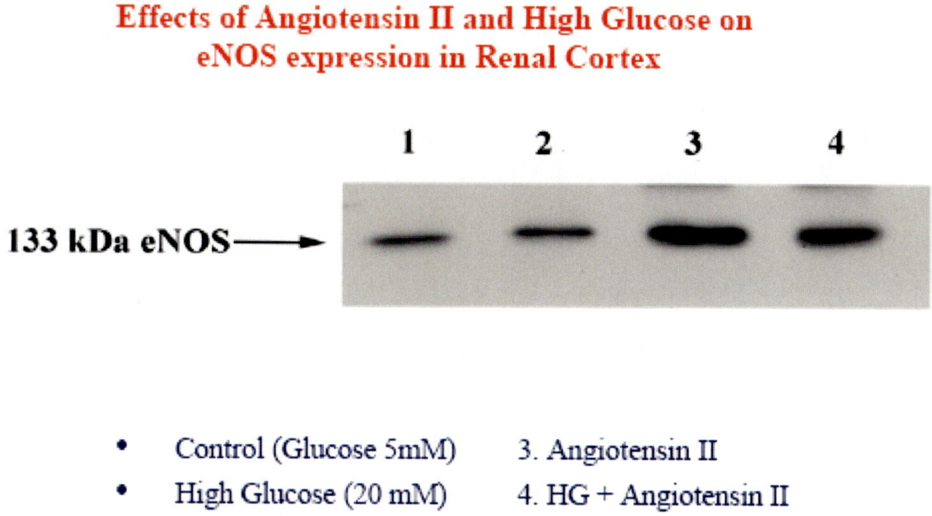

Figure 11. Effects of angiotensin on renal eNOS expression in the presence of high glucose.

Angiotensin II Subtype 2 (AT_2) Receptors Mediate Nitric Oxide Synthesis in Cultured Mesangial Cells

Since intrarenal angiotensin II (AII) and nitric oxide (NO) have been independently shown to impact on the progression of chronic renal failure, we hypothesized that AII exerts its effects at least in part by modulating the intrarenal NO levels. To verify this hypothesis, we examined AII effects on NO synthesis in cultured murine mesangial cells (MMC) and specifically the role of AII receptors (AT_1 and AT_2) in mediating these effects. MMC were grown in DMEM media with minimal nitrate and NO synthesis was induced with bacterial lipopolysaccharide (LPS) from E. coli 026:B6 at 10μg/ml and γ-interferon (γ-I) at 50 U/ml and NO was assayed by measuring the nitrites and nitrates (NOx) accumulated in MMC using an NO analyzer. AII (at 10^{-8} M concentrations) significantly inhibited (17.5 ± 1.45 μM, vs. control 44.43± 3.21 μM, $p<0.01$,n=12) while AT_1 blockade with losartan or LOS (at 10^{-5} M concentrations) reversed the ANG effects on NOx (49.5 ± 3.89 μM, n=12). Micromolar concentrations of PD 123,319 (DTFA or ditriflouroacetate), a selective AT_2 receptor antagonist significantly augmented the AII mediated NO inhibition (9.84 ± 1.23 μM, n=12, $p<0.05$ vs. 17.5 ± 1.45 μM). Since reversal of AII effects on NO by losartan could be due to either AT_1 blockade or due to unopposed stimulation of AT_2 receptors by AII, we examined the effects of combined AT_1 and AT_2 blockade on NOx in the presence of AII. These studies showed that blockade of both receptors significantly inhibited NOx (23.6± 1.71 μM, vs. control 42.8 ± 3.47 μM, n=12, $p<0.05$), suggesting a strong inhibitory effect of AT_2 blockade that exceeds the stimulatory effect of AT_1 blockade on NO synthesis.

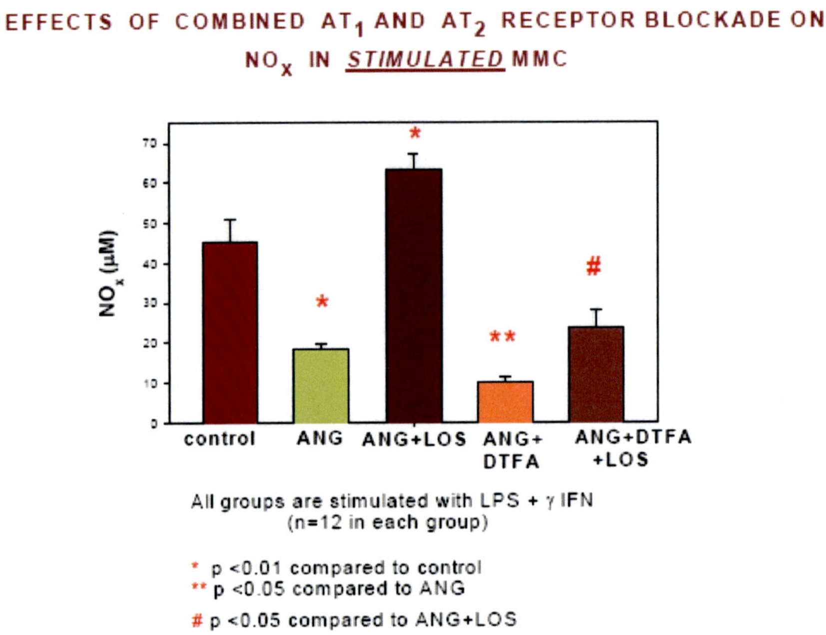

Figure 12. Angiotensin receptors and mesangial NO generation.

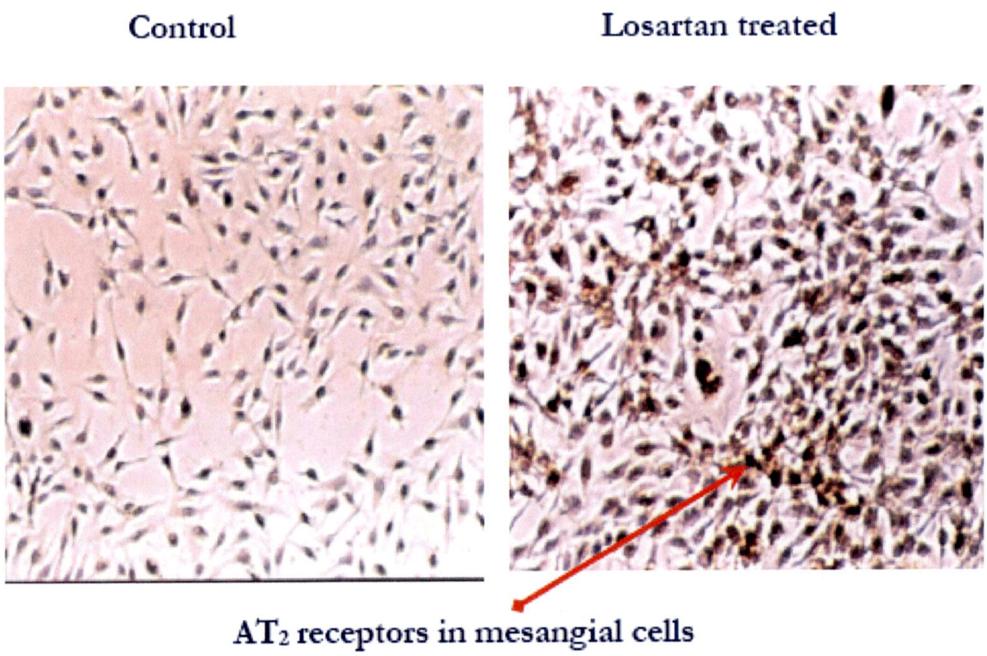

Figure 13. Expression of AT2 receptors in MMC.

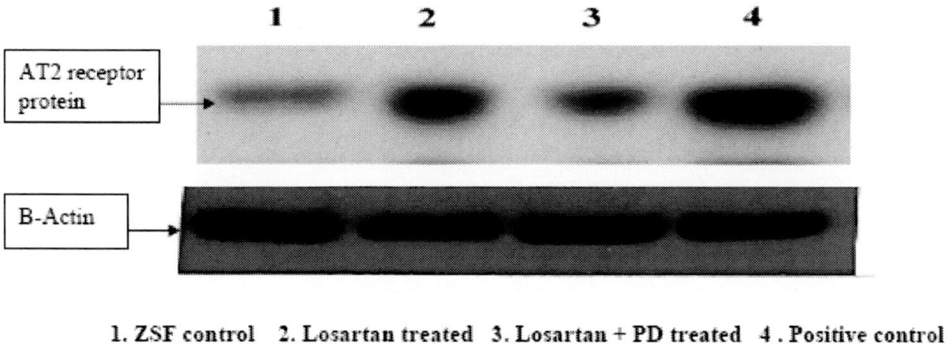

Figure 14. AT_2 receptor expression in ZSF rat kidney .

We conclude that AT_2 stimulation mediates NO synthesis that is supplementary to the effects of AT_1 blockade. This is the first demonstration of existence and activity of mesangial AT_2 receptors. Furthermore, these data indicate that maximal therapeutic effects of angiotensin blockade may be achieved by a combination of AT_1 blockade and AT_2 agonism rather than AT_1 blockade alone.

Renal Nitric Oxide Generation with Angiotensin Receptor Blockade is Associated with Increased Renal Expression of Angiotensin II Type 2 Receptors in ZSF Rats

Clinical and experimental studies have demonstrated that AT_1 receptor blockers protected kidneys from diabetic and nondiabetic injury. However the effects of AT_1 receptor blockade on renal nitric oxide (NO) production and the relation of such effects to renal expression of AT_2 receptors is currently unclear. We hypothesised that AT_1 receptor blockade results in increased renal NO production, and that such an effect is related to increased renal AT_2 receptor expression. To test our hypothesis, we studied ZSF rats, (a rat model for type 2 diabetes) for 12 weeks, from 8th to 20th week of life and administered losartan, an AT_1 receptor antagonist at 40 mg/L of drinking water. A subgroup of losartan treated rats received PD123319, an AT_2 receptor antagonist, administered subcutaneously at 30 mg/Kg body wt daily for 7 days during the 12th week of the study, to examine the effects on renal NO generation. Untreated obese ZSF rats served as controls. Renal NO generation was determined by measuring stable metabolites of NO (nitrite plus nitrate or NO_x) in the urine with an NO analyzer that uses chemiluminescence technique. Systemic blood pressure was measured by tail cuff plethysmography. Renal expression of AT_2 receptors was evaluated by immunoblotting (using specific AT_2 antibody), from cortical homogenates of kidneys harvested from ZSF rats at 20 weeks.

	Control	Losartan group	Losartan +PD
Urine NOx (uM/Kg BW)	5.6 ±1.1	9.4±2.3*	6.7±1.6
Systolic BP mm/Hg	179±23	138±14*	164±21
Diastolic BP mm/Hg	103±17	89±10*	98±12

PD= PD123319, N=12 in each group, *<0.01 vs. control

Our results showed that losartan treatment lowered blood pressure, proteinuria and serum creatinine and these expected results were associated with an increase in urinary NO_x levels. Furthermore renal expression of AT_2 receptors was increased in losartan treated rats compared to untreated ZSF rats at 20 weeks. However the increase in renal NO levels observed in losartan treated rats was significantly reduced in rats administered PD123319. We conclude that losartan mediated increase in renal NO generation may be primarily mediated through AT_2 receptors.

Enhanced Insulin Secretion Due to Angiotensin Inhibition Is Associated with Increased NO Synthesis in Pancreatic Islet B Cells

Recent large scale clinical trials have observed that angiotensin converting enzyme (ACE) inhibition as well as angiotensin receptor blockade (ARB) to treat hypertension is associated with prevention of new onset diabetes. Although recent literature suggests that angiotensin inhibition improves glucose uptake and metabolism in peripheral tissues, it is currently unknown if angiotensin blockade affects pancreatic insulin release. We therefore examined the effect of ACE inhibitors and ARBs on insulin secretion from pancreatic islets and examined the relationship of such an effect with local nitric oxide (NO) release. Pancreatic islet β cells derived from hamster were grown Kaighn's modified F-12K nutrient medium subcultured in serum free medium with minimal nitrate.Experimental groups were incubated in captopril (1 mM) or losartan (10 mM) seperately and and also along with a NO synthase blocker, NG-monomethyl L- arginine (L-NMMA) at 1mM for 18 hrs. Insulin secretion was assayed by a commercial insulin ELISA kit using a guinea-pig anti-insulin antibody and measuring absorbance at 492 nm. Nitric oxide release was assayed by measuring the end-products of NO, nitrites and nitrates (NOx) in the media by an NO analyzer using chemiluminescence technique.

Insulin and NO Release from Pancreatic Islet Beta Cells Exposed to Captopril and Losartan

The results indicated that exposure of islet β cells to captopril or losartan was associated increased insulin release as well as increased NO release. Co-incubation with L-NMMA abolished the augmentation of insulin secretion from islet cells seen with exposure to losartan and captopril. These data suggest that ACE inhibition and ARB enhance insulin release from pancreas by an NO-dependent mechanism and this effect may contribute to improved carbohydrate metabolism seen with blockade of renin-angiotensin axis [87].

Exptl Group	Control	Captopril	Losartan	Capto+L-NMMA	Losartan+L-NMMA
NOx (nM/L)	29.3±6.7	87.1±17.9*	59.5±11.4 #	26.4±8.4	22.8±5.8
Insulin (ng/ml)	6.7±1.3	19.3±4.3*	16.9±3.6*	5.8±1.1	7.1±1.5

P<0.01 vs. control, # P<0.05 vs. control

Conclusion

From what has been described above, it is clear that Ang II and NO are important counterbalancing forces in vascular biology. While Ang II is traditionally a well-noted vasoconstrictor, and NO is firmly established as a vasodilator, the two substances interact in an important self-regulatory loop (see figure 8). Ang II (by both stimulation of its receptors and the induction of shear stress) causes an increase in NO synthase activity via both stimulatory and anti-inhibitory mechanisms. Conversely, Ang II also generates reactive oxygen species that are capable of neutralizing NO activities and preventing further accumulation of NO. Upon accumulation, NO also inhibits the renin-angiotensin pathway at numerous steps, and thereby attempts to prevent further damage to the endothelium, as well as buffer further upregulation of NO. Thus, in normal physiological conditions, NO and Ang II work to balance the effect each has on the vasculature. However, in pathogenic states such as diabetes, a disequilibrium between the two results and lead to vascular damage including diabetic nephropathy. The plethora of the research data discussed above illustrates how complex these interactions could be and how the net directional change in nitric oxide and angiotensin II result from a balance of several opposing influences operating concurrently [88]. It is clear that such interactions decide not only alterations in macrocirculatory and microcirculatory hemodynamics and consequences thereof but also several non-hemodynamic tissue processes such as inflammation, tissue repair and remodeling. Inhibitors to the renin-angiotensin system have been used with great efficacy in preventing such pathologies, and they will certainly continue to be used. It is important that we continue to research to understand how these systems work, so that it will someday be possible to better understand the pathophysiology and discover innovative therapeutic strategies..

References

[1] Hall J.E. Historical Perspective of the Renin-Angiotensin System. Molecular *Biotechnology*, May 2003, Volume 24, Issue 1, pages 27-39

[2] Basso N; Terragno N. A. History About the Discovery of the Renin-Angiotensin System. *Hypertension.* 2001; Volume 38: pages 1246-1249

[3] Skeggs LT, Marsh W.H, Kahn J.R, Letntz K, Shumway N.P. The preparation, purification, and Amino Acid Sequence of a Polypeptide Renin Substrate. *The Journal of Experimental Medicine*, 1954, Volume 99, pages 275-282,

[4] UC Brewster, MA Perazella. The renin-angiotensin-aldosterone system and the kidey: effects on kidney disease. *The American Journal of Medicine*, 2004 Volume 116, Issue 4, pages 263-272.

[5] Danser A.H.J; Deinum J. Renin, Prorenin and the Putative (Pro)renin Receptor. *Hypertension*. 2005; Volume 46, pages 1069-1076

[6] Suzuki F, Hayakawa H , Nakagawa T, Nasir M, Ebihara A, Iwasawa A, Ishida Y, Nakamura Y. Human Prorenin Has "Gate and Handle" Regions for Its Non-proteolytic Activation. *J. Biol. Chem*, 2003; Volume 278, Issue 25,pages 22217-22222

[7] Takahashi H, Ichihara A, Kaneshiro Y, Inomata K,, Sakoda M, Takemitsu T, Nishiyama A, Itoh H. Regression of Nephropathy Developed in Diabetes by (Pro)renin Receptor Blockade. *J Am Soc Nephrol* 2007; Volume 18, pages 2054-2061

[8] Ichihara A, Matsuhiko Hayashi M, Kaneshiro K, Suzuki F, Nakagawa T, Tada Y, Koura Y, Nishiyama A, Okada H, Ishida Y, Inagami T, Saruta T. Inhibition of diabetic nephropathy by a decoy peptide corresponding to the "handle" region for nonproteolytic activation of prorenin. 2004, *J. Clin. Invest.* Volume 114, pages 1128-1135

[9] Saris J.J, Derkx F, de Bruin R, Dekkers D, Lamers J, Saxena P, Schalekamp M, Danser A. High-affinity prorenin binding to cardiac man-6-P/IGF-II receptors precedes proteolytic activation to renin, *Am J Physiol*, 2001, Volume 280, pages 1706–1715.

[10] Van Kesteren CAM, Danser AHJ, Derkx FHM, Dekkers DHW, Lamers JMJ, Saxena PR, Schalekamp MADH. Mannose 6-phosphate receptormediated internalization and activation of prorenin by cardiac cells, *Hypertension*, 1997, Volume 30, pages 1389 – 1396.

[11] Admiraal PJJ, van Kesteren CAM, Danser AHJ, Derkx FHM, Sluiter W, Schalekamp MADH. Uptake and proteolytic activation of prorenin by cultured human endothelial cells, *J Hypertens*, 1999 Volume 17, pages 621– 629.

[12] Groskopf JC, Syu LJ, Saltiel AR, Linzer DIH. Proliferin induces endothelial cell chemotaxis through a G protein- coupled, mitogen-activated protein kinase-dependent pathway, *Endocrinology*. 1997, Volume 138, pages 2835–2840.

[13] Di Bacco A, Gill G. The secreted glycoprotein CREG inhibits cell growth dependent on the mannose-6-phosphate/insulin-like growth factor II receptor, *Oncogene*, 2003, Volume 22, pages 5436 –5445

[14] Takahashi S, Inoue H, Miyake Y. The human gene for renin-binding protein. *J Biol Chem*. 1992, Volume 267, pages 13007–13013

[15] Schmitz C, Gotthardt M, Hinderlich S, Leheste JR, Gross V, Vorum H, Christensen EI, Luft FC, Takahashi S, Willnow TE. Normal blood pressure and plasma renin activity in mice lacking the renin-binding protein, a cellular renin inhibitor. *J Biol Chem*. 2000, Volume 275, Pages 15357–15362

[16] Christiane Rüster and Gunter Wolf. Renin-Angiotensin-Aldosterone System and Progression of Renal Disease *J Am Soc Nephrol*, 2006, Volume 17, Pages 2985–2991.

[17] Takahashi S, Inoue H, Miyake Y. The human gene for renin-binding protein. *J Biol Chem.* 1992, Volume 267, pages 13007–13013

[18] Michael A. Crackower, et al Angiotensin-converting enzyme 2 is an essential regulator of heart function. *Nature*, 2002, Volume 417, Pages 822-828

[19] Ingelfinger, JR. ACE2: A new target for prevention of diabetic nephropathy? *J. Am Soc Nephrol*, 2006 Nov; Volume 17, pages 2957-2959

[20] Donoghue M, Hsieh F, Baronas E, Godbout K, Gosselin M, Stagliano N, Donovan M, Woolf B, Robison K, Jeyaseelan R, Breitbart RE, Acton S: A novel angiotensin-converting enzyme-related carboxypeptidase (ACE2) converts angiotensin I to angiotensin 1-9. *Circ Rev* 2000, Volume 87, pages E1–E9

[21] 21. Li N, Zimpelmann J, Cheng K, John A, Wilkins K, Kevin D, Burns K.D, The Role of Angiotensin Converting Enzyme 2 in the Generation of Angiotensin 1-7 by Rat Proximal Tubules *Am J Physiol Renal Physiol*, October 5 2004, pages 1-38

[22] Unger T, Culman J, Gohlke P. Angiotensin II receptor blockade and end-organ protection. Pharmacological rationale and evidence, 1998, *J Hypertens.* Volume 16 (suppl), pages S3–S9

[23] Duncan JA, Scholey JW, Miller JA. Angiotensin II type 1-receptor gene polymorphisms in humans: physiology and pathophysiology of the genotypes. *Curr Opin Nephrol Hypertens.* 2001; Volume 10, pages 111–116.

[24] Nishiyama A, Seth DM, Navar LG: Renal interstitial fluid concentrations of angiotensins I and II in anesthetized rats. *Hypertension*, 2002, Volume 39, Pages 129–134.

[25] Beutler KT, Masilamani S, Turban S, Nielsen J, Brooks HL, Ageloff S, Fenton RA, Packer RK, Knepper MA: Long-term regulation of ENaC expression in kidney by angiotensin II. *Hypertension*, 2003, Volume 41, pages 1143–1150.

[26] Nishiyama A, Seth DM, Navar LG: Renal interstitial fluid I and angiotensin II concentrations during local angiotensin converting enzyme inhibition. *J Am Soc Nephrol*, 2002, Volume 13, pages 2207–2212.

[27] Zhuo JL, Imig JD, Hammond TG, Orengo S, Benes E, Navar LG: Ang II accumulation in rat renal endosomes during Ang II-induced hypertension. Role of AT1 receptor. *Hypertension*, 2002 Volume 39, pages 116–121

[28] Moncada S, Higgs A. The L-Arginine-Nitric Oxide Pathway. *NEJM*, 1993, Volume 329(27), pages 2002-2012

[29] Kone, BC. Nitric oxide synthesis in the kidney: isoforms, biosynthesis, and functions in health. Semin Nephrol 2004; Volume 24, page 299

[30] Bachmann, S, Mundel P, Nitric oxide in the kidney: synthesis, localization, and function. *Am. J. Kidney Dis.*1994, Volume 24, pages 112-129

[31] Feng Wu, Frank Park, Allen W. Cowley Jr., and David L. Mattson. Quantification of nitric oxide synthase activity in microdissected segments of the rat kidney. *Am J Renal Physiol,* 1999, Volume 276, pages F874-881.

[32] Kurtz A, Wagner C. Role of nitric oxide in the control of renin secretion. *Am J Renal Physiol*, 1998, Volume 275, pages F849-F862.

[33] Martin H, et al. Parallel regulation of constitutive NO synthase and renin at JGA of rat kidney under various stimuli. *Am J Physiol Renal Physiol*, 1995, Volume 269, pages 793 - 805

[34] Beierwaltes, W. H. Nitric oxide participates in calcium-mediated regulation of renin release. *Hypertension Dallas*, 1994, Volume 23, Suppl. I, pages 140-144.

[35] Chaki, S. and T. Inagami (1993) New signaling mechanism of angiotensin-ii in neuroblastoma neuro-2a cells - activation of soluble guanylyl cyclase via nitric-oxide synthesis. *Molecular Pharmacology,* 43, 603-608.

[36] Zarahn, E. D., X. Ye, A. M. Ades, L. P. Reagan and S. J. Fluharty (1992) Angiotensin-induced cyclic-gmp production is mediated by multiple receptor subtypes and nitric-oxide in n1e-115 neuroblastoma-cells. *Journal of Neurochemistry,* 58, 1960-1963.

[37] Furuta, H., D. F. Guo and T. Inagami (1992a) Molecular-cloning and sequencing of the gene encoding human angiotensin-ii type-1 receptor. *Biochemical and Biophysical Research Communications,* 183, 8-13.

[38] Furuta, H., D. F. Guo, N. Iwai and T. Inagami (1992b) nucleotide-sequence of a gene encoding human angiotensin-ii type-1 receptor. *Faseb Journal,* 6, A1578-A1578.

[39] Navar L, Rossival L. Contribution of the renin-angiotensin system to the control of intrarenal hemodynamics. *Kidney Int.* 1984 Jun;25(6):857-68.

[40] Ernsberger, P., J. Zhou, T. H. Damon and J. G. Douglas (1992) Angiotensin-ii receptor subtypes in cultured rat renal mesangial cells. *American Journal of Physiology,* 263, F411-F416.

[41] Zwart, A. S., E. A. Davis and R. E. Widdop (1998) Modulation of AT(1) receptor-mediated contraction of rat uterine artery by AT(2) receptors. *British Journal of Pharmacology,* 125, 1429-1436.

[42] Sasamura, H., L. Hein, J. E. Krieger, R. E. Pratt, B. K. Kobilka and V. J. Dzau (1992) Cloning, characterization, and expression of 2 angiotensin receptor (at-1) isoforms from the mouse genome. *Biochemical and Biophysical Research Communications,* 185, 253-259.

[43] MacTaggart, T. E., M. Ito, O. Smithies and S. W. M. John (1997) Mouse angiotensin receptor genes Agtr1a and Agtr1b map to chromosomes 13 and 3. *Mammalian Genome,* 8, 294-295.

[44] Ito, M., M. I. Oliverio, P. J. Mannon, C. F. Best, N. Maeda, O. Smithies and T. M. Coffman (1995) Regulation of blood-pressure by the type 1a angiotensin-ii receptor gene. *Proceedings of the National Academy of Sciences of the United States of America,* 92, 3521-3525.

[45] Oliverio, M. I., H. S. Kim, M. Ito, T. Le, L. Audoly, C. F. Best, S. Hiller, K. Kluckman, N. Maeda, O. Smithies and T. M. Coffman (1998) Reduced growth, abnormal kidney structure, and type 2 (AT(2)) angiotensin receptor-mediated blood pressure regulation in mice lacking both AT(1A) and AT(1B) receptors for angiotensin II. *Proceedings of the National Academy of Sciences of the United States of America,* 95, 15496-15501.

[46] Niimura, F., P. A. Labosky, J. Kakuchi, S. Okubo, H. Yoshida, T. Oikawa, T. Ichiki, A. J. Naftilan, A. Fogo, T. Inagami, B. L. M. Hogan and I. Ichikawa (1995) Gene targeting in mice reveals a requirement for angiotensin in the development and maintenance of

kidney morphology and growth factor regulation. *Journal of Clinical Investigation,* 96, 2947-2954.

[47] Krege, J. H., S. W. M. John, L. L. Langenbach, J. B. Hodgin, J. R. Hagaman, E. S. Bachman, J. C. Jennette, D. A. Obrien and O. Smithies (1995) Male-female differences in fertility and blood-pressure in ace-deficient mice. *Nature,* 375, 146-148.

[48] Siragy, H. M. and R. M. Carey (1996) The subtype-2 (AT(2)) angiotensin receptor regulates renal cyclic guanosine 3',5'-monophosphate and AT(1) receptor-mediated prostaglandin E(2) production in conscious rats. *Journal of Clinical Investigation,* 97, 1978-1982.

[49] Siragy, H. M. and R. M. Carey (1997) The subtype 2 (AT(2)) angiotensin receptor mediates renal production of nitric oxide in conscious rats. *Journal of Clinical Investigation,* 100, 264-269.

[50] Gohlke, P., W. Linz, B. A. Scholkens, G. Wiemer and T. Unger (1996) Cardiac and vascular effects of long-term losartan treatment in stroke-prone spontaneously hypertensive rats. *Hypertension,* 28, 397-402.

[51] Hein, L., G. S. Barsh, R. E. Pratt, V. J. Dzau and B. K. Kobilka (1995) Behavioral and cardiovascular effects of disrupting the angiotensin-ii type-2 receptor gene in mice. *Nature,* 377, 744-747.

[52] Ichiki, T., P. A. Labosky, C. Shiota, S. Okuyama, Y. Imagawa, A. Fogo, F. Niimura, I. Ichikawa, B. L. M. Hogan and T. Inagami (1995) Effects on blood-pressure and exploratory-behavior of mice lacking angiotensin-ii type-2 receptor. *Nature,* 377, 748-750.

[53] Siragy, H. M., T. Inagami, T. Ichiki and R. M. Carey (1999) Sustained hypersensitivity to angiotensin II and its mechanism in mice lacking the subtype-2 (AT(2)) angiotensin receptor. *Proceedings of the National Academy of Sciences of the United States of America,* 96, 6506-6510

[54] Alderton, W. K., C. E. Cooper and R. G. Knowles (2001) Nitric oxide synthases: structure, function and inhibition. *Biochemical Journal,* 357, 593-615.

[55] Dudzinski, D. M. and T. Michel (2007) Life history of eNOS: Partners and pathways. *Cardiovascular Research,* 75, 247-260.

[56] Ritter, O., K. Schuh, M. Brede, N. Rothlein, N. Burkard, L. Hein and L. Neyses (2002) AT(2)-Receptor activation regulates myocardial eNOS expression via the calcineurin-NF-AT pathway. *Faseb Journal,* 16, 283-+.

[57] Li, J. U., X. M. Zhao, X. M. Li, K. M. Lerea and S. C. Olson (2007) Angiotensin II type 2 receptor-dependent increases in nitric oxide synthase expression in the pulmonary endothelium is mediated via a G alpha(i3)/Ras/Raf/MAPK pathway. *American Journal of Physiology-Cell Physiology,* 292, C2185-C2196.

[58] Mount, P. F., B. E. Kemp and D. A. Power (2007) Regulation of endothelial and myocardial NO synthesis by multi-site eNOS phosphorylation. *Journal of Molecular and Cellular Cardiology,* 42, 271-279.

[59] Dimmeler, S., I. Fleming, B. Fisslthaler, C. Hermann, R. Busse and A. M. Zeiher (1999) Activation of nitric oxide synthase in endothelial cells by Akt-dependent phosphorylation. *Nature,* 399, 601-605.

[60] Fulton, D., J. P. Gratton, T. J. McCabe, J. Fontana, Y. Fujio, K. Walsh, T. F. Franke, A. Papapetropoulos and W. C. Sessa (1999) Regulation of endothelium-derived nitric oxide production by the protein kinase Akt. *Nature,* 399, 597-601.

[61] Hiyoshi, H., K. Yayama, M. Takano and H. Okamoto (2005) Angiotensin type 2 receptor-mediated phosphorylation of eNOS in the aortas of mice with 2-kidney, 1-clip hypertension. *Hypertension,* 45, 967-973.

[62] Yayama, K., H. Hiyoshi, D. Imazu and H. Okamoto (2006) Angiotensin II stimulates endothelial NO synthase phosphorylation in thoracic aorta of mice with abdominal aortic banding via type 2 receptor. *Hypertension,* 48, 958-964.

[63] Boo, Y. C. and H. Jo (2003) Flow-dependent regulation of endothelial nitric oxide synthase: role of protein kinases. *American Journal of Physiology-Cell Physiology,* 285, C499-C508.

[64] Boulanger, C. M., L. Caputo and B. I. Levy (1995) Endothelial at(1)-mediated release of nitric-oxide decreases angiotensin-ii contractions in rat carotid-artery. *Hypertension,* 26, 752-757.

[65] Pueyo, M. E., J. F. Arnal, J. Rami and J. B. Michel (1998) Angiotensin II stimulates the production of NO and peroxynitrite in endothelial cells. *American Journal of Physiology-Cell Physiology,* 43, C214-C220.

[66] Pueyo ME, Chen Y, D'Angelo G, Michel JB. Regulation of vascular endothelial growth factor expression by cAMP in rat aortic smooth muscle cells. *Exp. Cell. Res.* 1998 Feb 1;238(2):354-8.

[67] Thorup, C., M. Kornfeld, M. S. Goligorsky and L. C. Moore (1999) AT(1) receptor inhibition blunts angiotensin II-stimulated nitric oxide release in renal arteries. *Journal of the American Society of Nephrology,* 10, S220-S224

[68] Patzak, A., R. Mrowka, E. Storch, B. Hocher and P. B. Persson (2001) Interaction of angiotensin II and nitric oxide in isolated perfused afferent arterioles of mice. *Journal of the American Society of Nephrology,* 12, 1122-1127.

[69] Hennington, B. S., H. M. Zhang, M. T. Miller, J. P. Granger and J. F. Reckelhoff (1998) Angiotensin II stimulates synthesis of endothelial nitric oxide synthase. *Hypertension,* 31, 283-288.

[70] Ju, H., V. J. Venema, M. B. Marrero and R. C. Venema (1998) Inhibitory interactions of the bradykinin B2 receptor with endothelial nitric-oxide synthase. *Journal of Biological Chemistry,* 273, 24025-24029.

[71] Griendling, K. K., C. A. Minieri, J. D. Ollerenshaw and R. W. Alexander (1994) Angiotensin-ii stimulates nadh and nadph oxidase activity in cultured vascular smooth-muscle cells. *Circulation Research,* 74, 1141-1148.

[72] Beckman, J. S. and W. H. Koppenol (1996) Nitric oxide, superoxide, and peroxynitrite: The good, the bad, and the ugly. *American Journal of Physiology-Cell Physiology,* 40, C1424-C1437.

[73] Sigmon, D. H., O. A. Carretero and W. H. Beierwaltes (1992) Endothelium-derived relaxing factor regulates renin release invivo. *American Journal of Physiology,* 263, F256-F261.

[74] Kataoka, H., F. Otsuka, T. Ogura, T. Yamauchi, M. Kishida, M. Takahashi, Y. Mimura and H. Makino (2001) The role of nitric oxide and the renin-angiotensin system in salt-restricted Dahl rats. *American Journal of Hypertension,* 14, 276-285.

[75] Siragy, H. M., T. Inagami and R. M. Carey (2007) NO and cGMP mediate angiotensin AT(2) receptor-induced renal renin inhibition in young rats. *American Journal of Physiology-Regulatory Integrative and Comparative Physiology,* 293, R1461-R1467.

[76] Michel, J. B., Y. C. Xu, S. Blot, M. Philippe and G. Chatellier (1996) Improved survival in rats administered N-G-nitro L-arginine methyl ester due to converting enzyme inhibition. *Journal of Cardiovascular Pharmacology,* 28, 142-148.

[77] Persson, K. and R. G. G. Andersson (1999) Nitric oxide modulates captopril-mediated angiotensin-converting enzyme inhibition in porcine iliac arteries. *European Journal of Pharmacology,* 385, 21-27.

[78] Ackermann, A., M. S. Fernandez-Alfonso, R. S. de Rojas, T. Ortega, M. Paul and C. Gonzalez (1998) Modulation of angiotensin-converting enzyme by nitric oxide. *British Journal of Pharmacology,* 124, 291-298.

[79] Ichiki, T., M. Usui, M. Kato, Y. Funakoshi, K. Ito, K. Egashira and A. Takeshita (1998) Downregulation of angiotensin II type 1 receptor gene transcription by nitric oxide. *Hypertension,* 31, 342-348.

[80] Vandermeersch, S., V. Stefanovic, A. Hus-Citharel, R. Ardaillou, J. C. Dussaule and D. Chansel (2003) AT(1) receptor expression in glomeruli from NO-deficient rats. *Nephron Experimental Nephrology,* 95, E119-E128.

[81] Leclerc, P. C., P. M. Lanctot, M. Auger-Messier, E. Escher, R. Leduc and G. Guillemette (2006) S-nitrosylation of cysteine 289 of the AT(1) receptor decreases its binding affinity for angiotensin II. *British Journal of Pharmacology,* 148, 306-313.

[82] Onozato, M. L., A. Tojo, A. Goto, T. Fujita and C. S. Wilcox (2002) Oxidative stress and nitric oxide synthase in rat diabetic nephropathy: Effects of ACEI and ARB. *Kidney International,* 61, 186-194.

[83] Hink, U., H. G. Li, H. Mollnau, M. Oelze, E. Matheis, M. Hartmann, M. Skatchkov, F. Thaiss, R. A. K. Stahl, A. Warnholtz, T. Meinertz, K. Griendling, D. G. Harrison, U. Forstermann and T. Munzel (2001) Mechanisms underlying endothelial dysfunction in diabetes mellitus. *Circulation Research,* 88, E14-E22.

[84] Cosentino, F., K. Hishikawa, Z. S. Katusic and T. F. Luscher (1997) High glucose increases nitric oxide synthase expression and superoxide anion generation in human aortic endothelial cells. *Circulation,* 96, 25-28.

[85] Prabhakar SS Effects of Acute Angiotensin II Infusion on Nitric Oxide Synthesis and Nitric Oxide Synthase Isoforms in the Rat Kidney. *J Am Soc Neph* 15, p210A, 2004.

[86] Prabhakar SS. Angiotensin II Stimulates Intrarenal eNOS Expression and Nitric Oxide Generation in Anesthetized Rats Even in the Presence of Hyperglycemia. *J. Am. Soc. Neph.* 15, p 467A,2004.

[87] Prabhakar SS Enhanced insulin secretion due to angiotensin inhibition is associated with increased nitric oxide synthesis in pancreatic islet beta cells. *J Am. Soc Neph* (14) 2003

[88] Prabhakar SS. Regulatory and functional of interaction of vasoactive factors in the kidney and extracellular pH. Perspectives in Basic science. *Kidney International* 66 (5), 1742-1754, 2004

In: Angiotensin Research Progress
Editors: H. Miura and Y. Sasaki
ISBN: 978-1-60456-537-9

Chapter III

The Activity of Angiotensin II in the Human Placenta

Gershon Holcberg [1,3]***, Alaa Amash*** [2,3]***, Olga Sapir*** [1,3]***, Eyal Sheiner*** [1,3] ***and Mahmoud Huleihel*** [2,3]

[1] Department of Obstetrics and Gynecology, Soroka University Medical Center, Beer Sheva, Israel

[2] The Shraga Segal Department of Microbiology and Immunology, Beer Sheva, Israel

[3] Faculty of Health Sciences, Ben-Gurion University of the Negev, Beer Sheva, Israel

Abstract

The renin-angiotensin system (RAS) is one of main regulators of blood pressure and electrolyte balance during pregnancy. All the components of the RAS have been reported to be expressed in the utero-placental unit. Angiotensin (Ang) II is increasingly recognized as a growth factor, both in its own right and through interactions with other growth factors. Ang II synthesized in the placenta may serve as an autocrine/paracrine modulator of placental function. In addition to its role in regulation of blood pressure during normal pregnancy, Ang II is suggested to be involved in regulation of placentation through regulation of trophoblast invasion, angiogenesis and placental vascular development. These key effects of Ang II, in addition to its capacity to induce inflammatory response, suggest that increased vascular sensitivity to Ang II may have an important role in the increase in vascular resistance and elevated blood pressure seen pregnancy complications such as intra-uterin growth retardation (IUGR) and preeclampsia. Magnesium sulfate ($MgSO_4$) is used in pregnancy as either prophylaxis/treatment for eclamptic seizures or for preterm labor tocolysis. Our group has shown that $MgSO_4$ may attenuate Ang II induced placental vasoconstriction, and may affect Ang II induced placental cytokine production. This review summarizes the current understanding of the physiological aspects of Ang II in the utero-placental unit and its role in the pathogenesis of pregnancy disorders such as preeclampsia. Furthermore, the possible interactions between Ang II and $MgSO_4$ in the placental unit are also discussed.

Abbreviations

ACE	angiotensin-converting enzyme;
Ang	angiotensin;
ANP	atrial natriuretic peptide;
AT1	angiotensin II type 1 receptor;
AT1-AA	agonistic autoantibody to the angiotensin II type 1 receptor;
AT2	angiotensin II type 2 receptor;
B2	bradykinin B2 receptor;
BNP	brain natriuretic peptide;
Flt-1	fms-like tyrosine kinase-1;
HIF	hypoxia-inducible factor;
ICAM-1	intercellular adhesion molecule-1;
IL	interleukin;
IUGR	intrauterine growth retardation;
$MgSO_4$	magnesium sulfate;
NAD(P)H	nicotinamide adenine dinucleotide phosphate;
NF-κB	nuclear factor-kappa B;
NO	nitric oxide;
PA	plasminogen activator;
PAH	pregnancy-associated hypertension;
PAI	plasminogen activator inhibitor;
PDGF-β	platelet-derived growth factor-beta;
RAS	renin angiotensin system;
ROS	reactive oxygen species;
TGF-β	transforming growth factor-beta;
TNF-α	tumor necrosis factor-alpha;
VEGF	vascular endothelial growth factor;
VEGF-R	vascular endothelial growth factor receptor.

1. Introduction

The renin-angiotensin system (RAS) is a signaling cascade that plays a key role in regulating blood pressure and electrolyte balance. The reaction between renin, an aspartyl protease and its substrate, angiotensinogen, is the initial and rate-limiting step of this enzymatic cascade that generates the decapeptide angiotensin (Ang) I. Ang I is further processed to the most potent vasoactive octapeptide Ang II by angiotensin-converting enzyme (ACE) [1].

Traditionally, the RAS has been considered primarily as a circulating system involved in the regulation of blood pressure and salt and fluid homeostasis. In addition to this classic view of the RAS, accumulating evidence indicates that the components of the RAS are synthesized in many tissues, such as brain, heart, ovary, and placenta, and that Ang II levels can be controlled locally, independent of circulating Ang II [2-3]. These local RASs may function independently or in concert with the circulating RAS and participate in a paracrine and autocrine regulation of various aspects of tissue function.

Ang II is responsible for most of the physiological and pathophysiological effects of the RAS. It is not just a potent vasopressor which raises systemic blood pressure but it also mediates a broad array of physiological and pathophysiological effects by binding to specific cell membrane receptors. Two distinct types of G-protein-coupled receptors for Ang II have been cloned and characterized in humans and rodents. The angiotensin type 1 (AT1) receptor is predominantly expressed in the kidneys, adrenal glands, vascular smooth muscle cells and the heart. The regulatory actions of Ang II on blood pressure and salt/water balance have been attributed to this receptor [4-6]. The AT2 receptor is present at high density during fetal development, while in the adult, significant AT2 receptor expression occurs only in the adrenal medulla, uterus, ovary, vascular endothelium, adrenal glands and certain areas in the brain [7-8]. The AT2 receptor is thought to counterbalance effects mediated by the AT1 receptor: it appears to induce vasodilatation and may be involved in the control of cell proliferation, differentiation and angiogenesis [5,7-8].

2. Angiotensin II in the Utero-Placental Unit

The expression and localization of all RAS components in the utero-placental unite are well studied and defined. Renin is presented by all parts of the human utero-placental unit, with highest renin levels in the fetal membranes and the decidua [9]. The expression of renin mRNA by human placenta indicates the potential local renin synthesis [10]. Angiotensinogen peptide and mRNA were also found in the human placenta [10-12]. Moreover, expression and high activity of ACE were reported in the human uterus, placenta and fetal membranes [13-14].

The examination of ACE localization in the placenta directly revealed that ACE mRNA is localized primarily in vessel fraction of the placenta, especially in venous endothelial cells of stem villous of the placenta [15]. The ACE activity in vessel fractions is significantly higher than those in other fractions of the placenta. Moreover, the ACE activity of umbilical venous endothelial cells is significantly higher than that of umbilical arterial endothelial cells, suggesting that the placental and umbilical venous endothelial cells would be the main site for the conversion of Ang I to Ang II.

Ang II receptors of both AT1 and AT2 were demonstrated in placentas and fetal membranes of humans and other species [16]. The AT1 receptors are the predominant receptors in the human placenta [17-18], and are localized in the cytotrophoblast and syncytiotrophoblasts in the placental villi, in extravillous trophoblast, and around the blood vessels of the placental villi [18-19].

The presence of cathepsin D and chymase in the uterus indicates that alternative pathways for Ang II generation may exist independent from renin and ACE [20-21]. The placenta is known to produce large amounts of enzymes capable of Ang II degradation [22], such as aminopeptidase A produced in the fetal placenta which can degrade locally generated Ang II and thus may act as an effective barrier for Ang II between the fetoplacental and the maternal intervillous circulations [23].

The conditions for Ang II formation exist and Ang II receptors are present throughout the human utero-placental unit, indicating the presence of a functional local RAS. These findings also suggest that Ang II synthesized in the placenta may serve as an autocrine/paracrine modulator of placental function.

3. Physiological Aspects of Angiotensin II in the Utero-Placental Unit

3.1. Regulation of the Utero-Placental Vascular Resistance and Blood Flow

The primary function of the placenta is to act as an interface between mother and fetus that allows, and even promotes, fetal growth and development, and to contribute to the maternal cardiovascular adaptations of pregnancy. Appropriate perfusion of the placenta is necessary for its critical endocrine and exchange functions. The regulation of vascular resistance and blood flow in the utero-placental unit is complex [24]. As this circulation is not innervated, vascular tone and resistance to blood flow must be largely controlled by autocrine or paracrine factors [25]. One of the postulated mechanisms by which the placenta may influence maternal vascular tone is the placental RAS.

In human pregnancy, on the maternal side of utero-placental unit, there are two major renin-producing circulations, namely, renal circulation and utero-placental circulation. Experimentally, in pregnant rabbits, a reciprocal relationship between uterine and renal renin synthesis has been demonstrated; i.e., when uterine renin synthesis is increased, renal renin synthesis is reciprocally decreased [26]. A complete local tissue-based RAS in the utero-placental unit may regulate the regional maternal intervillous blood flow [27].

The maternal circulating RAS and utero-placental tissue RAS has been thought to support maternal placental flow by raising maternal arterial pressure or changing placental vascular resistance. Previously, our group has examined the vasoactive effect of Ang II on uncomplicated human perfused placenta, and showed that bolus injections of Ang II (10^{-9}–10^{-5} M) resulted in a significant increase in perfusion pressure in a dose-dependent manner as compared with basal pressure, indicating that Ang II is a possible regulator of placental blood flow during pregnancy (figure 1) [28].

The placenta or uterus may alter maternal circulating RAS. During normal pregnancy, plasma renin concentration, renin activity, and Ang II levels are elevated. However, the vascular responsiveness to Ang II appears to be reduced [29-30]. The systemic vasculature is thus refractory to Ang II in normal pregnancy, i.e., requires a higher Ang II infusion rate (almost twice as high compared with nongravid at the peak refractory state) for the same degree of vascular response [31-32]. This vascular refractoriness to Ang II in pregnancy has been ascribed, in part, to progesterone and prostacycline [33]. Angiotensinogen levels also increase in pregnancy, but the exact timing of this change is not well defined. Aldosterone levels are increased, and this may contribute to sodium retention and resultant obligatory water retention, which is one of the mechanisms of volume expansion in pregnancy [34].

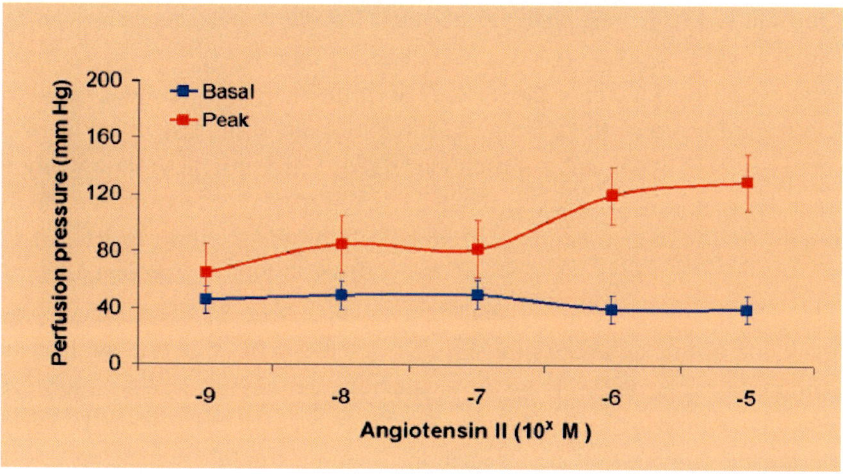

Figure 1. Pressor response to Ang II in an isolated placental cotyledon from a normal placenta.

The control of vascular tone and blood pressure achieved by Ang II is probably mediated by vascular endothelial synthesis of prostaglandins. The vascular response to Ang II has been shown to be diminished by the use of prostaglandin inhibitors, indomethacin and aspirin [35-36]. During normal pregnancy there is an increased production of the potent vasodilator prostacyclin, and the vasoconstrictor thromboxane A_2, with a delicate balance which favors prostacyclin [37]. Our group has also shown that the vasoconstrictor effect of Ang II on feto-placental vasculator is attenuated by atrial natriuretic peptide (ANP) and brain natriuretic peptide (BNP) [38].

3.2. Implantation and Placentation

Ang II is increasingly recognized as a growth factor, independently and through interactions with other growth factors. In addition to its well-known effects on the cardiovascular system and on salt and water homeostasis, Ang II through its AT1 receptor also mediates processes that are likely to be important for successful placentation, including cellular proliferation and growth [39] and angiogenesis [40-41]. The AT1 receptor is present in the placenta throughout pregnancy [42] and has been localized to cytotrophoblast, syncytiotrophoblast, extravillous trophoblast and perivascular cells [43].

Trophoblast invasion, implantation and placental development depend, in part, on the controlled production of plasmin from plasminogen, a process regulated by plasminogen activators (PAs) and plasminogen activator inhibitors (PAIs) [44-45]. Ang II, through the AT1 receptor, has been shown to regulate plasminogen activation through stimulation of PAI-1 synthesis and secretion in human trophoblasts [46]. Moreover, activation of the RAS may induce multiple growth factors that have been implicated in fibrosis, such as transforming growth factor-β (TGF-β) and platelet-derived growth factor-β (PDGF-β) [47-49]. Induction of these growth factors is mediated by Ang II action on the AT1 receptor [50]. TGF-β1 inhibits trophoblast invasion [51], proliferation [52], migration [53], and limits degradation of the extracellular matrix by extravillous trophoblast [52]. Since TGF-β1 is produced predominantly by the maternal decidua, it may provide a mechanism to prevent over-invasion.

Furthermore, all RAS components also exist in the maternal placenta, i.e., the decidua. The decidua has been shown to be a major source of renin in the gravid human uterus [54]. Angiotensinogen, renin, ACE, and AT1 are expressed in and around the spiral arteries of the first-trimester decidua [55]. Moreover, renin and angiotensinogen expression has been also described in prolactin-producing human decidual cells from the third trimester of pregnancy [56]. The molecular evidence of a tissue-based RAS in the decidua may have several implications: The local spiral artery RAS may play a role in the pregnancy-associated vascular modeling of the spiral arteries [57]. The synthesis of renin and angiotensinogen in the prolactin-producing endocrine cells of the decidua has implications for the regulation of this RAS [56]. Transformation of uterine-lining endometrial stromal cells to the decidual cells of pregnancy and expression and secretion of prolactin are regulated by estrogen, progesterone, and several other paracrine factors [58]. The sex steroids also regulate hepatic angiotensinogen expression [59]. Therefore, decidual RAS may be regulated by estrogen and progesterone and by other factors involved in decidualization [56]. Progesterone treatment of decidual cells or endometrial stromal cells in vitro increases active renin secretion [60-61]. Uterine renin secretion is a constitutive secretion, with a response time of hours for various stimuli compared with renal renin, which has a regulated secretion, cells having storage granules and a response time of few minutes [62].

Additional important growth factor in placental development is the vascular endothelial growth factor (VEGF) family and their receptors. VEGF is expressed in the placenta by villous trophoblasts in the first trimester, by extravillous trophoblasts at term and by both fetal and maternal macrophages [63]. There are two receptors for VEGF, VEGF-R1, also known as fms-like tyrosine kinase (or Flt-1) and kinase domain receptor (VEGF-R2). Ang II induces angiogenesis in several in vivo models [40,64-66], including the chroioallantoic membrane assay [66] and the corneal pocket model of the rabbit [40]. Ang II infusion increases capillary density in several animal models [40,65-66]. Ang II has several stimulatory effects on angiogenesis including the induction of VEGF [67-68], hypoxia-inducible factor (HIF)-1α [69], and VEGFR-2 [70] expression. The proangiogenic effects of Ang II appear to be mediated by the AT1 receptor [65,71-72]. Ang II can also act at the AT2 receptor and activation of the AT2 receptor has been reported to have an inhibitory effect on angiogenesis. Ang II has been reported to inhibit VEGF mediated tube formation by endothelial cells through AT2 activation [73].

Towards the end of gestation, continued placental vascular development is terminated, a process accomplished in part through the production of antiangiogenic factors. One such antiangiogenic factor is a soluble form of VEGFR-1 (sFlt-1) that binds to circulating VEGF and placental growth factor, thereby inhibiting their angiogenic activities. sFlt-1, a vascular growth factor antagonist produced by the placenta, is released into the maternal circulation [74]. Infusion of Ang II has been shown to increase circulating levels of sFlt-1 in pregnant mice, demonstrating that Ang II is a regulator of sFlt-1 secretion in vivo. Furthermore, Ang II stimulated sFlt-1 production by human villous explants and cultured trophoblasts but not by endothelial cells, suggesting that trophoblasts are the primary source of sFlt-1 during pregnancy.

In summary, Ang II is one of the key factors regulating the process of placentation through regulation of trophoblast invasion, angiogenesis and placental vascular development.

4. The Role of Angiotensin II in Preeclampsia

The fact that the RAS is one of the main regulators of vascular resistance and blood flow during pregnancy suggests that disruptions in this system may play a critical role in pregnancy disorders. Several indirect lines of evidence indicate that the up-regulation of RAS in the placenta might be important for the pathogenesis of some clinical disorders, such as pregnancy-associated hypertension (PAH) and preeclampsia [75-77].

Preeclampsia is a multisystem disorder generally appearing after the 20th week of gestation characterized by hypertension, proteinuria, vascular abnormalities, and possibly intrauterine growth retardation (IUGR) [78-80]. It affects approximately 3–5% of all pregnancies and considered to be one of the leading causes for maternal mortality and a major contributor to maternal and perinatal morbidity. Nevertheless, the mechanisms responsible for the pathogenesis of preeclampsia are poorly understood. The only effective treatment is delivery of the fetus and placenta, often resulting in serious complications of prematurity for the neonate.

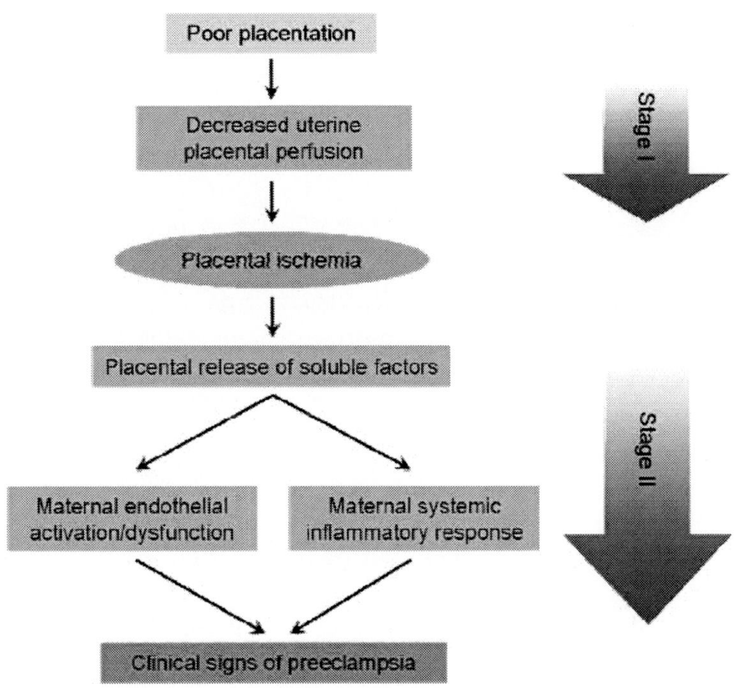

Figure 2. The two-stage model for preeclampsia.

Preeclampsia is currently believed to be a two-stage disease [79,81-82]. The first stage is characterized by shallow cytotrophoblast invasion of maternal spiral arterioles, resulting in placental insufficiency. The hypoxic placenta releases soluble factors into the maternal circulation, which induce systemic endothelial dysfunction. This causes the second stage of the disease: the maternal syndrome. During this stage, hypertension and proteinuria, the clinical signs of preeclampsia, are manifested [83] (figure 2).

4.1. Circulating Angiotensin in Preeclampsia

The RAS has been implicated in the pathogenesis of preeclampsia [27]. While renin, angiotensinogen, Ang II, and aldosterone are increased in the peripheral blood in normal pregnancy [84], in patients with preeclampsia, plasma renin activity and aldosterone are paradoxically suppressed with relatively higher levels of aldosterone for the given level of renin [85-86], suggesting increased adrenal sensitivity to Ang II. Moreover, normal pregnancy is associated with decreased vascular responsiveness to Ang II [31], and preeclampsia is associated with increased sensitivity to Ang II that may develop before the clinical manifestations of the disease [32]. In addition, recent genetic studies have suggested that elevated expression of maternal Ang in human decidual spiral arteries may promote impaired remodeling of these vessels and reduce utero-placental blood flow, potentially initiating for the cascade preceding preeclampsia [12,57] and IUGR [87].

The increased renal sensitivity to Ang II in preeclampsia may be a result of the reduced utero-placental perfusion, and it may be achieved through reductions in nitric oxide (NO) or prostacyclin synthesis or by enhanced formation of thromboxane. Consistent with this suggestion are studies indicating enhanced vascular responsiveness to Ang II in vessels from animals or humans with preeclampsia [29]. Furthermore, previous studies have found that, unlike normal conditions, the preglomerular vessels of the renal circulation become extremely sensitive to the vasoconstrictor actions of Ang II when the renal synthesis of NO or prostacyclin is reduced or when thromboxane synthesis is elevated [88]. Increased vascular Ang II responsiveness during preeclampsia, however, does not prove that Ang II is an important endogenous mediator of the vasoconstriction or hypertension in experimental models of preeclampsia because increased responsiveness may only reflect low endogenous Ang II formation. Thus, the importance of increased Ang II in the regulation of renal function and blood pressure during preeclampsia remains unclear.

4.2 Placental Angiotensin in Preeclampsia

The capacity of the placenta to produce Ang II seems to be also increased in preeclampsia. Recently, increased ACE expression and activity have been reported in placental tissues as well as in umbilical venous plasma samples from pregnancies complicated with preeclampsia [77]. Previously, it has been shown that the level of Ang II in umbilical venous plasma was higher than that in umbilical arterial plasma in preeclamptic placentas and uncomplicated placentas, and that the level of Ang II in maternal peripheral venous plasma was much higher in preeclampsia than in uncomplicated pregnancies [13,89-90]. The level of Ang II in umbilical venous plasma in preeclampsia was also increased as compared to uncomplicated pregnancy, the steeper gradient was noted between umbilical venous and arterial plasma [77]. These data suggest that ACE activity in preeclamptic placenta would be elevated. On the contrary, Kalenga et al did not find any significant differences in placental ACE and Ang II levels between uncomplicated pregnancies and preeclampsia [14]. This controversy may be explained by differences in the investigation methodology used by each group.

4.3. Role of Angiotensin II Receptors

Recent studies in preeclamptic women have revealed several exciting findings regarding the role that Ang II receptors may play in the pathophysiology of preeclampsia. AT1 receptors were shown to form heterodimers with the bradykinin B2 receptor and results in enhanced Ang II sensitivity [91-92]. Further, the AT1-B2 heterodimers are present in greater abundance in preeclamptic women, suggesting that this heterodimerization may play a part in the long observed increased Ang II sensitivity in preeclampsia [91-92].

Furthermore, increased circulating levels of an agonistic autoantibody to the Ang II type 1 receptor (AT1-AA) were demonstrated in preeclamptic women [93-94]. The AT1-AA appear to be responsible for a variety of effects in several different tissues ranging from increased intracellular Ca^{++} mobilization to monocyte activation and stimulation of interleukin (IL)-6 production from mesangial cells [95-98]. While these findings potentially implicate AT1 as a central mediator of several pathways in preeclampsia, both the specific mechanisms that lead to excess production and the mechanisms whereby AT1-AA increases blood pressure during pregnancy remain unclear. Consequently, this has become an area of intense interest.

The AT2 receptors appear to function as an antagonist of AT1 [99]. Increased responsiveness to Ang II has been demonstrated in the uterine vasculature during AT2 receptor blockade [100]. Moreover, enhanced Ang II reactivity has been observed in the uterine vasculature with decreased expression of AT2 in sheep exhibiting preeclampsia-like syndrome after Ang II infusion [101]. Recently, Judson et al., using immunohistochemical techniques, reported that while in normal patients, the level of AT2 receptors is higher than the level of AT1 receptors in patients with preeclampsia this balance is disrupted and the level of AT1 receptors is higher than the level of AT2 receptors [102]. However, more investigations are needed in order to determine whether the expression of AT2 receptors is altered in preeclampsia.

4.4. Angiotensin and Endothelial Dysfunction

Endothelial dysfunction plays a central role in the pathogenesis of the maternal syndrome in preeclampsia. Dysfunctional endothelial cells produce altered quantities of vasoactive mediators, which lead to a tip in the balance towards vasoconstriction [103]. Multiple interconnected pathways including oxidative stress, cytokine release, and a generalized intra-vascular inflammatory response, were reported to be involved in this process. This culminates in increased endothelial cell permeability, lipid peroxidation [104], platelet activation [105], activation of the coagulation cascade, oxidative stress [106], and a shift in the balance of vasoactive mediators favoring vasoconstriction.

There is accumulating evidence to indicate that Ang II is also capable of inducing inflammatory response in the vascular wall [107]. Characteristic features of inflammation include changes in vascular permeability, recruitment of leukocytes from the circulation to the interstitial tissue, and tissue repair, which involves cell growth and fibrosis [108]. Ang II has been shown to participate in all these processes [107].

Recently, it has been suggested that Ang II, via AT1 receptors, enhances the production of reactive oxygen species (ROS) through stimulation of nicotinamide adenine dinucleotide phosphate [NAD(P)H] oxidase in the vascular wall [109]. Increased oxidative stress contributes to endothelial dysfunction and to vascular inflammation by stimulating the redox-sensitive transcription factor, nuclear factor-kappa B (NF-κB), and by upregulating adhesion molecules, cytokines, and chemokines [107].

Cytokines are messengers for the regulation of the inflammatory cascade, and tumor necrosis factor-α (TNF-α), IL-1β, IL-6 and IL-8 are the main pro-inflammatory cytokines [110]. It has been reported that plasma levels of TNF-α and IL-1β are increased in preeclampsia, whereas conflicting reports on IL-6 and IL-8 show either no difference or increased concentrations [111-115]. Pro-inflammatory cytokines can induce functional and structural alterations, including oxidative damage or interference with the mechanism of concentration/relaxation leading to alterations in vascular integrity, tone and coagulation [110]. Several lines of evidence support the hypothesis that the ischemic placenta in preeclampsia contributes to endothelial cell activation/dysfunction of the maternal circulation by enhancing the synthesis of cytokines such as TNF-α and IL-6 [116].

TNF-α enhances the formation of number of endothelial cell substances such as endothelin and reduces acetylcholine-induced vasodilatation [116]. It has been shown to directly induce oxidative damage as it destabilizes electron flow in mitochondria, resulting in release of oxidizing free radicals and formation of lipid peroxides. Lipid peroxides and oxygen radicals can damage endothelial cells as they are highly reactive compounds. Recently, Ang II has been shown to stimulate the production of cytokines such as TNF-α and IL-6 via the AT1 receptor [117], whereas Ang II receptor blocking attenuated the Ang II stimulated production of these cytokines [118]. Previously, we have shown that stimulation of normal human perfused placenta with Ang II resulted in increased secretion of TNF-α (figure 3) [28]. These findings suggest that inflammatory cytokines play an important role in mediating the effect of Ang II on the vascular system. On the other hand, TNF-α and IL-1β were shown to act coordinately to increase AT1 receptor density on non-myocytes in the post-MI heart [119].

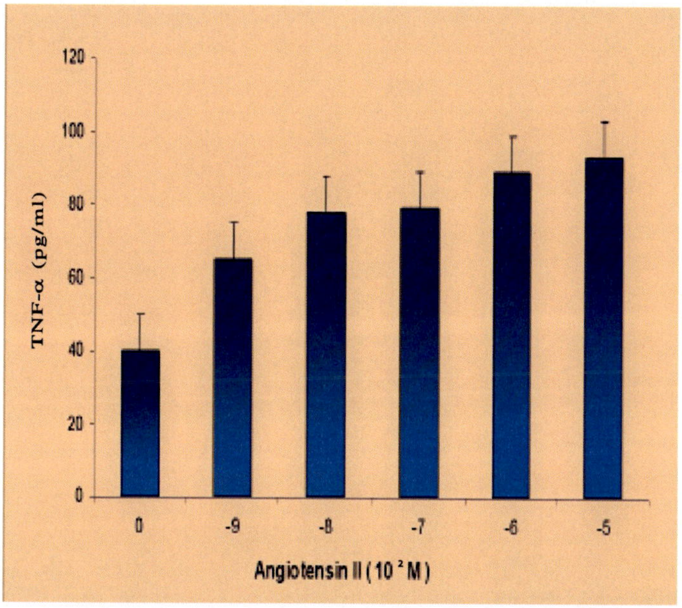

Figure 3. Secretion of TNF-α by normal human placenta before and after Ang II injection (10^{-9} – 10^{-5} M).

5. Magnesium Sulfate and Regulation of Angiotensin II Activity

Magnesium sulfate ($MgSO_4$) is used in pregnancy as prophylaxis and treatment for eclamptic seizures or for preterm labor tocolysis [120]. Magnesium has been shown to reach fetal blood and the amniotic cavity promptly in significant levels following maternal treatment [121-122]. Potential beneficial effects, such as neuro-protection during periods of perinatal hypoxic-ischemia and an association with a decreased risk of cerebral palsy, are being actively studied. Magnesium may improve neonatal outcome through its action on placental vessels. It has been suggested that magnesium may protect against infection-mediated damage by reducing cytokines and bacterial toxin synthesis [123-125].

Previously, Kovac et al. reported an increased feto-placental perfusion pressure in placentas from preeclamptic pregnancies, which decreases with exposure to $MgSO_4$ [126]. Moreover, they showed that feto-placental vascular response to Ang II, either in normotensive or in preeclamptic placenta, was not affected by $MgSO_4$. However, recently, using a placental perfusion system previously described by Schneider and Huch [127] and modified by Holcberg et al. (figure 4) [128], we have shown that $MgSO_4$ selectively attenuates the vasoconstrictor effect of Ang II (figure 5) and endothelin-1, but not of thromboxane, in the feto-placental vasculature of human normotensive placenta (figure 5) [128]. The mechanism of action of $MgSO_4$ is thought to be secondary to a competitive inhibition of calcium and its effect on smooth muscle contraction [129].

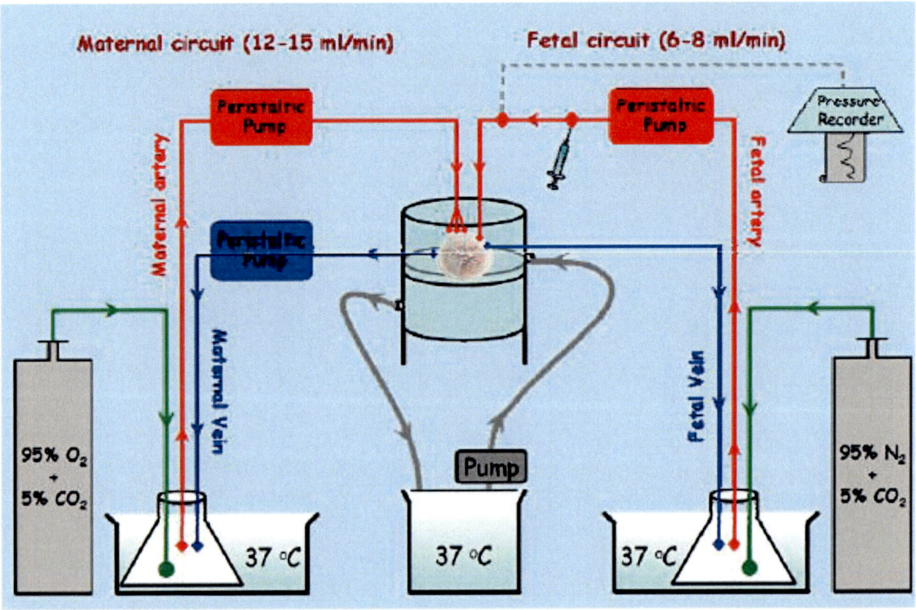

Figure 4. The placental perfusion system as previously described by Schnider and Huch, and modified by Holcberg et al.,.

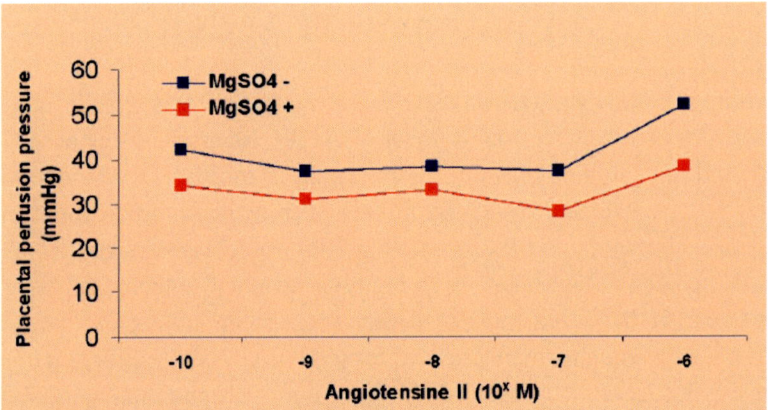

Figure 5. Effect of MgSO4 on Ang II (10-10-10-6M) induced vasoconstriction in perfused human normotensive placenta.

Recent findings suggest that $MgSO_4$ may function as an anti-inflammatory agent in certain pathological conditions such as preterm labor [130], hypertension, vascular disease and stroke [131]. Rochelson et al. suggested that $MgSO_4$ may inhibit endothelial cell activation, as measured by levels of IL-8 and endothelial cell-associated intercellular adhesion molecule-1 (ICAM-1) via NF-κB, and that $MgSO_4$ treatment may function as an anti-inflammatory agent in preterm labor [132]. Since NFκB is one of the main transcription factors mediating Ang II inflammatory response, it is possible to suggest that $MgSO_4$ may be a possible regulator for this response via inhibition of NFκB. Consisting with these suggestion, recently, we have recently shown that $MgSO_4$ attenuates Ang II induced IL-1β secretion by human normotensive perfused placenta (figure 6) [128]. These data suggest that

$MgSO_4$ may be a possible regulator not only of Ang II vasoconstrictor effect on the vascular smooth muscle cells, but also of the inflammatory effect of Ang II on vascular endothelial cells (figure 7).

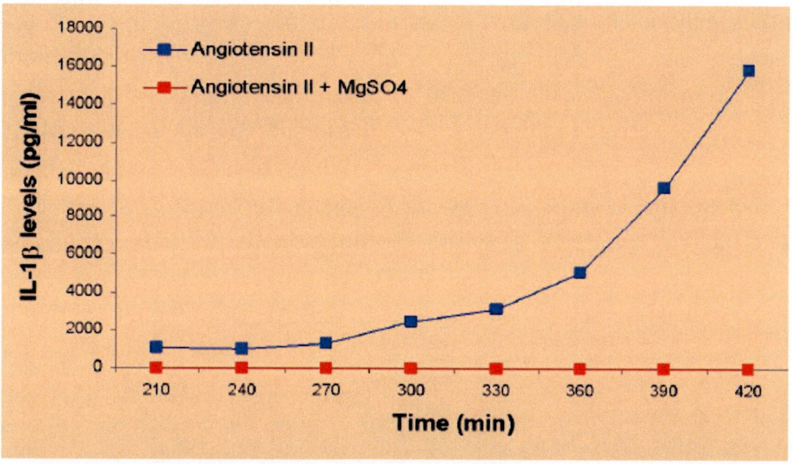

Figure 6. Effect of $MgSO_4$ on Ang II-induced IL-1β secretion by perfused human normotensive placenta.

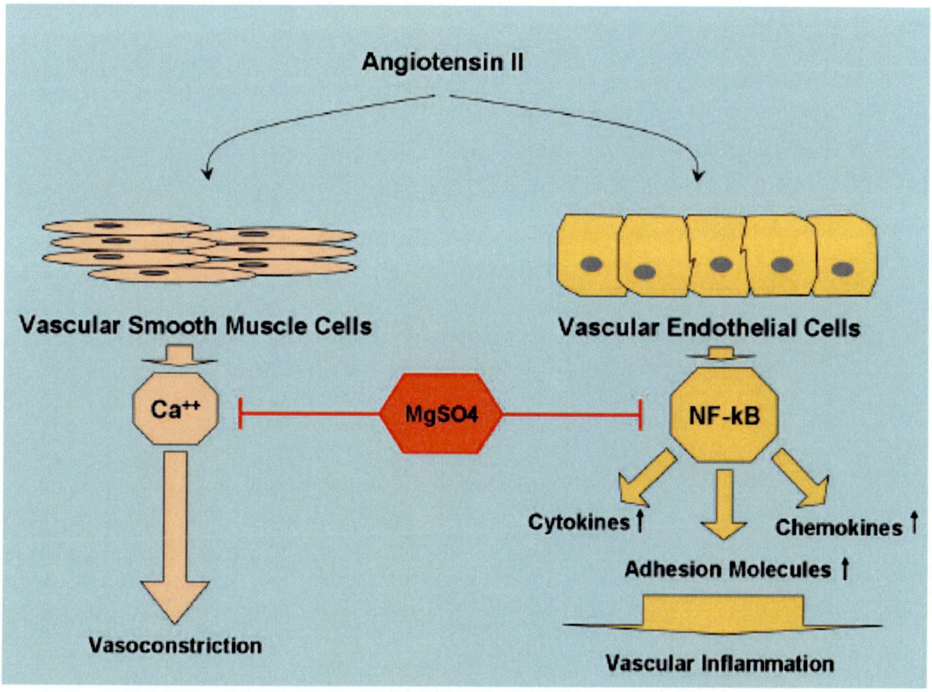

Figure 7. Possible interaction between MgSO4 and Ang II.

Recently, our group has shown that $MgSO_4$, in addition to its placental vascular vasodilatation effect, may regulate the capacity of human placenta to secrete inflammatory cytokines such as TNF-α and IL-6 [132]. We showed that $MgSO_4$ is capable of increasing TNF-α and IL-6 secretion from the maternal compartment of the human placenta, and that Ang II significantly attenuates this effect (figure 8A and B). However, in the fetal placental compartment, the interaction of $MgSO_4$ with Ang II leads to a significant increase in TNF-α and IL-6 levels, and $MgSO_4$ alone has no significant effect on TNF-α or IL-6 secretion into the fetal compartment (figure 9A and B). This may reflect a different response or function of the fetal and maternal compartments of the placenta to these agents. These findings suggest also that $MgSO_4$ may be associated with a significant increase in TNF-α, and IL-6 secretion by the human placenta, and bolus injections of Ang II differently affect secretion of these cytokines into the maternal versus the fetal compartments. The fact that increased levels of TNF-α and IL-6 were secreted into the fetal compartment, after exposure to $MgSO_4$ and Ang II, suggests the possibility of harmful effects of TNF-α to the fetal brain, which may affect late stages of development. Thus, it is possible to suggest that $MgSO_4$, at least partially, could be involved in adverse fetal outcome and the decision for $MgSO_4$ administration during pregnancy should be carefully considered. Further investigation is needed in order to determine this possible interaction between $MgSO_4$ and Ang II in the feto-placental unit.

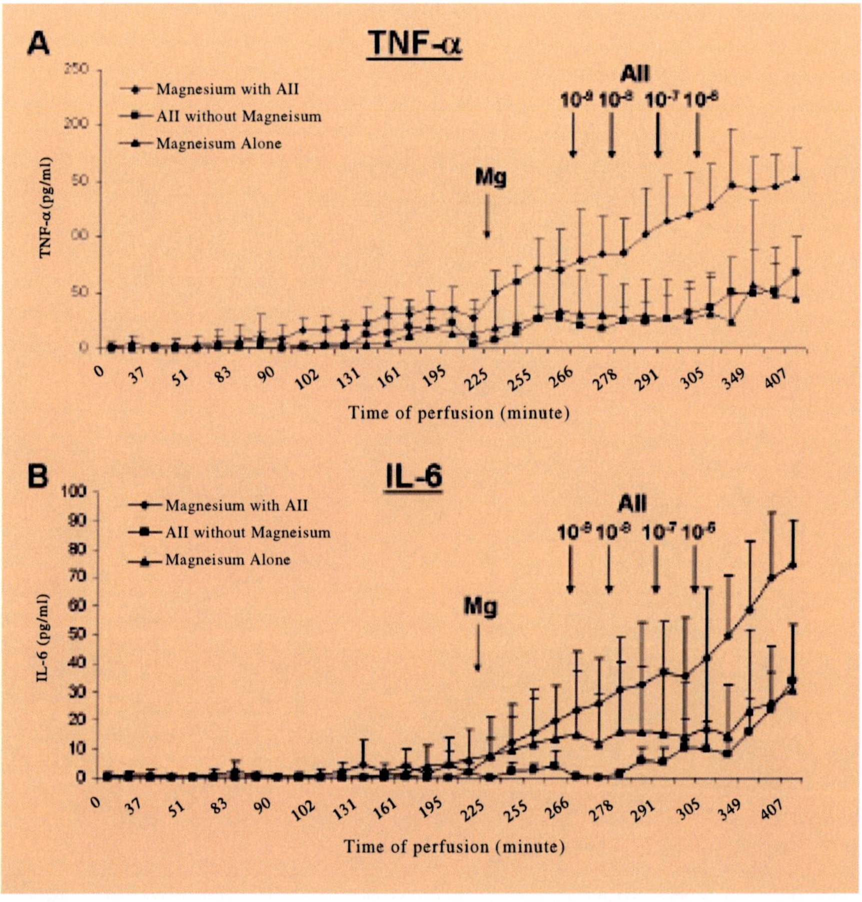

Figure 8. Secretion levels of TNF-α (A) and IL-6 (B) into the maternal compartment of normotensive human term placenta in presence of Ang II alone (four sequential bolus injections into the fetal circuit [10^{-9}-10^{-6} M], at 20 min intervals), $MgSO_4$ alone (6–7mg% in constant infusion into the maternal circuit), and both Ang II and $MgSO_4$ together.

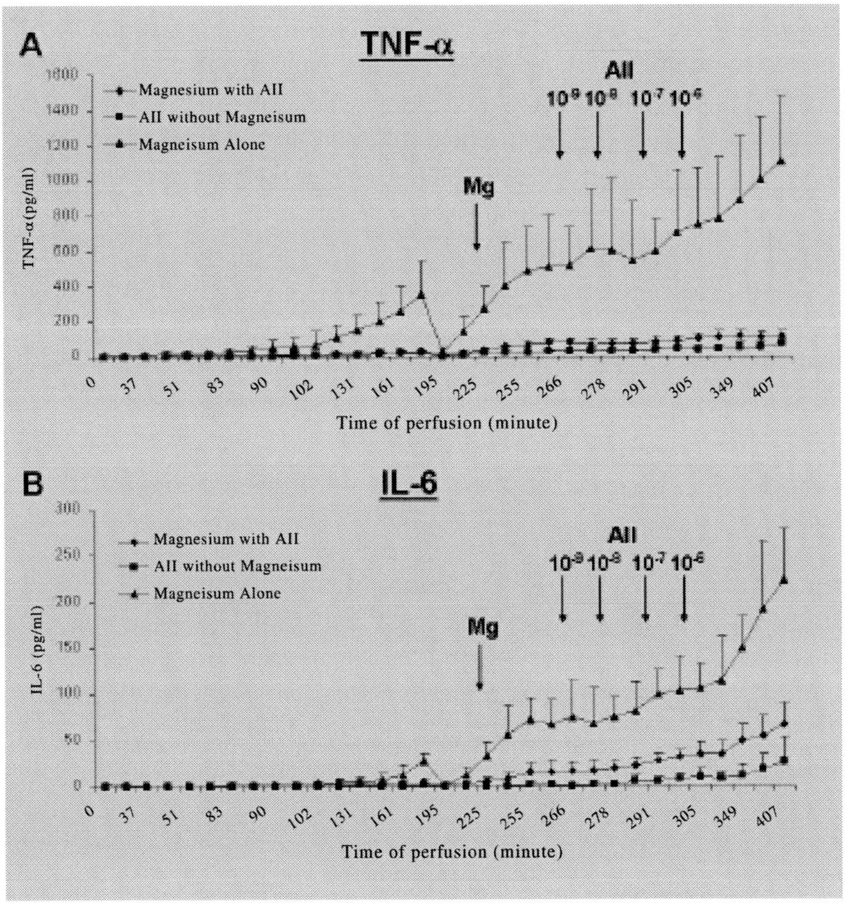

Figure 9. Secretion levels of TNF-α (A) and IL-6 (B) into the fetal compartment of normotensive human term placenta in presence of Ang II alone (four sequential bolus injections into the fetal circuit [10^{-9}-10^{-6} M], at 20 min intervals), $MgSO_4$ alone (6–7mg% in constant infusion into the maternal circuit), and both Ang II and $MgSO_4$ together.

Conclusion

The conditions for Ang II formation exist and Ang II receptors are present throughout the human utero-placental unit, indicating the presence of a functional local RAS in this organ. Ang II synthesized in the placenta may serve as an autocrine/paracrine modulator of placental function. During normal pregnancy, Ang II levels are elevated although the vascular responsiveness to Ang II appears to be reduced. Ang II is one of the key factors regulating the process of placentation through regulation of trophoblast invasion, angiogenesis and placental vascular development.

Up-regulation of placental RAS might be important for the pathogenesis of some clinical disorders, such as preeclampsia. Increased vascular responsiveness to Ang II, as well as increased placental production of Ang II, were reported in preeclampsia. Moreover, increased AT1 receptor activity was reported in preeclampsia, which may be a result of increased heterodimerization of AT1 with the bradykinin B2 receptor and increased circulating levels of an agonistic autoantibody to the AT1 receptor. Ang II, via AT1 receptor, increases oxidative stress and contributes endothelial dysfunction and to vascular inflammation by stimulating the redox-sensitive transcription factor NF-κB, and by upregulating adhesion molecules, cytokines, and chemokines. $MgSO_4$ may be a possible regulator not only of Ang II vasoconstrictor effect on the vascular smooth muscle cells, but also of the inflammatory effect of Ang II on vascular endothelial cells.

In conclusion, Ang II has significant physiological aspects in reproduction in general, and in placental development and regulation of the utero-placental perfusion in particular. Further studies are still needed in order to assess more accurately the possible interactions between Ang II and $MgSO_4$ in general and the utero-placental unit in particular.

References

[1] August, P., Lenz, T., Ales, K. L., Druzin, M. L., Edersheim, T. G., Hutson, J. M., Muller, F. B., Laragh, J. H.,and Sealey, J. E. (1990). Longitudinal study of the renin-angiotensin-aldosterone system in hypertensive pregnant women: deviations related to the development of superimposed preeclampsia. *Am. J. Obstet. Gynecol.* 163, 1612–1621.

[2] Poisner, A. M. (1998). The human placental renin-angiotensin system. *Front. Neuroendocrinol.* 19, 232-252.

[3] Hagemann, A., Nielsen, A. H.,and Poulsen, K. (1994). *Exp. Clin. Endocrinol.* 102, 252-261

[4] Dinh, D.T. (2001). Angiotensin receptors: distribution, signaling and function. *Clin. Science,* 100, 481–492.

[5] Brede, M.and Hein, L. (2001). Transgenic mouse models of angiotensin receptor subtype function in the cardiovascular system. *Regul. Pept.* 96, 125–132.

[6] Ito, M., Oliverio, M.I., Mannon, P.J., Best, C.F., Maeda, N., Smithies, O.and Coffman, T.M. (1995). Regulation of blood pressure by the type 1A angiotensin II receptor gene. *Proc. Natl. Acad. Sci. U.S.A.* 92, 3521–3525.

[7] Turner, A.J.and Hooper, N.M. (2002). The angiotensin-converting enzyme gene family: genomics and pharmacology. *Trends Pharmacol. Sci.* 23, 177–183.

[8] Meffert, S., Stoll, M., Steckelings, U.M., Bottari, S.P.and Unger, T. (1996). The angiotensin II AT2 receptor inhibits proliferation and promotes differentiation in PC12W cells. *Mol. Cell. Endocrinol.* 22, 59–67.

[9] Skinner, S.L., Lumbers, E.R.and Symonds E.M. (1968). Renin concentration in human fetal and maternal tissues. *Am. J. Obstet Gynecol*, 101, 529–533.

[10] Paul, M., Wagner, J.and Dzau, V.J. (1993). Gene expression of the reninangiotensin system in human tissues. Quantitative analysis by the polymerase chain reaction. *J. Clin. Investig*, 91, 2058–2064.

[11] Tewksbury, D.A. (1996). Quantification of five forms of high molecular weight angiotensinogen from human placenta. *Am. J. Hypertension*, 9, 1029–1034.

[12] Morgan, T., Craven, C., Nelson, L., Lalouel, J.M.and Ward, K. (1997). Angiotensinogen T235 expression is elevated in decidual spiral arteries. *J. Clin Investig*, 100, 1406–1415.
[13] Yagami, H., Kurauchi, O., Murata, Y., Okamoto, T., Mizutani, S.and Tomoda, Y. (1994). Expression of angiotensin-converting enzyme in human placenta and its physiologic role in the fetal circulation. *Obstet. Gynecol*, 84, 453–457.
[14] Kalenga, M.K., Thomas, K., de Gasparo, M.and de Hertogh, R. (1996). Determination of renin, angiotensin converting enzyme and angiotensin II levels in human placenta, chorion and amnion from women with pregnancy induced hypertension. *Clin. Endocrinol*, 44, 429–433.
[15] Ito, M., Itakura, A., Ohno, Y., Nomura, M., Senga, T., Nagasaka, T.and Mizutani, S. (2002). Possible activation of the renin–angiotensin system in the feto-placental unit in preeclampsia. *J. Clin. Endocrinol. Metab*. 87, 1871– 1878.
[16] Kalenga, M.K., de Gasparo, M., Thomas, K.and de Hertogh, R. (1996). Review article. Angiotensin II and its different receptor subtypes in placenta and foetal membranes. *Placenta*, 17, 103–110.
[17] Kalenga, M.K., de Gasparo, M., de Hertogh, R., Whitebread, S., Vankrieken, L.and Thomas, K. (1991). Angiotensin II receptors in the human placenta are type AT1. *Reproduction, Nutrition, Developement*, 31, 257–267.
[18] Knock, G.A., Sullivan, M.H.F., McCarthy, A., Elder, M.G., Polak, J.M.and Wharton, J. (1994). Angiotensin II (AT1) vascular binding sites in human placentae from normal-term, preeclamptic and growth retarted pregnancies. *J. Pharmacol. Exp. Therapeutics*, 271, 1007–1015.
[19] Li, X., Shams, M., Zhu, J., Khalig, A., Wilkes, M., Whittle, M., Barnes, N.and Ahmed, A. (1998). Cellular localization of AT1 receptor mRNA and protein in normal placenta and its reduced expression in intrauterine growth restriction. *J. Clin. Investig*, 101, 442–454.
[20] Sapolsky, A.I.and Woessner, J.F. (1972). Multiple forms of cathepsin D from bovine uterus. *J. Biol. Chem*, 247, 2069–2076.
[21] Urata, H., Strobel, F.and Ganten, D. (1994). Wide spread tissue distribution of human chymase. *J. Hypertension*, 12, s17–s22.
[22] Mitzutani, S., Goto, K., Safwat, M., Itakura, A., Kurauchi, O., Kikkawa, F.and Tomoda, Y. (1994). Initiating enzymes of angiotensin II degradation in human placenta and pregnancy serum. *Medical Science Research*, 22, 537–544.
[23] Hariyama, Y., Itakura, A., Okamura, M., Ito, M., Murata, Y., Nagasaka, T., Nakazato, H.and Mizutani, S. (2000). Placental aminopeptidase A as a possible barrier of angiotensin II between mother and fetus. *Placenta*, 21: 621– 627.
[24] Poston, L. (1997). The control of blood flow to the placenta. Exp Physiol, 82, 377–387.
[25] Boura, A.L., Walters, W.A., Read, M.A.and Leitch, I.M. (1994) Autocoids and control of human placental blood flow. *Clin. Exp. Pharmacol. Physiol*, 21, 737–748.
[26] Dzau, V.J., Gonzalez, D., Ellison, K., Churchill, S.and Emmett, N. (1987). Characterization of purified rabbit uterine renin: influence of pregnancy on uterine inactive renin. *Endocrinology*, 120: 358–364.

[27] Nielsen, A.H., Schauser, K.H.and Poulsen, K. (2000). Current topic: the uteroplacental renin-angiotensin system. *Placenta*, 21, 468–477.

[28] Holcberg, G., Huleihel, M., Sapir, O., Katz, M., Tsadkin, M., Furman, B., Mazor, M.and Myatt, L. (2001). Increased production of tumor necrosis factor-alpha TNF-alpha by IUGR human placentae. *Eur. J. Obstet. Gynecol. Reprod. Biol*, 94, 69-72.

[29] August, P., and Lindheimer, M.D. (1995). *Hypertension: Pathophysiology of preeclampsia.* (2nd ed). New York, NY: Raven Press.

[30] Shah,D.M. (2005). Role of the renin-angiotensin system in the pathogenesis of preeclampsia. *Am. J. Physiol. Renal. Physiol*, 288, F614-F625.

[31] Abdul-Karim, R.and Assali, N. (1961). Pressor response to angiotonin in pregnant and nonpregnant women. *Am. J. Obstet. Gynecol.* 82, 246–251.

[32] Gant, N.F., Daley, G.L., Chand, S., Whalley, P.J.and MacDonald, P.C. (1973). A study of angiotensin II pressor response throughout primigravid pregnancy. *J. Clin. Invest*, 52, 2682–2689.

[33] Gant, N.F., Worley, R.J., Everett, R.B.and MacDonald, P.C. (1980). Control of vascular responsiveness during human pregnancy. *Kidney Int*, 18, 253–258.

[34] Brown, M.A., Gallery, E.D., Ross, M.R.and Esber, R.P. (1988). Sodium excretion in normal and hypertensive pregnancy: a prospective study. *Am. J. Obstet. Gynecol*, 159, 297–307.

[35] Everett, R.B., Worley, R.J., MacDonald, P.C., Gant, N.F. (1978). Effect of prostaglandin synthetase inhibitors on pressor response to angiotensin II in human pregnancy. *J. Clin. Endocrinol. Metab*, 46, 1007-10.

[36] Sanchez-Ramos, L., O'Sullivan, M.J., Garrido-Calderone, J. (1987). Effect of low-dose aspirin on angiotensin II pressor response in human pregnancy. *Am. J. Obstet. Gynecol*, 156, 193-4.

[37] Friedman, S.A. (1988). Preeclampsia: A review of the role of prostaglandins. *Obstet. Gynecol*, 71, 122-37.

[38] Holcberg, G., Kossenjans, W., Brewer, A., Miodovnik, M.and Myatt, L. (1995). The action of two natriuretic peptides (atrial natriuretic peptide and brain natriuretic peptide) in the human placental vasculature. *Am. J. Obstet. Gynecol*, 172, 71-7.

[39] Dinh, D.T., Frauman, A.G., Johnston, C.I. and Fabiani, M.E. (2001). Angiotensin receptors: distribution, signalling and function. *Clin. Sci*, 100, 481–492.

[40] Fernandez, L.A., Twickler, J.and Mead, A. (1985). Neovascularization produced by angiotensin II. *J. Lab. Clin. Med*, 105, 141–145.

[41] Le Noble, F.A.C., Kekking, J.W.M., Van Staaten, H.W.M., Slaaf, D.W.and Struyker Boudier, H.A. (1991). Angiotensin II stimulates angiogenesis in the chorio-allantoic membrane of the chick embryo. *Eur. J. Pharmacol*, 195, 305–306.

[42] Cooper, A., Robinson, G., Vinson, G., Cheung, W.and Broughton Pipkin, F. (1999). The localization and expression of the renin-angiotensin system in the human placenta throughout pregnancy. *Placenta*, 20, 467–474.

[43] Li, X., Shams, M., Zhu, J., Khaliq, A., Wilkes, M., Whittle, M., Barnes, N.and Ahmed, A. (1998). Cellular localization of AT1 sub 1 receptor mRNA and protein in normal placenta and its reduced expression in intrauterine growth restriction: angiotensin II stimulates the release of vasorelaxants. *J. Clin. Invest*, 101, 442–454.

[44] Strickland, S., Reich, E.and Sherman, M. I. (1976). Plasminogen activator in early embryogenesis: enzyme production by trophoblast and parietal endoderm. *Cell*, 9, 231-240.

[45] Strickland, S.and Beers, W.H. (1976). Studies on the role of plasminogen activator in ovulation. In vitro response of granulosa cells to gonadotropins, cyclic nucleotides, and prostaglandins. *J. Biol. Chem*, 251, 5694-5702.

[46] Xia, Y., Wen, H.Y.and Kellems, R.E. (2002). Angiotensin II Inhibits Human Trophoblast Invasion through AT1 Receptor Activation. J. Biol. Chem, 277, 24601-8.

[47] Wolf, G. (1998). Angiotensin II is involved in the progression of renal disease: importance of non-hemodynamic mechanisms. *Nephrologie*, 19, 451-456.

[48] Matsusaka, T., Hymes, J.,and Ichikawa, I. (1996). Angiotensin in progressive renal diseases: theory and practice. J. Am. Soc. Nephrol, 7, 2025-2043

[49] Ketteler, M., Noble, N.A., and Border, W.A. (1995). Transforming growth factor-beta and angiotensin II: the missing link from glomerular hyperfiltration to glomerulosclerosis? *Annu. Rev. Physiol*, 57, 279-295.

[50] Hahn, A.W., Regenass, S., Kern, F., Buhler, F.R.,and Resink, T.J. (1993). Expression of soluble and insoluble fibronectin in rat aorta: effects of angiotensin II and endothelin-1. Biochem. *Biophys. Res. Commun*, 192, 189-197.

[51] Graham, C.and Lala, P. (1991). Mechanism of control of trophoblast invasion in situ. J. Cell Physiol, 148, 228–234.

[52] Graham, C. (1997). Effect of transforming growth factor-beta on the plasminogen activator system in cultured first trimester trophoblast cells. Placenta, 18, 137–143.

[53] Irving, J.and Lala, P.K. (1995). Functional role of cell surface integrins on human trophoblast cell migration: regulation by TGF-beta, IGF-II, and IGFBP-I. *Exp. Cell Res,* 217, 419–427.

[54] Shaw, K.J., Do, Y.S., Kjos, S., Anderson, P.W., Shinagawa, T., Dubeau, L.and Hsueh, W.A. (1989). Human decidua is a major source of renin. J Clin Invest, 83, 2085–2092.

[55] Morgan, T., Craven, C.and Ward, K. (1998). Human spiral artery renin-angiotensin system. *Hypertension*, 32, 683–687.

[56] Li, C., Ansari, R., Yu, Z.and Shah, D. (2000). Definitive molecular evidence of renin-angiotensin system in human uterine decidual cells. *Hypertension*, 36, 159–164.

[57] Morgan, T., Craven, C., Lalouel, J.M.and Ward, K. (1999). Angiotensinogen Thr235 variant is associated with abnormal physiologic change of the uterine spiral arteries in first-trimester decidua. *Am. J. Obstet. Gynecol*, 180, 95–102.

[58] Zoumakis, E., Margioris, A.N., Stournaras, C., Dermitzaki, E., Angelakis, E., Makrigiannakis, A., Koumantakis, E. and Gravanis, A. (2000). Corticotrophin-releasing hormone (CRH) interacts with inflammatory prostaglandins and interleukins and affects the decidualization of human endometrial stroma. *Mol. Hum. Reprod*, 6, 344–351.

[59] Oelkers, W.K. (1996). Effects of estrogens and progestogens on the renin-aldosterone system and blood pressure. *Steroids*, 61, 166–171.

[60] Shah, D.M., Higuchi, K., Inagami, T.and Osteen, K.G. (1991). Effect of progesterone on renin secretion in endometrial stromal, chorionic trophoblast, and mesenchymal monolayer cultures. *Am. J. Obstet. Gynecol*, 164, 1145–1150.

[61] Yan, J.S., Guo, L.H., Liu, J.and Wang, H. (1999). Modulation of the secretion of active renin in human decidual cells by progesterone. *Sheng Li Xue Bao*, 51, 211–218.

[62] Smith, D.L., Law, R.E., Shaw, K.J., Do, Y.S., Hsueh, W.A.and Morris, B.J. (1994). Proximal 6.2 kb of 5'-flanking DNA is insufficient for human renin promoter activity in renin-synthesizing chorio-decidual cells. *Biochim. Biophys. Acta*, 1219, 465–474.

[63] Sharkey, A.M., Charnock-Jones, D.S., Boocock, C.A., Brown, K.D.and Smith, S.K. (1993). Expression of mRNA for vascular endothelial growth factor in human placenta. *J. Reprod. Fertil*, 99, 609–615.

[64] Emanueli, C., Salis, M.B., Stacca, T., Pinna, A., Gaspa, L.and Madeddu, P. (2002). Angiotensin AT1 receptor signaling modulates reparative angiogenesis induced by limb ischemia. Br. J. Pharmacol, 135, 87–92.

[65] Munzemaier, D.H.and Greene, A.S. (1996). Opposing actions of antgiotensin II on microvascular growth and arterial blood pressure. *Hypertension*, 27, 760–5.

[66] Le Noble, F.A., schreurs, N.H., van Straaten, H.W., Slaaf, D.W., Smits, J.F., Rogg, H.and Struijker-Boudier, H.A. (1993). Evidence for a novel angiotensin II receptor involved in angiogenesis in chick embryo chorioallantoic membrane. *Am. J. Physiol*, 264, R460–5.

[67] Chua, C.C., Hamdy, R.C.and Chua, B.H. (1998). Upregulation of vascular endothelial growth factor by anigiotensin II in rat heart endothelial cells. *Biochem. Biophys. Acta*, 1401, 187–194.

[68] Williams, B., Baker, A.Q., Gallacher, B. and Lodwick, D. (1995). Angiotensin II increases vascular permeability factor gene expression by human vascular smooth muscle cells. *Hypertension*, 25, 913–7.

[69] Page, E.L., Robitaille, G.A., Poussegur, J.and Richard, D.E. (2002). Induction of hypoxiainducible factor-1a by transcriptional and translational mechanisms. *J. Biol. Chem*, 277: 48403–9.

[70] Otani, A., Takagi, H., Suzuma, K.and Honda, Y. (1998). Angiotensin II potentiates vascular endothelial growth factor-induced angiogenic activity in retinal microcapillary endothelial cells. *Circ Res*, 82, 619–28.

[71] Amaral, S.L., Linderman, J.R., Morse, M.M.and Greene, A.S. (2001). Angiogenesis induced by electrical stimulation is mediated by angiotensin II and VEGF. *Microcirculation*, 8, 57–67.

[72] Stoll, M., Steckelings, U.M., Paul, M., Bottari, S.P., Metzger, R.and Unger, T. (1995). The angiotensin AT2-receptor mediates inhibition of cell proliferation in coronary endothelial cells. *J. Clin. Invest*, 95, 651–7.

[73] Benndorf, R., Böger, R.H., Ergün, S., Steenpass, A., Wieland, T. (2003). Angiotensin II type 2 receptor inhibits vascular endothelial growth factor-induced migration and in vitro tube formation of human endothelial cells. *Circ. Res*, 93, 438–47.

[74] Clark, D.E., Smith, S.K., He, Y., Day, K.A., Licence, D.R., Corps, A.N., Lammoglia, R.and Charnock-Jones, D.S. (1998). A vascular endothelial growth factor antagonist is produced by the human placenta and released into the maternal circulation. *Biol. Reprod*, 59, 1540–1548.

[75] Brar, H.S., Kjos, S.L., Dougherty, W., Do, Y.S., Tam, H.B.and Hsueh, W.A. (1987). Increased fetoplacental active renin production in pregnancy-induced hypertension. *Am. J. Obstet. Gynecol*, 157, 363–367
[76] Leung, P.S., Tsai, S.J., Wallukat, G., Leung, T.N., Lau, T.K. (2001). The upregulation of angiotensin II receptor AT1 in human preeclamptic placenta. *Mol. Cell Endocrinol*, 184, 95–102.
[77] Ito, M., Itakura, A., Ohno, Y., Nomura, M., Senga, T., Nagasaka, T., Mizutani, S. (2002). Possible activation of the renin-angiotensin system in the feto-placental unit in preeclampsia. *J. Clin. Endocrinol. Metab*, 87, 1871-1878.
[78] Redman, C.W.and Sargent I.L. (2005). Latest advances in understanding preeclampsia. *Science*, 308, 1592-1594.
[79] Roberts, J.M. (2000). Preeclampsia: what we know and what we do not know. *Semin Perinatol*, 24, 24–28.
[80] Granger, J.P., Alexander, B.T., Llinas, M.T., Bennett, W.A.and Khalil, R.A. (2001). Pathophysiology of hypertension during preeclampsia linking placental ischemia with endothelial dysfunction. *Hypertension*, 38, 718-722.
[81] Roberts, J.M.and Cooper, D.W. (2001). Pathogenesis and genetics of pre-eclampsia. *Lancet*, 357, 53–56.
[82] Roberts, J.M. (1998). Endothelial dysfunction in preeclampsia. Semin. Reprod. Endocrinol, 16, 5–15.
[83] Ferris, T.F. (1991). Pregnancy, preeclampsia, and the endothelial cell. *N. Engl. J. Med*, 325, 1439–1440.
[84] Skinner, S.L., Lumbers, E.R.and Symonds, E.M. (1972). Analysis of changes in the renin-angiotensin system during pregnancy. *Clin. Sci*, 42, 479–488.
[85] Brown, M.A., Zammit, V.C., Mitar, D.A.and Whitworth, J.A. (1992). Renin-aldosterone relationships in pregnancy-induced hypertension. *Am. J. Hypertens*, 5, 366–371.
[86] Symonds, E.M., Broughton Pipkin, F.and Craven, D.J. (1975). Changes in the renin-angiotensin system in primigravidae with hypertensive disease of pregnancy. *Br. J. Obstet. Gynaecol*, 82, 643–650.
[87] Zhang, X.Q., Varner, M., Dizon-Townson, D., Song, F.and Ward, K. (2003). A molecular variant of angiotensinogen is associated with idiopathic intrauterine growth restriction. *Obstet. Gynecol*, 101, 237–242.
[88] Granger, J.P.and Alexander, B.T. (2000). Abnormal pressure natriuresis in hypertension: Role of nitric oxide. *Acta Physiologica Scandinavia*, 168, 161–168.
[89] Broughton Pipkin, F.and Symonds, E.M. (1977). Factors affecting factor angiotensin II concentrations in the human infant at birth. *Clin. Sci. Mol. Med*, 52, 449-456.
[90] Symonds, E.M. (1979). The placental and the renin-angiotensin system. *J. Reprod. Med*, 23, 129-133.
[91] AbdAlla, S., Lother, H., el Massiery, A.and Quitterer, U. (2001). Increased AT(1) receptor heterodimers in preeclampsia mediate enhanced angiotensin II responsiveness. *Nat. Med* 7, 1003-1009.
[92] AbdAlla, S., Lother, H.and Quitterer, U. (2000). AT1-receptor heterodimers show enhanced G-protein activation and altered receptor sequestration. *Nature*, 407, 94- 98.

[93] Wallukat, G., Homuth, V., Fischer, T., Lindschau, C., Horstkamp, B., Jupner, A., Baur, E., Nissen, E., Vetter, K., Neichel, D., Dudenhausen, J.W., Haller, H.and Luft, F.C. (1999). Patients with preeclampsia develop agonistic autoantibodies against the angiotensin AT1 receptor. *J. Clin. Invest*, 103, 945-952.

[94] Wallukat, G., Neichel, D., Nissen, E., Homuth, V.and Luft, F.C. Agonistic autoantibodies directed against the angiotensin II AT1 receptor in patients with preeclampsia. Can. *J. Physiol. Pharmacol*, 81, 79-83.

[95] Bobst, S.M., Day, M.C., Gilstrap, L.C., III, Xia, Y.and Kellems, R.E. (2005). Maternal autoantibodies from preeclamptic patients activate angiotensin receptors on human mesangial cells and induce interleukin-6 and plasminogen activator inhibitor-1 secretion. *Am. J. Hypertens*, 18, 330-336.

[96] Stepan, H., Faber, R., Wessel, N., Wallukat, G., Schultheiss, H.P.and Walther, T. (2006). Relation between Circulating Angiotensin II Type 1 Receptor Agonistic Autoantibodies and Soluble fms-Like Tyrosine Kinase 1 in the Pathogenesis of Preeclampsia. *J. Clin. Endocrinol. Metab*, 91, 2424-2427.

[97] Walther, T., Wallukat, G., Jank, A., Bartel, S., Schultheiss, H.P., Faber, R.and Stepan, H. (2005). Angiotensin II Type 1 Receptor Agonistic Antibodies Reflect Fundamental Alterations in the Uteroplacental Vasculature. *Hypertension*, 46, 1275-1279.

[98] Xia, Y., Wen, H., Bobst, S., Day, M.C.and Kellems, R.E. (2003). Maternal autoantibodies from preeclamptic patients activate angiotensin receptors on human trophoblast cells. *J. Soc. Gynecol. Investig*, 10, 82-93.

[99] AbdAlla, S., Lother, H., Abdel-tawab, A.M.and Quitterer, U. (2001). The angiotensin II AT2 receptor is an AT1 receptor antagonist. J. Biol. Chem, 276, 39721–39726.

[100] McMullen, J.R., Gibson, K.J., Lumbers, E.R., Burrell, J.H.and Wu, J. (1999). Interactions between AT1 and AT2 receptors in uterine arteries from pregnant ewes. *Eur. J. Pharmacol*, 378, 195–202.

[101] McMullen, J.R., Gibson, K.J., Lumbers, E.R.and Burrell, J.H. (2001). Selective down-regulation of AT2 receptors in uterine arteries from pregnant ewes given 24-h intravenous infusions of angiotensin II. *Regul Pept*, 99, 119–129.

[102] Judson, J.P., Nadarajah, V.D., Bong, Y.C., Subramaniam, K., Sivalingam, N. (2006). A preliminary finding: immunohistochemical localisation and distribution of placental angiotensin II receptor subtypes in normal and preeclamptic pregnancies. *Med. J. Malaysia*, 61, 173-80.

[103] Baumwell, S.and Karumanchi, S.A. (2007). Pre-eclampsia: clinical manifestations and molecular mechanisms. *Nephron Clin. Pract*, 106, c72-81.

[104] Hubel, C.A., McLaughlin, M.K., Evans, R.W., Hauth, B.A., Sims, C.J.and Roberts, J.M. (1996). Fasting serum triglycerides, free fatty acids, and malondialdehyde are increased in preeclampsia, are positively correlated, and decrease within 48 hours post partum. *Am. J. Obstet Gynecol*. 174, 975-982.

[105] Kolben, M., Lopens, A., Blaser, J., Huber, A., Frank, M., Wilhelm, O., Wilhelm, S., Schneider, K.T., Ulm, K., Tschesche, H., et al. (1995). Measuring the concentration of various plasma and placenta extract proteolytic and vascular factors in pregnant patients with HELLP syndrome, pre-/eclampsia and highly pathologic Doppler flow values. *Gynaekol Geburtsh Rundsch*, 35(suppl 1), 126-131.

[106] Davidge, S.T. (1998). Oxidative stress and altered endothelial cell function in preeclampsia. *Semin. Reprod. Endocrinol*, 16, 65-73.

[107] Cheng, Z.J., Vapaatalo, H., Mervaala, E. (2005). Angiotensin II and vascular inflammation. *Med. Sci. Monit*, 11, RA194-205.

[108] Touyz, R.M. (2005). Molecular and cellular mechanisms in vascular injury in hypertension: role of angiotensin II. *Curr. Opin. Nephrol. Hypertens*, 14, 125–131.

[109] Dandona, P., Kumar, V., Aljada, A., Ghanim, H., Syed, T., Hofmayer, D., Mohanty, P., Tripathy, D.and Garg, R. (2003). Angiotensin II receptor blocker valsartan suppresses reactive oxygen species generation in leukocytes, nuclear factor-kappa B, in mononuclear cells of normal subjects: evidence of an antiinflammatory action. *J. Clin .Endocrino. Metab*, 88: 4496–501

[110] Luppi, P., Deloia, J.A. (2006). Monocytes of preeclamptic women spontaneously synthesize pro-inflammatory cytokines, *Clin. Immunol.* 118, 268–275.

[111] Williams, M.A., Farrand, A., Mittendorf, R., Sorensen, T.K., Zingheim, R.W., O'Reilly, G.C., King, I.B., Zebelman, A.M.and Luthy, D.A. (1999). Maternal second trimester serum tumor necrosis factor-alpha-soluble receptor p55 (sTNFp55) and subsequent risk of preeclampsia. *Am. J. Epidemiol*, 149: 323-9.

[112] Kupferminc, M.J., Peaceman, A.M., Wigton, T.R., Tamura, R.K., Rehnberg, K.A.and Socol, M.L. (1994). Immunoreactive tumor necrosis factor-alpha is elevated in maternal plasma but undetected in amniotic fluid in the second trimester. *Am. J. Obstet. Gynecol*, 171(4), 976-9.

[113] Greer, I.A., Lyall, F., Perera, T., Boswell, F.and Macara, L.M. (1994). Increased concentrations of cytokines interleukin-6 and interleukin-1 receptor antagonist in plasma of women with preeclampsia: a mechanism for endothelial dysfunction? *Obstet. Gynecol.* 84(6): 937-40.

[114] Al-Othman, S., Omu, A.E., Diejomaoh, F.M., Al-Yatama, M.and Al-Qattan, F. (2001). Differential levels of interleukin 6 in maternal and cord sera and placenta in women with pre-eclampsia. *Gynecol. Obstet. Invest*, 52, 60–65.

[115] Conrad, K.P., Miles, T.M.and Benyo, D.F. (1998). Circulating levels of immunoreactive cytokines in women with preeclampsia. *Am. J. Reprod. Immunol*, 40, 102–111.

[116] Conrad, K.P.and Benyo, D.F. (1997). Placental cytokines and the pathogenesis of preeclampsia. *Am J Reprod Immun.* 37, 240 –249.

[117] Brasier, A.R.and Recinos, A. 3rd, Eledrisi, M.S. Vascular inflammation and the renin-angiotensin system. *Arterioscler Thromb Vasc Biol*, 22 :1257-66.

[118] Manabe, S., Okura, T., Watanabe, S., Fukuoka, T.and Higaki, J. (2005). Effects of angiotensin II receptor blockade with valsartan on pro-inflammatory cytokines in patients with essential hypertension. *J. Cardiovasc. Pharmacol*, 46, 735-9.

[119] Gurantz, D., Cowling, R.T., Varki, N., Frikovsky, E., Moore, C.D.and Greenberg, B.H. (2005). IL-1beta and TNF-alpha upregulate angiotensin II type 1 (AT1) receptors on cardiac fibroblasts and are associated with increased AT1 density in the post-MI heart. *J. Mol. Cell Cardiol*, 38: 505-15.

[120] Pritchard, J.A. (1979). The use of magnesium sulfate in preeclampsia-eclampsia. *J. Reprod. Med*, 23, 107–114.

[121] Hankins, G.D., Hammond, T.L.and Yeomans, E.R. (1991). Amniotic cavity accumulation of magnesium with prolonged magnesium sulfate tocolysis. *J. Reprod. Med*, 36, 446–449.

[122] Hallak, M., Berry, S.M., Madincea, F., Romero, R., Evans, M.I.and Cotton, D.B. (1993). Fetal serum and amniotic fluid magnesium concentrations with maternal treatment. *Obstetr. Gynecol*, 81, 185–188.

[123] Lipsitz, P.J.and English, I.C. (1967). Hypermagnesemia in the newborn infant. *Pediatrics*, 40, 856–862.

[124] Nelson, K.B.and Grether, J.K. (1995). Can magnesium sulfate reduce the risk of cerebral palsy in very low birthweight infants? *Pediatrics*, 95, 263–269.

[125] Hauth, J.C., Goldenberg, R.L., Nelson, K.G., et al., (1995). Reduction of cerebral palsy with maternal magnesium sulfate treatment in newborns weighing 500-1000 g. *Am. J. Obstetr. Gynecol*, 172, 419, Abstract #581.

[126] Kovac, C.M., Howard, B.C., Pierce, B.T., Hoeldtke, N.J., Calhoun, B.C.and Napolitano, P.G. (2003). Fetoplacental vascular tone is modified by magnesium sulfate in the preeclamptic ex vivo human placental cotyledon. *Am. J. Obstet. Gynecol*, 189, 839-42.

[127] Schneider, H., Huch, A., (1985). Dual in vitro perfusion of an isolated lobe of human placenta: method and instrumentation. *Contrib. Gynecol. Obstetr.* 13, 40–47.

[128] Holcberg, G., Sapir, O., Hallak, M., Alaa, A., Shorok, H.Y., David, Y., Katz, M.and Huleihel, M. (2004). Selective vasodilator effect of magnesium sulfate in human placenta. *Am. J. Reprod. Immunol*, 51, 192-7.

[129] Lewis, D.F. (2005). Magnesium sulfate: the first-line tocolytic. *Obstet. Gynecol. Clin. North Am.* 32, 485–500.

[130] Rochelson, B., Dowling, O., Schwartz, N.and Metz, C.N. (2007). Magnesium sulfate suppresses inflammatory responses by human umbilical vein endothelial cells (HuVECs) through the NFkappaB pathway. *J. Reprod. Immunol.* 73, 101-7.

[131] Altura, B.M., Kostellow, A.B., Zhang, A., Li, W., Morrill, G.A., Gupta, R.K.and Altura, B.T. (2003). Expression of the nuclear factor-kappaB and proto-oncogenes c-fos and c-jun are induced by low extracellular Mg2+ in aortic and cerebral vascular smooth muscle cells: possible links to hypertension, atherogenesis, and stroke. *Am. J. Hypertens.* 16: 701-7.

[132] Holcberg, G., Amash, A., Sapir, O., Hallak, M., Sheiner, E., Ducler, D., Katz, M.and Huleihel, M. (2006). Different effects of magnesium sulfate and angiotensin II on the capacity of the fetal and maternal compartments of normal human placenta to secrete TNF-alpha and IL-6. *J. Reprod. Immunol.* 69: 115-25.

In: Angiotensin Research Progress
Editors: H. Miura and Y. Sasaki

ISBN: 978-1-60456-537-9

Chapter IV

How Is the Insulin-Regulated Aminopeptidase Involved in the Physiological and *In Vitro* Effects of Angiotensin IV?

Demaegdt Heidi*
Department of Molecular and Biochemical Pharmacology, Vrije Universiteit Brussel, Pleinlaan 2, 1050 Brussels, Belgium

Abstract

Evidence is accumulating that Angiotensin IV (Ang IV or Ang II (3-8)), a major metabolite of Ang II, exerts physiological effects on its own. Ang IV accomplishes cardiovascular, renal and, more importantly, central effects. In this context, it was shown that i.c.v. administration of Ang IV increases memory and learning in different in vivo studies. The exact mechanism by which Ang IV exerts this beneficial effect is not yet understood. Some of the effects of Ang IV occur at nanomolar concentration and are not blocked by the classical AT_1 and AT_2 receptor antagonists. This, together with the discovery of binding sites with high affinity for $[^{125}I]$Ang IV in the central, vascular and renal system, has led to the concept of a novel angiotensin receptor subtype: the AT_4 receptor. Recently, this receptor was identified as the insulin regulated aminopeptidase (IRAP), a Zn^{2+} dependent gluzincin aminopeptidase. IRAP is known for it's co-localization with the GLUT4 glucose transporter in specialized intracellular vesicles. To regulate glucose homeostasis, they translocate quickly to the cell surface in the presence of insulin. Ang IV has been shown to bind IRAP with high affinity and to inhibit its catalytic activity, albeit with lower potency. This discrepancy can be attributed to the mandatory presence of chelators in the binding studies. This implies that high affinity binding of $[^{125}I]$Ang IV only takes place to the IRAP apo-enzyme. These properties of IRAP have led to three hypotheses concerning the way Ang IV exerts its effects.

* Tel: (+32) 2 629 19 46; Fax: (+32) 2 629 13 58; email: hdemaegd@vub.ac.be

Some of IRAP's in vitro substrates are known, among them we have oxytocin and vasopressin. These peptides have been recognised to have a positive effect on memory and learning. By inhibiting the catalytic activity of IRAP, Ang IV is supposed to prolong the half-life of these compounds, resulting in a positive effect on memory and learning.

Secondly, IRAP could be capable of triggering intracellular events and thus function as a receptor.

Finally, increased glucose uptake in neurons has been shown to improve memory and learning. It has therefore been proposed that Ang IV interferes in the recycling of IRAP and GLUT4 in such a way that it prolongs the residence time of GLUT4 at the cell surface and thus, glucose uptake.

This review focuses on the molecular and physiological aspects of the Ang IV-IRAP interaction with special attention to recent findings and the above-mentioned hypotheses.

Ang II and Its Metabolite Ang IV

The renin-angiotensin system (RAS) is widely recognized as the main signaling system that controls arterial blood pressure, sodium balance and body fluid volumes. Its major peptide, Angiotensin II (Ang II) is well known to cause hypertension and to stimulate cardiac remodelling. G-protein-coupled receptors of the AT_1 subtype play a major role in these processes while the AT_2 receptors, the other major subtype, are primordially expressed in foetal tissues and involved in the inhibition of cell proliferation, apoptosis and neuronal differentiation (Saavedra, 1999; Allen et al., 2000; de Gasparo et al., 2000). Ang II has long been considered to represent the end product of the RAS, but there is accumulating evidence that some of its fragments such as Ang III (deletion of the N-terminal amino acid) and Ang IV (deletion of the two N-terminal amino acids) have central, cardiovascular and renal actions as well (Ardaillou, 1997; Chansel and Ardaillou, 1998; Mustafa et al., 2001a). Ang III is still a potent AT_1 and AT_2 receptor agonist. On the other hand, Ang IV possesses only very low potency for these receptors (Harding et al., 1994, Kramar et al., 1998). Yet, most of its biological effects are already observed at nanomolar concentrations and are not blocked by the classical non-peptide AT_1 and AT_2 receptor antagonists such as losartan and PD123177. Moreover, high affinity binding sites for [^{125}I]-Ang IV have been detected to be present in the central nervous, vascular and renal systems (Harding et al., 1994; Swanson et al., 1992; Moeller et al., 1996b; Coleman et al., 1998a; Zhang et al., 1998; Chai et al., 2000; Von Bohlen und Halbach, 2003; Thomas and Mendelsohn, 2003; Fernando et al., 2005). These findings provided a solid argument in favour of the existence of the so-called "AT_4 receptors" (de Gasparo et al., 1995; 2000; Harding et al., 1994). The peptide LVV-hemorphin 7 (LVV-H7) is quite abundant in the brain and it is now also widely recognized that, along with Ang IV, it acts as an endogenous AT_4 receptor agonist (Moeller et al., 1997).

In Vivo Effects of Ang IV and LVV-H7

Central Effects

When directly administered to the brain, Ang IV and LVV-H7 have been found to elicit a number of physiological effects (extensively reviewed by von Bohlen und Halbach, 2003; Chai et al., 2004; Stragier et al., 2007).

Most importantly, direct administration of Ang IV, its analogue Norleucine[1]-Ang IV (Nle[1]-Ang IV) or LVV-H7 into the cerebral ventricles (i.c.v.) has been found to facilitate memory retention and retrieval in rats in passive and conditioned avoidance and Morris water maze paradigms (Braszko et al., 1988a, b, 2004; Wright et al., 1993, 1995, 1996a, 1999; Tchekalarova et al., 2001, Lee et al., 2004). All these effects can be blocked by the putative AT_4 antagonist Divalinal-Ang IV. The exact mechanism by which such "AT_4 receptor agonists" are able to stimulate memory recall and learning is not yet fully understood, but some pertinent information in this respect is starting to emerge. Interesting breakthroughs are briefly enumerated below. In the rat model of scopolamine-induced amnesia, Nle[1]-Ang IV and LVV-H7 are able to reverse the performance deficits in the Morris water maze test (Pederson et al., 1998, Albiston et al., 2004). As scopolamine is a competitive antagonist for the M_1 muscarinic acetylcholine receptor, the counteracting effect of the AT_4 receptor agonists could interact with the release of acetylcholine (Pederson et al., 1998). The ability of Ang IV and LVV-H7 to potentiate potassium-evoked release of acetylcholine from rat hippocampal slices (Lee et al., 2001) lends further support to this hypothesis. Ang IV and LVV-H7 also significantly enhance long-term potentiation (LTP) in the dentate gyrus and the CA1 field of the hippocampus both *in vitro* (Kramár et al., 2001) and *in vivo* (Wayner et al., 2001). This effect could be blocked by the "AT_4 receptor antagonist" divalinal-Ang IV. At the cellular level, Ang IV has been found to restore astrocyte adhesion, growth and morphology, to prolong the survival and to inhibit apoptosis of hippocampal cells in angiotensinogen knock-out (KO) mice (Kakinuma et al., 1997; 1998). Ang IV inhibits neurite outgrowth in cultured embryonic chicken sympathetic neurons and may therefore play an important role in neurite modelling and neuronal development (Møeller et al., 1996a). Finally, i.c.v. administration of Ang IV induces c-Fos expression in hippocampal pyramidal cells (Roberts et al., 1995).

Next to its memory promoting properties, i.c.v. administered Ang IV has been shown to be anticonvulsant in the acute pilocarpine model for epilepsy in rats (Stragier *et al.*, 2006). Ang IV also dose-dependently attenuates pentylenetetrazol (PTZ, a non-competitive GABA antagonist) induced seizures in mice (Tchekalarova *et al.*, 2001). The anti-epileptogenic properties of Ang IV are further evidenced by its ability to suppress the seizure expression not only directly during kindling procedure but it can also block the development of an epileptic-like state in mice (Tchekalarova *et al.*, 2005a, b). Moreover, Ang IV and LVV-H7 are also able to influence the neurotransmission in the striatum. Local administration of both compounds produces an increase of the extracellular dopamine concentration in rat striatum (Stragier et al., 2004). This may pave the way to new therapeutic strategies to aleviate extrapyramidal symptoms in patients suffering from Parkinson's disease since they result from a depletion of dopamine in the striatum (Riederer and Wuketich, 1976).

Like Ang II and Ang III, Ang IV also increases blood pressure (BP) after i.c.v. administration in anaesthesized rats (Wright et al., 1996b; Yang et al., 2008). BP is also abnormally high in a transgenic mouse model in where Ang IV is chronicly elevated in the brain (Lochard et al, 2004). However, these centrally mediated pressor responses can be blocked by AT_1 receptor-selective antagonists and, hence, appear to involve AT_1 receptor stimulation (Yang et al., 2008).

Vascular and Renal Effects

With respect to its peripheral effects, Ang IV is certainly active at the vascular level since its internal carotid or renal infusion produces changes in the cerebral and renal blood flow, respectively. However, the implication of Ang IV receptors is still matter of debate. Some researchers observed an increased cerebral and renal cortical blood flow and, because of the inhibitory effect of Divalinal-Ang IV thereon, they proclaimed the involvement of AT_4 receptors (Harding et al., 1994; Kramar et al., 1998; Coleman et al., 1998a, b). Other groups, however, observed a reduction in renal blood flow by Ang IV and found that this response is mediated by AT_1 receptor stimulation (Gardiner et al., 1993; Fitzgerald et al., 1999, Van Rodijnen et al., 2002; Handa, 2006; Yang et al., 2008).

Biochemical Effects of Ang IV

The "AT_4 receptor" concept sprouted from the refractoriness of several in vivo effects of Ang IV to classical AT_1 and AT_2 receptor antagonists and from the existence of high affinity Ang IV binding sites (i.e. sites labeled by [^{125}I]-Ang IV and/or its analogues) (de Gasparo et al., 1995; 2000). In agreement with this hypothesis, Ang IV has been reported to trigger biochemical events within different isolated cell types. Ang IV was especially found to stimulate a transient increase in the cytosolic calcium level ($[Ca^{2+}]_i$) in cells of pulmonary, cardiac and renal origin. In porcine pulmonary arterial endothelial cells this increase was found to trigger the production of NO· radicals and cyclic GMP. None of these effects could be blocked by AT_1 receptor antagonists (Dulin et al., 1995; Handa et al., 1999, 2001; Patel et al., 1998; Hill-Kapturczak et al., 1999; Mustafa et al., 2001a).

Ang IV was also found to stimulate p38 kinase activity and to increase the level of phosphorylated Erk-2 and $[Na^+]_i$ in human proximal tubule epithelial cells (Handa et al., 2001a), to stimulate c-Fos expression in the hippocampus and piriform cortex (Roberts et al., 1995) and to activate the nuclear transcription factor-κB (NF-κB) in vascular smooth muscle cells (Esteban et al., 2005). In addition, Ang IV and LVV-H7 stimulate DNA synthesis in the human neuroblastoma SK-N-MC cell line (Mustafa et al., 2001b), in the uterine epithelium and in the adrenal cortex (Pawlikowski et al., 1999; 2001). Ang IV has also been shown to activate signaling molecules involved in lung endothelial cell proliferation (Li et al, 2002; Lu et al., 2005). This included an increased activity and/or phosphorylation of PI3 kinase, Akt, Erk1/2, p70 ribosomal S6 kinase and the eukaryotic RNA translation initiation factor 4EBP1. Unfortunately, the potential interference of AT_1 receptor stimulation has only been addressed/ ruled out in some of these studies. Recently, Vinh et al. (2007) showed that chronic Ang IV treatment in apolipoprotein E-deficient mice, that exhibit endothelial dysfunction, produced a vasoprotective effect mediated by improved NO bioavailability attenuated by both Divalinal-Ang IV and an AT_2 receptor antagonist.

It is also important to notice that Ang IV failed to trigger a cellular response in a number of studies. In neuroblastoma NG108-16 cells, LVV-H7, but not Ang IV was capable of stimulating DNA synthesis (Moeller et al., 1999). In bovine aortic endothelial cells (BAEC), Ang IV was unable to modify the concentration of the classical second messengers cAMP, $[Ca^{2+}]_i$, inositol phosphates, nitric oxide or arachidonic acid derivatives (Briand et al., 1998). Ang IV failed to trigger $[Ca^{2+}]_i$ transients and Erk phosphorylation in IRAP- expressing recombinant HEK293 cells (Albiston et al., 2007). In the same vein, we were also unable to record AT_4 Ang IV-mediated $[Ca^{2+}]_i$ transients in CHOA12 and MDBK cells (Demaegdt, unpublished results).

Finally, some responses cannot always be reproduced at will. A plausible explanation for the above mentioned controversial results is that the effects of Ang IV are dependent on cell type but also on culture conditions. Indeed, in about half of the experiments, we found that Ang IV was able to stimulate DNA synthesis in CHO-K1 cells and that this effect was not blocked by the potent AT_1 receptor antagonist candesartan (figure 1). In the remaining experiments, Ang IV failed to reproduce this response. At present, there is no conclusive explanation for the elusiveness of the "AT_4 receptor" activity. More detailed research focused on the role of cell type, culture conditions, cell cycle,... could help solving this issue.

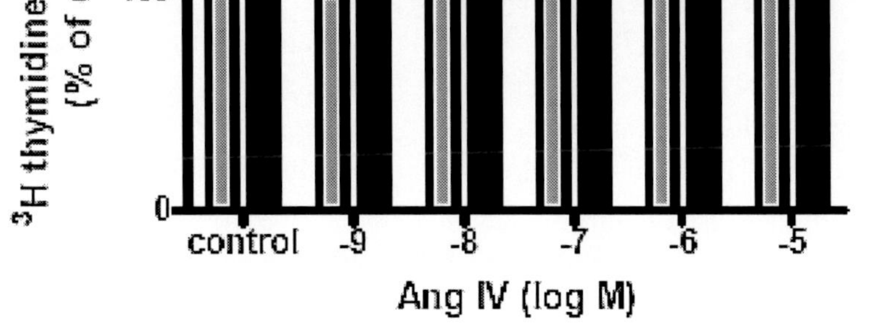

Figure 1. ^{3}H-thymidine incorporation in CHO-K1 cells. Cells were put serum free 40 hours before the experiment. 24 hours before measuring, Ang IV in the absence or presence of 10^{-7} M candesartan (CV) was added and 4 hours before measurement ^{3}H-thymidine was added to each well (1 µCi/well). Cells were washed with cold PBS and treated 3 times with 10 % TCA for 5 min. Remaining radioactivity was measured in a beta counter. The incorporation was expressed as a percentage of the control (stimulation with buffer). Data are the mean +- SEM of 4 experiments performed in triplicate.

Identification of AT_4 Receptor as IRAP Based on Binding Studies

The assumption that an AT_4 receptor is implicated in the physiological effects of Ang IV is compatible with the highly conserved regional distribution of high affinity Ang IV binding sites in the brains of sheep, monkey, guinea pig and humans (Miller-Wing et al., 1993; Moeller et al., 1995; Moeller et al., 1996b; Chai et al., 2000). Binding was found to be most prominent in structures that are important to cognitive processing and sensory and motor functions, including neocortex, hippocampus, basal ganglia, dentate gyrus, amygdale, thalamus, cerebellum and spinal cord (Miller-wing et al., 1993; Wright et al., 1995, Fernando et al., 2005).

Experiments with [^{125}I]benzoylphenyl-Ala-Ang IV, a labeled photoaffinity analogue of Ang IV, revealed that its high affinity binding sites in bovine tissues exist as complexes of three different peptides with molecular weights of 165 kDa, 50 to 60 kDa and 70 to 80 kDa, respectively (Zhang et al. 1998,1999). In agreement, [^{125}I]Nle1-BzPhe6-Gly7-Ang IV was also shown to crosslink to a 180 kDa peptide in membranes from the SK-N-MC human neuroblastoma cell line (Mustafa et al., 2001b). This peptide has now been identified as a membrane-associated aminopeptidase. This enzyme is known under no less than 13 different denominations, including insulin-regulated aminopeptidase (IRAP), placental leucine aminopeptidase (P-LAP), oxytocinase (Otase) (Albiston et al., 2001) and cysteine aminopeptidase (EC 3.4.11.3) (Tsujimoto et al., 1992). Experiments with IRAP-expressing recombinant HEK293 cells reveal that this peptide indeed contains a high affinity binding site for [^{125}I]Ang IV and its analogue [^{125}I]Nle1-Ang IV (Albiston et al., 2001; Demaegdt et al., 2006). The correspondence between the high affinity Ang IV binding sites and IRAP was further evidenced by the similar regional distribution of IRAP mRNA (in situ hybridization), IRAP positive immunoreactivity (immunohistochemistry) and [^{125}I]Nle1-Ang IV binding (autoradiography) in mouse brain slices (Albiston et al., 2001).

The properties of IRAP have previously been reviewed by Keller (2003) and Albiston et al. (2003, 2007). In short, IRAP has been cloned and sequenced (Keller et al., 1995; Rogi et al., 1996; Laustsen et al., 1997) and was found to be homologous to aminopeptidase A (AP-A), aminopeptidase N (AP-N), and other Zn^{2+}-dependent aminopeptidases that make part of the M1 family of gluzincin aminopeptidases (Rogi et al., 1996). It is a type II integral membrane protein of 1025 amino acid residues, consisting of an acidic intracellular domain (110 AA, N-terminal) followed by a hydrophobic transmembrane segment (22 AA α-helix) and a 893 AA extracellular domain bearing the catalytic domain (Keller et al., 1995; Rogi et al., 1996). Typical for the members the M1 gluzincin aminopeptidase family, this domain consists of a GAMEN motif and the $HEXXH(X)_{18}E$ Zn^{2+}-binding motif (Keller et al., 1995).

"P-LAP"/"Otase", the major human placental protease, is regarded to be the human homologue of IRAP (Rogi et al., 1996; Laustsen et al., 1997; Herbst et al., 1997; Mizutani, 1998; Horio et al., 1999). In fact, it refers to the soluble extracellular part of IRAP that is proteolytically cleaved from its N-terminal region and secreted into serum (Yamahara et al., 2000). The fact that this process only takes place in the placenta explains why P-LAP/Otase is only detected in serum during pregnancy. Interestingly, its serum levels have been found to increase with advancing gestation until just before the onset of labor (Mizutani et al., 1996; Yamahara et al., 2000).

The denomination "insulin regulated aminopeptidase" refers to the fact that, in the absence of insulin, it is mainly co-localized with the glucose transporter GLUT4 in specialized vesicles in adipocytes and skeletal muscle cells. In the presence of insulin, these vesicles translocate 10 times faster to the cell surface. The resulting higher cell surface concentration of GLUT4 is then responsible for an increased uptake of glucose (Czech and Corvera, 1999). This translocation is a major mechanism by which insulin regulates glucose homeostasis (Kandror et al., 1994; Keller et al., 1995; Ross et al., 1996; 1997; Garza and Birnbaum, 2000). Rather than being simply co-translocated with GLUT4, IRAP seems to play an active, and even essential role in this process. The impaired GLUT4 translocation and glucose uptake in IRAP knockdown 3T3-L1 adipocytes (Yeh et al., 2007) pleads in favor of this hypothesis. However, it should be noted that GLUT4 translocation and glucose uptake remain intact in adipocytes isolated from IRAP knock-out mice, possibly through the development of compensatory mechanisms (Keller et al., 2002).

Several studies have also shed light on the co-localisation of IRAP and GLUT4 in several brain regions. In some regions of rodent brain, the distribution of GLUT4 strongly resembles that of IRAP; both are expressed in neurons in hippocampus, cerebral cortex, septum, hypothalamus, cerebellum and in most motor and motor-associated nuclei (El Messari et al., 1998; Fernando et al., 2005; Matsumoto et al., 2001; Choeiri et al., 2002; Sankar et al., 2002). Furthermore, it was demonstrated that GLUT4 and IRAP are co-localized in the pyramidal cells in CA1-CA3 region of the hippocampus (Fernando et al., 2005). However, it is not yet known whether both are also able to co-translocate to the surface of these neurons and whether such process is controlled by insulin or other stimuli.

However, IRAP is not necessarily always associated with GLUT4. Indeed, this enzyme is also found to be present in tissues in where GLUT4 is not readily detectable; i.e. in heart, kidney, brain, spleen, lung and testis (Keller et al., 1995; Matsumoto et al., 2001). In this respect, we were also able to detect high levels of endogenous IRAP in Chinese hamster ovary (CHO-K1) cells despite the fact that they are apparently devoid of GLUT4 (Demaegdt et al., 2004a, Demaegdt et al., 2006; Shibasaki et al., 1992). In those cells, insulin was repeatedly shown to trigger the translocation of recombinant IRAP (Johnson et al., 1998; Lim et al., 2001) and, recently, we have demonstrated that this process also takes place in the case of the endogenously expressed IRAP (see Applications for binding studies).

Comparison between "AT_4 Receptor" Binding Studies and IRAP Activity Measurements

The pharmacological profile of the "AT_4 receptor" (i.e. defined as high affinity Ang IV binding sites) has been studied in great detail. In this respect, binding studies revealed that the first three amino acid residues of the hexapeptide Ang IV are critical for their high affinity binding (Sardinia *et al.*, 1993). At the N-termius, a primary α-amine and an L-conformation for the first amino acid α-carbon are required (Sardinia *et al.*, 1994). An activated aromatic ring in the side chain of amino acid residue 2 and a hydrophobic amino acid in position 3 are important as well (Krishnan *et al.*, 1999). To confirm the proposal that the "AT_4 receptor" corresponds to IRAP, extensive binding studies were undertaken to determine the binding profile of Ang IV, LVV-H7 and their fragments for this enzyme expressed in recombinant HEK293 cells (Lee et al., 2003; Lew et al., 2003) and endogenous IRAP in CHO-K1 cells (Demaegdt et al., 2004a). These studies revealed that, similar to the earlier observations with the "AT_4 receptor", C-terminal deletions of Ang IV and LVV-H7 only result in modest changes in their affinity for IRAP, whereas deletion of the first three N-terminal residues abolish the binding.

The IRAP catalytic activity can easily be determined based on the cleavage of the substrates leucine-p-nitroanilide (L-Leu-pNA) and leucine-β-naphthylamide (L-Leu-βNA) into L-leucine and p-nitroaniline or β-naphthylamine respectively, which display a characteristic light absorption maximum at 405 nm or fluorogenic excitation and emission wavelengths of 340 and 410 nm (Laustsen et al., 2001; Herbst et al., 1997). Although L-Leu-pNA and L-Leu-βNA are fairly non-selective substrates, a large portion of the catalytic activity present in membranes from recombinant HEK293 cells and CHO-K1 cells could indeed be inhibited by Ang IV and related "AT_4 receptor" ligands such as LVV-H7, divalinal-Ang IV and Nle^1-Ang IV (Albiston et al., 2001; Lee et al., 2003; Lew et al., 2003; Demaegdt et al., 2004a; Demaegdt et al., 2006). This inhibition took place in a competitive fashion, indicating that these ligands have access to at least part of the enzyme's catalytic site. These ligands inhibited the IRAP enzyme activity only with moderate potency (i.e. Ki values between 0.1 and 2.3 μM). Whereas one would expect these potencies to be the same as those obtained in the related [^{125}I]Ang IV competition binding experiments, this was clearly not the case. Instead, Ang IV and some of its analogues displayed up to about 20 fold lower potency in the enzyme assays (figure 2). The cause for this discrepancy was found when comparing both experimental conditions. In fact, a particularity of the radioligand binding assays with IRAP (Lew et al., 2003, Demaegdt et al., 2004a, b) as well as of nearly all the previous binding assays dealing with "AT_4 receptors" is that they are carried out in the presence of high concentrations of metal chelators, usually EDTA in combination with 1,10-phenantrolin (Lew et al., 2003, Demaegdt et al., 2004). Whereas high affinity [^{125}I]Ang IV binding could only be observed in the presence of these chelators, the catalytic activity of IRAP appeared to be completely abolished under those conditions (Demaegdt et al., 2004b; Laeremans et al. 2005). Based on the fact that IRAP is a metalloprotease, these studies clearly established that high affinity [^{125}I]Ang IV binding is only observed when it takes place to the IRAP apo-enzyme (i.e. the enzyme from which the catalytic Zn^{2+} has been removed by the chelators).

The necessity to perform [^{125}I]-Ang IV binding studies in the presence of metal chelators could merely reflect the ability of these chelators to prevent the degradation of these ligands by IRAP and/or related metalloproteases. As binding studies are only carried out in the presence of minute quantities of radioligand (about 1 nM in the case of [^{125}I]Ang IV), metal chelators should be necessary to prevent its rapid breakdown and, hence, to preserve a detectable amount of binding. In agreement with this hypothesis, substantial degradation of Ang IV was found to take place in the presence of CHO-K1 cell membranes (Axen et al., 2006) and rat kidney homogenates (Krebs et al., 1996). In this respect, Ang IV is known to be degraded into smaller peptide fragments by mainly the metalloproteases AP-N and dipeptidyl aminopeptidase (DAP) (Chansel et al., 1998; Ardaillou et al., 1998) and possibly also by other proteases. These authors showed that the degradadtion of Ang IV by rat mesangial cells was completely blocked by adding inhibitors of AP-N and DAP. In a recent study, the implication of AP-N was also illustrated by the observation that Ang IV induces an increase in dopamine release in rat striatum and that this effect is markedly boosted by the AP-N-selective inhibitor 7B (Stragier et al., 2007). 7B has lately also been shown to almost completely prevent the degradation of Ang IV in the presence of CHO-K1 cell membranes (Vanderheyden et al., personal communication). Based on the assumption that metal chelators enable [^{125}I]Ang IV binding only by virtue of their ability to prevent its degradation, it was speculated that the same result should also be obtained by replacing those chelators by 7B. Yet, still no high affinity binding of [^{125}I]Ang IV could be detected in the presence of 7B instead of the chelators (Vanderheyden et al., personal communication). It is therefore more likely that the apo- and catalytically active forms of IRAP display different pharmacological profiles. Most important is that the IRAP apo-enzyme displays higher affinity for Ang IV and some of its analogues than its Zn^{2+} - containing catalytic form. This hypothesis is also compatible with the preference of Ang II and Ang (4-8) for the catalytically active form of IRAP (figure 2) and with the differences in potency of stable compounds like 7B, divalinal-Ang IV and L-methionine for both forms of the enzyme (table 1, figure 1). While it is well established that Ang IV binds the catalytic site of IRAP, the possibility arises that, in the presence of chelators, its binding occurs to a completely different region of the protein (Caron et al., 2003).

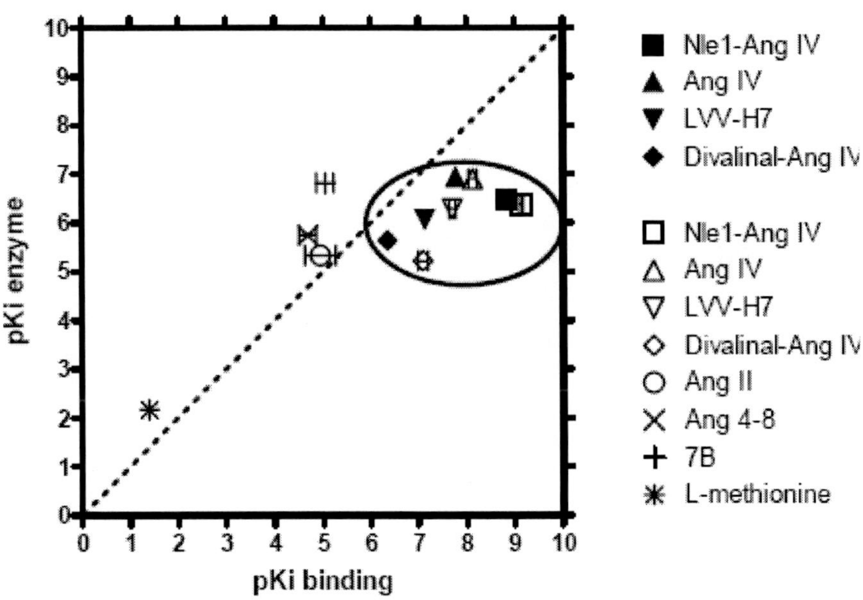

Figure 2. Comparison of pK_i values calculated from [^{125}I]Ang IV competition binding studies as well as inhibition of enzyme activity. Closed symbols refer to results with recombinant HEK293 cells expressing human IRAP (Lew et al., 2003). All other symbols refer to results with endogenous IRAP in CHO-K1 cells (Demaegdt et al., 2004a, 2006 and unpublished data).Encircled compounds are known as AT_4 ligands.

Table 1. Pk_i values of competition binding with [^{125}I]Ang IV and inhibition of IRAP enzyme activity in CHO-K1 cell membranes for stable AT_4 ligands

	Binding	Enzyme activity
Divalinal-Ang IV	7.10 ± 0.10	5.22 ± 0.06
7B	5.05 ± 0.18**	6.82 ± 0.09
L-methionine	< 1.4*	2.17 ± 0.02*

* Values from Demaegdt et al., 2006
** Values from Vanderheyden et al., unpublished results.

Detailed studies with chelators unveiled that EDTA and 1,10-phenantrolin are able to remove the catalytic Zn^{2+} from IRAP in a synergistic fashion (Demaegdt et al., 2006b, Laeremans et al, 2005). This shed light on the important role of a hitherto undetected regulatory divalent cation binding site on this enzyme. A mechanism was proposed in where 1,10-phenantrolin is responsible for the removal of catalytic Zn^{2+} and that, by binding to the regulatory site, EDTA facilitates the access of 1,10-phenantrolin to the catalytic site by promoting a conformational change of the enzyme (Demaegdt et al., 2006b, Laeremans et al., 2005). Interestingly, the same mechanism was also shown to be operative in the case of AP-N (Vanderheyden et al., 2007). We should therefore not be surprised to learn that other members of the M1 family of gluzincin aminopeptidases also contain a similar regulatory divalent cation binding site.

From the physiological point of view, it is obviously only relevant to compare the *in vivo* and *in vitro* behaviour of "AT_4 receptor" ligands when the latter assays are performed with the native enzyme. Presently, this condition is only satisfied by the IRAP enzyme activity measurements (Demaegdt et al., 2006a). This does not mean, however, that [^{125}I]Ang IV binding should be totally dismissed. On the contrary, this approach can be of great help for applications that rely on the specific detection of IRAP.

Applications for Binding Studies

Selective IRAP Detection

The catalytic activity of the related AP-N and AP-A enzymes can also be inhibited by Ang IV or LVV-H7, albeit with somewhat lower potency when compared to IRAP (Garreau et al., 1998; Chansel et al., 1998, Goto et al., 2006). Recent comparative studies with recombinant IRAP and AP-N confirm that their pharmacological profile is quite distinct: i.e. the catalytic activity of human IRAP is preferentially inhibited by Ang IV, some of its novel synthetic analogues and LVV-H7 while the catalytic activity of human AP-N is preferentially inhibited by 7B, L-methionine and substance P (Demaegdt et al., 2006, Axen et al., 2006; 2007; Lukaszuk et al., 2008; Stragier et al., 2007). However, the still limited selectivity of the hitherto tested compounds (i.e. no more than a 100-fold difference in inhibitory potency) hampers the positive identification of IRAP by catalytic activity measurements when AP-N is also present. This appears to be the case in tissues like brain and kidney (Stragier et al., 2007,

Yang et al., 2008) as well as in some cell lines like J774, a mouse monocyte cell line (Demaegdt et al., unpublished data).

Alternatively, IRAP could also be detected by using commercial antibodies directed against its extracellular C-terminal region. Yet, we as well as others (Prof. Chai SY, University of Melbourne, Australia; Prof. M. Ruiz-Ortega, Universidad Autónoma Madrid, Spain, personal communications) have only met limited success in using this approach with a variety of cell systems.

Fortunately, and in contrast to earlier suggestions based on circumstantial evidence (Garreau et al., 1998), experiments with recombinant AP-N reveal that neither the active nor the apo-form of this enzyme possess sufficiently high affinity for $[^{125}I]$Ang IV to permit their detection by radioligand binding. At least compared to AP-N, only the IRAP-apo-enzyme can be detected by $[^{125}I]$Ang IV binding. The very similar pharmacological profiles of the $[^{125}I]$Ang binding sites in all hitherto published studies on tissue extracts and cell lines further suggests that such binding does not take place to other IRAP-related enzymes either. Hence, radioligand binding represents an easy and selective tool for the detection of IRAP.

Translocation Studies

IRAP and the insulin-dependent glucose transporter GLUT4 co-localize in specific intracellular vesicles (i.e. GLUT4 vesicles). These vesicles move slowly to the cell surface but their translocation is markedly enhanced by insulin, resulting in higher glucose uptake (Keller et al., 1995; Ross et al., 1996). Previous studies of the insulin mediated translocation of IRAP to the cell surface have been hampered by the laborious detection of the enzyme at the cell surface (biotinylation, precipitation, SDS-page and western blotting) (Nakamura et al., 2000; Garza and Birnbaum, 2000). Recently, we found that only cell surface IRAP will be labeled when intact cells are treated with $[^{125}I]$Ang IV or $[^{3}H]$Ang IV and that the internalization of this enzyme can be prevented by adding phenyl arsene oxide (PAO). By using this convenient detection method we were able to observe insulin-mediated translocation of IRAP to the surface of CHO-K1 and HEK293 cells (Demaegdt et al., manuscript in preparation). Hence, Ang IV radioligand binding represents a fast and easy detection method to study the translocation of IRAP.

Pregnancy

In contrast to the widespread tissue distribution of membrane-bound IRAP, the soluble form of the enzyme can only be detected in serum during pregnancy. When IRAP is produced in the placenta, it will be proteolytically cleaved at a location close to the cell surface, so that its soluble, extracellular part is secreted into maternal serum (Yamahara et al., 2000). Serum levels of IRAP increase with advancing gestation until just before the onset of labor, before diminishing postpartum. (Mizutani et al., 1998; Yamahara et al., 2000). Furthermore, based on a preliminary clinical study, it was concluded that a low concentration of soluble IRAP could predict preterm delivery (Rogi et al., 1996). This suggests that the concentration of IRAP in maternal serum can be used as a clinical marker. Until now, this concentration was only determined enzymatically; i.e. by measuring the degradation of the substrate L-Leu-pNA. (Kozaki et al. 2001) However, as this substrate lacks specificity towards IRAP, the possibility arises that the activity of other plasma-borne enzymes are measured as well. Here again, [^{125}I]Ang IV binding could provide a more accurate quantitation of the IRAP concentration in maternal serum.

Stability of AT_4 Ligands towards Metalloproteases

It is also possible to determine the stability of the peptide analogues towards their breakdown by metalloproteases in membrane homogenates by comparing their [^{125}I]Ang IV-competition binding curves obtained by the co-administration of both ligands (i.e. co-incubation) or after preliminar incubation of the peptide analogue with the membranes (i.e. pre-incubation). In the first case no breakdown by metalloproteases occurs since the assay is carried out in the presence of the chelators, EDTA and 1,10-phenantrolin, known to effectively prevent metalloprotease activity (Lowther et al., 2002). In the second case, the peptide analogues got the opportunity to be broken down for some time before the addition of radioligand and chelators. This method has already been proven useful in the search for new and stable analogues for Ang IV (Axen et al., 2006; 2007, Lukaszuk et al., 2008). After pre-incubation with CHO-K1 membranes, the Ang IV experienced clear rightward shift (Axen et al., 2006). This indicates that Ang IV is quickly metabolized. A similar conclusion was reached when using rat kidney homogenates (Krebs et al., 1996).

Ang IV analogues Nle1-Ang IV and Divalinal-Ang IV are occasionally claimed to be metabolically stable (Chai et al., 2000; Albiston et al., 2001). Using the method presented above, we have re-assessed this claim and found that, from both, only Divalinal-Ang IV resists metallopeptidase activity (figure 3). This is in agreement with the early claims by Sardinia et al. (1994) and Wright et al. (1999). Therefore, caution is needed when using Nle^1-Ang IV in functional studies.

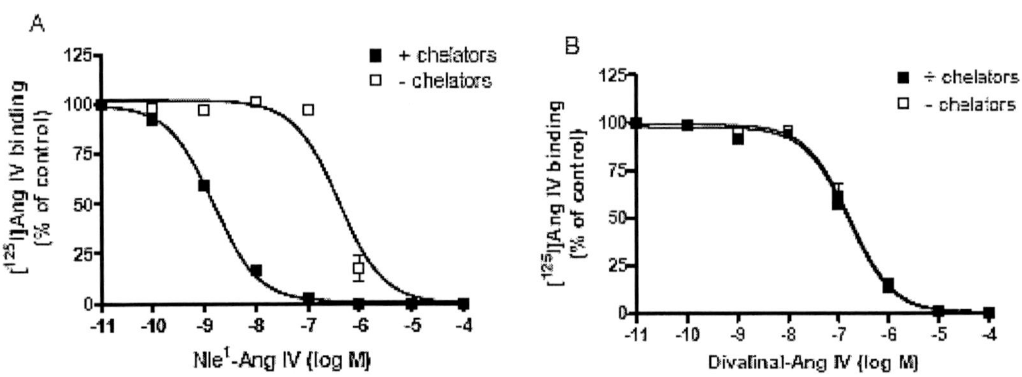

Figure 3. Stability of Nle^1-Ang IV and Divalinal-Ang IV (new preparations by A. Lukaszuk, Department of Organic Chemistry, Vrije Universiteit Brussel, Belgium) towards breakdown by metalloproteases present in CHO-K1 cell membranes. Membranes were preincubated for 40 min at 37°C with increasing concentrations of Nle^1-Ang IV (A) and Divalinal-Ang IV (B) in the presence (●) or absence (▲) of EDTA and 1.10-phenantrolin before 30 min incubation with 1 nM [^{125}I]Ang IV. Data refer to specific binding of [^{125}I]Ang IV (expressed as percent of control binding in the absence of unlabelled compound), calculated by subtracting non-specific binding in the presence of 10 μM unlabelled Ang IV from total binding. Data are the mean +-SEM of three independent experiments performed in duplicate.

Detection of Alternative High Affinity Ang IV Targets

Whereas pharmacological profiles of IRAP have already been extensively documented both by means of binding studies and by means of catalytic activity measurements (Lee et al., 2003, Lew et al., 2003, Demaegdt et al., 2004a, 2006, Mustafa et al., 2001b; Handa et al., 1999; Harding et al., 1994; Swanson et al., 1992; Hanesworth et al., 1993), only little attention has been focused on this aspect in the studies dealing with Ang IV- mediated responses. Pending a more detailed knowledge about the pharmacological profile of the cellular recognition sites that convey these responses, we cannot exclude that they are distinct from IRAP. As presently carried out, [^{125}I]Ang IV binding experiments may only offer little help for the detection of alternative non-IRAP high affinity targets. Indeed, binding to such targets could be shielded by a potential excess of IRAP apo-enzymes if the assay is carried out in the presence of divalent cation chelators. This "interference" could be avoided by omitting the chelators and by replacing the presently used metabolically unstable radioligands like [^{125}I]Ang IV and [^{125}I]Nle1-Ang IV by stable ones. As binding studies with such ligands could be performed in the absence of chelators, IRAP should no longer interfere with the detection of alternative targets with high affinity for Ang IV. The use of IRAP KO mice could also be very useful in functional and binding studies since the alternative targets should still be able to convey Ang IV- mediated responses.

Hypotheses on How Ang IV Enhances Cognition

Three potential mechanisms have been put forward to explain the role of IRAP in the increased cognitive performance after Ang IV administration (Chai et al., 2004). Yet, there is presently still no conclusive experimental evidence with respect to the molecular mechanism(s) involved.

Inhibition of IRAP Activity

The ability of Ang IV to inhibit the catalytic activity of IRAP has already been well documented in *in vitro* conditions (Albiston et al., 2001; Lew et al., 2003, Demaegdt et al., 2004). Substrates of IRAP that were found *in vitro* are somatostatin CCK-8, vasopressin and oxytocin (Herbst et al., 1997; Matsumoto et al., 2001; Tsujimoto et al., 1992). Recently, Wallis and coworkers (2007) found higher levels of endogenous vasopressin in the blood of IRAP KO mice when compared to controls. Moreover, vasopressin was not processed when injected in these mice. This constitutes the first demonstration that IRAP exerts also proteolytic activity *in vivo*.

The substrates cited above are known for their memory-facilitating or anticonvulsant properties (reviewed by Kovacs and de wied, 1994; Chai et al., 2004; Stragier et al., 2007). In short, vasopressin is known for facilitating memory consolidation and retrieval in the passive avoidance paradigm (de Wied et al., 1993; Kovacs and de Wied, 1994; Engelmann et al., 1996). Moreover, vasopressin reverses memory deficits induced by scopolamine (Fujiwari et al., 1997). Oxytocin might also be involved since it improves reference memory but not working memory in female mice (Tomizawa et al., 2003). However, the role of oxytocin on memory storage and retrieval is more controversial as for vasopressin (de Wied et al, 1993; Engelmann et al., 1996). Cholecystokinin-8 also facilitates memory consolidation in the inhibitory avoidance paradigm (Huston et al., 1998) and the two-trial memory task (Sebret et al., 1999). Decreased cholecystokinin-8 concentrations within areas such as the prefrontal cortex, amygdala and hippocampus are associated with learning and memory deficits (Virgo et al., 1995). Although somatostatin is not generally thought to be associated with cognitive function, somatostatin-14 also facilitates acquisition in spatial tasks (Lamirault et al., 2001) and depletion of endogenous somatostatin results in memory deficits that are reversed by somatostatin treatment (Schettini et al., 1988).

According to this hypothesis, any compound that inhibits the catalytic activity of IRAP increases the concentration of its natural substrates and, by this way, increases the cognitive performance (Albiston et al., 2001). For this hypothesis to hold true, the effects of Ang IV on learning and memory should be blocked by antagonizing the interaction between those natural substrates and their receptors. Although this has not been demonstrated so far with respect to cognition, it was recently shown that the anticonvulsant effect of Ang IV can be blocked by a somatostatin receptor antagonist. Taken together with the fact that somatostatin-14 possesses anticonvulsant properties via activation of the sst2 receptor (Vezzani and Hoyer, 1999; Cammalleri et al., 2004), this finding suggests that Ang IV can effectively induce a build-up of somatostatin by inhibiting the enzymatic activity of IRAP (Stragier et al., 2006). Nevertheless, additional work is needed to provide more direct proof for the physiological relevance of such mechanism. Theoretically, according to this hypothesis, Ang IV might also increase the more negative effects of the IRAP substrates. For example vasopressin can stimulate aggression, stress and anxiety (Caldwell et al., 2007). However, in literature, no evidence was given for odd behaviour in rodents treated with Ang IV or in IRAP KO mice.

A major caveat of this hypothesis is that it does not readily explain why Ang IV is already able to trigger biochemical effects in intact cells at nanomolar concentrations. This behaviour has been repeatedly shown and it is especially difficult to reconcile with the observation that Ang IV blocks the catalytic activity of IRAP with a pKi of about 7 (Demaegdt et al., 2004a, 2006, Lew et al., 2003). Moreover, since Ang IV itself is rapidly degraded, it seems unlikely that it can be maintained at high enough concentrations to significantly inhibit the catalytic activity of IRAP *in vivo* (Wallis et al., 2007).

IRAP as a Receptor

As second hypothesis, Vauquelin et al. (2002) have been the first to propose that IRAP could be capable of triggering intracellular events on top of its enzymatic activity. Indeed, the biochemical effects of Ang IV in intact cells (see Biochemical effects of Ang IV) provide circumstantial evidence that IRAP is capable of transferring extracellular information across the cell membrane in its own right; i.e. that it also acts as a classical receptor. Interestingly, while Ang IV caused an increase of the dopamine concentration in rat striatum, this effect could not be reproduced with the AP-N inhibitor 7B at a concentration at which it inhibited most of the IRAP catalytic activity (Stragier et al., 2007). Hence, this *in vivo* effect of Ang IV seems to be unconnected to the catalytic activity of IRAP.

It may be argued that the designation of IRAP as a receptor is difficult to reconcile with its enzymatic activity and, from the structural point of view, with the fact that it only possesses a single membrane spanning α-helix. However, homodimer formation is one of the characteristic features of the membrane-bound M1 metallopeptidase family (Shipp and Look, 1993) to which IRAP belongs. As a dimer, this enzyme is theoretically capable to convey information across the cell membrane in the same way as growth factors and cytokine receptors. Moreover, it was shown that the intracellular, 110 amino acid-long N-terminal domain of IRAP is highly conserved between species and that, like GLUT4, it contains 2 dileucine motifs (Keller et al., 1995; Ross et al. 1996) which are required for correct intracellular targeting (Waters et al., 1997; Subtil et al., 2000; Johnson et al., 2001). The N-terminal domain of IRAP has also been shown interact with cytosolic proteins like tankyrase, a poly(ADP-ribose)polymerase; acyl-coenzyme A dehydrogenase; p115 and Akt substrate 160 (reviewed by Albiston et al., 2007). While these proteins are thought to play a role in retention or sorting in the specialized (GLUT4) vesicles, it remains to be established whether IRAP is also capable to interact with proteins that are implicated in intracellular signalling.

In support of a receptor function for IRAP, it has also been reported that the structurally related AP-N (EC 3.4.11.2) and dipeptidylpeptidase IV (DP IV, EC 3.4.14.3) metalloproteases are also able to mediate intracellular signalling. Monoclonal antibodies against AP-N trigger both $[Ca^{2+}]_i$ transients and phosphorylation of the mitogen-activated protein kinases (Erk1/2, JNK and p38) in human monocytes (Navarrete Santos et al., 2000a,b). In the same line, synthetic inhibitors of the DP IV enzymatic activity are capable of inducing intracellular tyrosine phosphorylation, transient MAPK, p38 activation, secretion of TGF-β1 and suppression of DNA synthesis and cytokine production in human and murine T cells (Kahne et al., 1998; 2000).

This hypothesis could be conciliated with the high potency of Ang IV in some of its cellular effects, but this requires certain conditions to be met. For example, only part of the IRAP proteins need to be occupied to trigger a prominent response (i.e. when there is a large "receptor reserve"). Alternatively, because of its high affinity for Ang IV, the IRAP apo-enzyme could function as a receptor. In this respect, it was indeed shown that catalytic Zn^{2+} and free Zn^{2+} are in rapid equilibrium with each-other (Laeremans et al., 2005). Yet, pending more conclusive experimental evidence, the receptor hypothesis remains highly controversial (Wright and Harding, 2004; Albiston et al., 2007).

Ability of IRAP to Affect Glucose Uptake in Neurons

A well-established correlation exists between the cognitive function and enhanced glucose utilisation in the hippocampus (McNay et al., 2000). Moreover, exogenous glucose is known to enhance cognitive functions in animals and in humans (Kopf et al., 2001; Korol et al., 1998). Possibly, the effect of glucose on learning and memory results from its ability to affect neurotransmission. In this respect, glucose has been shown to enhance both acetylcholine synthesis and release (Kopf et al., 1998). Glucose may also account for the higher metabolic rate that arises during learning tasks (McNay et al., 2000). Finally, glucose stimulates the tuberous sclerosis complex-mammalian target of rapamycin (TSC-mTOR) pathway (Dash et al., 2006), a major intracellular cascade that controls the synthesis of several proteins that are involved in cellular growth processes by activation and inactivation of ribosomal S6 kinase and 4E binding protein 1 (Inoki et al., 2005). A disruption of the TSC-mTOR pathway impairs memory formation.

Based on these considerations, Chai et al. (2004) proposed that the binding of Ang IV to IRAP may interfere with IRAP recycling and, consequently, may enhance/prolong the exposure of GLUT4 at the cell surface. Therefore, the possibility arises that the beneficial effects of Ang IV on memory and learning involve an increased glucose uptake in hippocampal neurons. Although still little is known about the sub-cellular trafficking of IRAP and GLUT4 in the brain, there is circumstantial evidence that neuronal cells may contain a secretory vesicle compartment that is similar to the GLUT4 vesicles in the insulin responsive tissues (Thoidis and Kandror, 2001). Indeed, the location, structure and appearance of IRAP containing vesicles in the hippocampus seems to be consistent with those of neurosecretory vesicles (Fernando et al., in press). GLUT4 was also found to be targeted to IRAP containing vesicles and to the regulated neurosecretory pathway in the PC12 neuroendocrine cell line (Thoidis and Kandror 2001; Hudson et al., 1993). With respect to the physiological stimuli that induce the translocation of IRAP and/or GLUT4 to the plasma membrane we have recently observed an insulin- and dbCAMP- mediated increase in glucose uptake in the hippocampal P40H1 cell line (Demaegdt et al., unpublished observations). This paves the way to directly assess the effect of Ang IV and related "AT4 receptor" ligands on glucose transport in intact hippocampal cells.

Conclusion and Drug Perspectives

Conservatively speaking, we assume that Ang IV exerts its effect through IRAP by one of the three hypotheses mentioned above. The involved pathway most likely depends on cell type and cell conditions. Among others, this could explain why in some studies Divalinal-Ang IV behaves as an "AT4 receptor" antagonist (IRAP as a receptor) while it mimics the effects of Ang IV in other studies (IRAP as an enzyme) (Krebs et al., 1996; Handa et al., 2001). However, we cannot exclude the existence of alternative targets for Ang IV like AP-N or even structurally unrelated ones. Binding studies with a stable Ang IV analogue and/or the use of IRAP KO animals are likely to provide some more conclusive answers in hopefully a not to distant future.

Clearly, there is a need of stable analogues of Ang IV with low affinity for the traditional AT_1 and AT_2 receptors as well as for the AP-N enzyme. In this respect, Axen et al. (2006, 2007) recently produced compounds with higher stability and selectivity for IRAP by performing disulfide cyclizations of Ang IV as well as by modifying its C-terminal end. Similarly, a recently synthesized Ang IV analogue containing two beta amino acids was found to be highly stable (Lukaszuk et al., 2008). Besides the traditional focus on IRAP, it could be of great help to screen such newly developed compounds for their ability to trigger biochemical responses in intact cell systems as well as their ability to trigger *in vivo* physiological effects (i.e. agonistic vs. antagonistic properties). In view of the already significant effects of Ang IV in facilitating learning and memory, stable analogues may pave the way for the development of memory-enhancing drugs.

Acknowledgements

I am very grateful to my promotor, Prof. Georges Vauquelin (Head of Department of Molecular and Biochemical Pharmacology, Vrije Universiteit Brussel, Belgium) for constructive discussions.. My obligations also go to the Research Council of the 'Vrije Universiteit Brussel' (GOA-2002) and the 'Fonds voor Wetenschappelijk Onderzoek Vlaanderen' and the 'Geneeskundige Stichting Koningin Elisabeth' for their financial support. H.D. has a grant from the "Institute for the Promotion of Innovation through Science and Technology in Flanders" (I.W.T.-Vlaanderen).

References

Albiston, AL, McDowall, SG, Matsacos, D, Sim, P, Clune, E, Mustafa, T, Lee, J, Mendelsohn, FAO, Simpson, RG, Connolly, L, Chai, SY. Evidence that the Angiotensin IV (AT_4) receptor is the enzyme Insulin-regulated aminopeptidase. *J. Biol. Chem,* 2001 276, 48623-48626.

Albiston, AL, Pederson, ES, Burns, P, Purcell, B, Wright, JW, Harding, JW, Mendelsohn, FA, Weisinger, RS, Chai, SY. Attenuation of scopolamine-induced learning deficits by LVV-hemorphin-7 in rats in the passive avoidance and water maze paradigms. *Behav. Brain. Res,* 2004 154, 239-243.

Albiston, AL, Pham, V, Chai, SY. Lack of intra-cellular signaling by angiotensin IV in IRAP transfected cells. *Inter. Journal of peptide research and therapeutics*, 2007.

Albiston, AL, Peck, GR, Yeatman, HR, Fernando, R, Ye, S, Chai, SY. Therapeutic targeting of insulin-regulated aminopeptidase: Heads and tails? *Pharmacol. Ther*, 2007 116, 417-427.

Allen, AM, Zhuo, J, Mendelsohn, FA. Localization and function of angiotensin AT1 receptors. *Am. J. Hypertens*, 2000 13, 31-38.

Ardaillou, R. Active fragments of angiotensin II: enzymatic pathways of synthesis and biological effects. *Curr. Opin. Nephrol. Hypertens*, 1997 6, 28-34.

Ardaillou, R, Chansel, D. Chapter 4: Angiotensin II fragments: effects and pathways of synthesis. In: Hulfendahl, HR and Aurel, M, editors. Wenner-Gren Intnl Series, Volume 74. Renin Angiotensin. London: Porland Press Ltd; 1998; 47-55.

Axén, A, Lindeberg, G, Demaegdt, H, Vauquelin, G, Karlén, A, Hallberg, M. Cyclic insulin-regulated aminopeptidase (IRAP)/AT4 receptor ligands. *J. Pept. Sci,* 2006 12, 705-713.

Axén, A, Andersson, H, Lindeberg, G, Rönnholm, H, Kortesmaa, J, Demaegdt, H, Vauquelin, G, Karlén, A, Hallberg, M. Small potent ligands to the insulin-regulated aminopeptidase (IRAP)/AT4 receptor. *J. Pept. Sci*, 2007 13, 434-444.

Braszko, JJ, Kupryszewski, G, Witczuk, B, Wisniewski, K. Angiotensin II (3-8)-hexapeptide affects motor activity, performance of passive avoidance and a conditioned avoidance response in rats. *Neuroscience,* 1988a 27, 777-783.

Braszko, JJ and Wisniewski, K. Effective angiotensin II and saralasin on motor activity in the passive avoidance behavior of rats. *Peptides*, 1988b 9, 475-479.

Braszko, JJ. Involvement of D1 dopamine receptors in the cognitive effects of angiotensin IV and des-Phe6 angiotensin IV. *Peptides*, 2004 25, 1195-1203.

Briand, SI, Bellemare, JM, Bernier, SG, Guillemette, G. Study on the functionality and molecular properties of the AT4 receptor. *Endocr. Res*, 1998 24, 315-323.

Caldwell, HK, Lee, HJ, Macbeth, AH, Young, WS, Vasopressin: Behavioral roles of an "original" neuropeptide. *Prog. Neurobiol,* 2007 [Epub ahead of print].

Cammalleri, M, Cervia, D, Langenegger, D, Liu, Y, Dal Monte, M, Hoyer, D, Bagnoli, P, Somatostatin receptors differentially affect spontaneous epileptiform activity in mouse hippocampal slices. *Eur. J. Neurosci*, 2004 10, 2711-2721.

Caron, AZ, Arguin, G, Guillemette, G. Angiotensin IV interacts with a juxtamembrane site on AT(4)/IRAP suggesting an allosteric mechanism of enzyme modulation. *Regul. Pept*, 2003 113, 9-15.

Chai, SY, Bastias, MA, Clune, EF, Matsacos, DJ, Mustafa, T, Lee, JH, McDowall, SG, Mendelsohn, FA, Albiston, AL, Paxinos, G. Distribution of angiotensin IV binding sites (AT4 receptor) in the human forebrain, midbrain and pons as visualised by in vitro receptor autoradiography. *J. Chem. Neuroanat*, 2000 20, 339-348.

Chai, SY, Fernando, R, Peck, G, Mendelsogn, FA, Jenkins, TA, Albiston, AL. The angiotensin IV/AT4 receptor. *Cell Mol. Life Sci*, 2004 61, 2728-2737.

Chansel, D, Ardaillou, R. Active metabolites derived from angiotensin II. *Nephrology*, 1998 19, 427-432.

Chansel, D, Czekalski, S, Vandermeersch, S, Ruffet, E, Fournié-Zaluski, MC, Ardaillou, R. Characterization of angiotensin IV-degrading enzymes and receptors on rat mesangial cells. *Am. J. Physiol*, 1998 275, F535-542

Choeiri, C, Staines, W, Messier, C. Immunohistochemical localization and quantification of glucose transporters in the mouse brain. *Neuroscience*, 2002 111, 19-34.

Coleman, JKM, Krebs, LT, Hamilton, TA, Ong, B, Lawrence, KA, Sardinia, MF, Harding, JW, Wright, JW. Autoradiographic identification of kidney angiotensin IV binding sites and angiotensin IV-induced renal cortical blood flow changes in rats. *Peptides,* 1998a 19, 269-277.

Coleman, JK, Lee, JI, Miller, JM, Nuttall, AL. Changes in cochlear blood flow due to intra-arterial infusions of angiotensin II (3-8) (angiotensin IV) in guinea pigs. *Heart Res,* 1998b 119, 61-68.

Czech, MP and Corvera, S. Signaling Mechanisms That Regulate Glucose Transport. *J. Biol. Chem*, 1999 274, 1865-1868.

Dash, PK, Orsi, SA, Moore, AN. Spatial memory formation and memory-enhancing effect of glucose involves activation of the tuberous sclerosis complex-mammalian target of rapamycin pathway. *J. Neurosci*, 2006 26, 8048-8056.

de Gasparo, M, Husain, A, Alexander, W, Catt, KJ, Chiu, AT, Drew, M, Goodfriend, T, Harding, JW, Inagami, T, Timmermans, PB. Proposed update of angiotensin receptor nomenclature. *Hypertension*, 1995 25, 924-939.

de Gasparo, M, Catt, KJ, Inagami, T, Wright, JW, Unger, T. International Union of Pharmacology. XXIII. The Angiotensin II Receptors. *Pharmaco. Rev,* 2000 52, 415-472.

de Wied, D, Diamant, M, Fodor, M. Central nervous system effects of the neurohypophyseal hormones and related peptides. *Front Neuroendocrinol*, 1993 14, 251-302.

Demaegdt, H, Vanderheyden, PML, De Backer, JP, Mosselmans, S, Laeremans, H, Le, MT, Kersemans, V, Michotte, Y, Vauquelin, G. Endogenous cystinyl aminopeptidase in Chinese hamster ovary cells: characterization by [(125)I]Ang IV binding and catalytic activity. *Biochem. Pharmacol,* 2004a 68, 885-892.

Demaegdt, H, Laeremans, H, De Backer, JP, Mosselmans, S, Le, MT, Kersemans, V, Michotte, Y, Vauquelin, G, Vanderheyden, PML. Synergistic modulation of cystinyl aminopeptidase by divalent cation chelators. *Biochem. Pharmacol*, 2004b 68, 893-900.

Demaegdt, H, Lenaerts, PJ, Swales, J, De Backer, JP, Laeremans, H, Le, MT, Kersemans, K, Vogel, L, Michotte, Y, Vanderheyden, P, Vauquelin, G. AT4 receptor ligand interaction with cystinyl aminopeptidase and aminopeptidase N: [125I]Ang IV only binds to the cystinyl aminopeptidase apo-enzyme. *Eur. J. Pharmacol*, 2006 546, 19-27.

Dulin, NO, Madhun, ZT, Chang, C, Bert-Mattera, L, Dickens, D, Douglas, JG. Angiotensin IV receptors and signalling in opossum kidney cells. *Am. J. Physiol,* 1995 269, F644-F652.

El Messari, S, Leloup, C, Quignon, M, Brisorgueil, MJ, Penicaud, L, Arluison, M. Immunocytochemical localization of the insulin-responsive glucose transporter 4 (Glut4) in the rat central nervous system. *J. Comp. Neurol*, 1998 399, 492-512.

Engelmann, M, Wotjak, CT, Neumann, I, Ludwig, M, Landgraf, R. Behavioral consequences of intracerebral vasopressin an oxytocin: focus on learning and memory. *Neurosci. Biobehav. Rev,* 1996 20, 341-358.

Esteban, V, Ruperez, M, Sanchez-Lopez, E, Rodriguez-Vita, J, Lorenzo, O, Demaegdt, H, Vanderheyden, PML, Egido, J, Ruiz-Ortega, M. Angiotensin IV activates the nuclear transcription factor-kappaB and related proinflammatory genes in vascular smooth muscle cells. *Circ. Res,* 2005 96, 965-973.

Fernando, RN, Larm, J, Albiston, AL, Chai, SY. Distribution and cellular localization of insulin-regulated aminopeptidase in the rat central nervous system. *J. Comp. Neurol*, 2005 487, 372-390.

Fernando, RN, Luff, SE, Albiston, AL, Chai, SY. Sub-cellular localization of insulin-regulated membrane aminopeptidase, IRAP to vesicles in neurons. *J. Neurochem*, 2007 102, 967-76.

Fitzgerald, SM, Evans, RG, Bergstrom, G, Anderson, WP. Renal hemodynamic response to intrarenal infusion of ligands for the putative angiotensin IV receptor in anesthetized rats. *J. Cardiovasc. Pharmacol,* 1999 34, 206-211.

Fujiwara, M, Ohgami, Y, Inada, K, Iwasaki, K. Effect of active fragments of arginine-vasopressin on the disturbance of spatial cognition in rats. *Behav. Brain. Res,* 1997 83, 91-96.

Gardiner, SM, Kemp, PA, March, JE, Bennett, T. Regional haemodynamic effects of angiotensin II (3-8) in conscious rats. *Br. J. Pharmacol,* 1993 110, 159-162.

Garreau, I, Chansel, D, Vandermeersch, S, Fruitier, I, Piot, JM, Ardaillou, R. Hemorphins inhibit angiotensin IV binding and interact with aminopeptidase N. *Peptides*, 1998 19, 1339-1348.

Garza, LA.and Birnbaum, MJ. Insulin-responsive aminopeptidase trafficking in 3T3-L1 adipocytes. *J. Biol. Chem,* 2000 275, 2560-2567.

Goto, Y, Hattori, A, Ishii, Y, Mizutani, S, Tsujimoto, M. Enzymatic properties of human aminopeptidase A. Regulation of its enzymatic activity by calcium and angiotensin IV. *J. Biol. Chem,* 2006 281, 23503-13.

Handa, RK, Krebs, LT, Harding, JW, Handa, SE. Angiotensin IV AT4-receptor system in the rat kidney. *Am. J. Physiol*, 1998 274, F290-F299.

Handa, RK, Harding, JW, Simasko, SM. Characterization and function of the bovine kidney epithelial angiotensin receptor subtype 4 using angiotensin IV and divalinal angiotensin IV as receptor ligands. *J. Pharmacol. Exp. Ther,* 1999 291, 1242-1249.

Handa, RK. Characterization and Signalling of the AT(4) Receptor in Human Proximal Tubule Epithelial (HK-2) Cells. *J. Am. Soc. Nephrol,* 2001 12, 440-449.

Handa, RK. Biphasic actions of angiotensin IV on renal blood flow in the rat. *Regul. Pept,* 2006 136, 23-29.

Hanesworth, JM, Sardinia, MF, Krebs, LT, Hall, KL, Harding, JW. Elucidation of a specific binding site for angiotensin II(3-8), angiotensin IV, in mammalian heart membranes. *J. Pharmacol. Exp. Ther,* 1993 266, 1036-1042.

Harding, JW, Wright, JW, Swanson, GN, Hanesworth, JM, Krebs, LT. AT4 receptors: specificity and distribution. *Kidney Int,* 1994 46, 1510-1512.

Herbst, JJ, Ross, SA, Scott, HM, Bobin, SA, Morris, NJ, Lienhard, GE, Keller, SR. Insulin stimulates cell surface aminopeptidase activity toward vasopressin in adipocytes. *Am. J. Physiol,* 1997 272, E600-E606.

Hill-Kapturczak, N, Kapturczak, MH, Block, ER, Patel, JM, Malinski, T, Madsen, KM, Tisher, CC. Angiotensin II-stimulated nitric oxide release from porcine pulmonary endothelium is mediated by angiotensin IV. *J. Am. Soc. Nephrol,* 1999 10, 481-491.

Horio, J, Nomura, S, Okada, M, Katsumata, Y, Nakanishi, Y, Kumano, Y, Takami, S, Kinoshita, M, Tsujimoto, M, Nakazato, H, Mizutani, S. Structural organization of the 5'-end and chromosomal assignment of human placental leucine aminopeptidase/insulin-regulated membrane aminopeptidase gene. *Biochem. Biophys. Res. Commun*, 1999 262, 269-274.

Hudson, AW, Fingar, DC, Seidner, GA, Griffiths, G, Burke, B, Birnbaum, MJ. Targeting of the "insulin-responsive" glucose transporter (GLUT4) to the regulated secretory pathway in PC12 cells. *J. Cell .Biol,* 1993 122, 579-88.

Huston, JP, Schildein, S, Gerhardt, P, Privou, C, Fink, H, Hasenohrl, RU. Modulation of memory, reinforcement and anxiety parameters by intra-amygdala injection of cholecystokinin-fragments Boc-CCK-4 and CCK-8s. *Peptides,* 1998 19, 27-37.

Inoki, K, Corradetti, MN, Guan, KL. Dysregulation of the TSC-mTOR pathway in human disease. *Nat. Genet,* 2005 37, 19-24.

Johnson, AO, Subtil, A, Petrush, R, Kobylarz, K, Keller, SR, McGraw, TE. Identification of an insulin-responsive, slow endocytic recycling mechanism in Chinese hamster ovary cells. *J. Biol. Chem*, 1998 273, 17968-17977.

Johnson, AO, Lampson, MA, McGraw, TE. A di-leucine sequence and a cluster of acidic amino acids are required for dynamic retention in the endosomal recycling compartment of fibroblasts. *Mol. Biol. Cell*, 2001 12, 367-381.

Kahne, T, Neubert, K, Faust, J, Ansorge, S. Early phosphorylation events induced by DPIV/CD26-specific inhibitors. *Cell. Immunol*, 1998 189, 60-66.

Kahne, T, Reinhold, D, Neubert, K, Born, I, Faust, J, Ansorge, S. Signal transduction events induced or affected by inhibition of the catalytic activity of dipeptidyl peptidase IV (DP IV, CD26). *Adv. Exp. Med. Biol*, 2000 477, 131-137.

Kakinuma, Y, Hama, H, Sugiyama, F, Goto, K, Mltrakami, K, Fukamizu, A. Anti-apoptotic action of angiotensin fragments to neuronal cells from angiotensinogen knock-out mice. *Neurosci. Lett*, 1997 232, 167-170.

Kakinuma, Y. Impaired blood-brain barrier function in angiotensinogen-deficient mice. *Nature. Med*, 1998 4, 1078-1080.

Kandror, KV, Yu, L, Pilch, PF. The major protein of GLUT4-containing vesicles, gp160, has aminopeptidase activity. *J. Biol. Chem*, 1994 269, 30777-30778.

Keller, SR, Scott, HM, Mastick, CC, Aebersold, R, Lienhard, GE. Cloning and characterization of a novel insulin-regulated membrane aminopeptidase from Glut4 vesicles. *J. Biol. Chem*, 1995 270, 23612-23618.

Keller, SR, Davis, AC, Clairmont, KB. Mice Deficient in the Insulin-regulated Membrane Aminopeptidase Show Substantial Decreases in Glucose Transporter GLUT4 Levels but Maintain Normal Glucose Homeostasis. *J. Biol. Chem*, 2002 277, 17677-17686.

Keller, SR. The insulin-regulated aminopeptidase: a companion and regulator of GLUT4. *Front Biosci*, 2003 8, 410-420.

Korol, DL, Gold, PE. Glucose, memory, and aging. *Am. J. Clin. Nutr*, 1998 67, 764S-771S.

Kopf, SR, Buchholzer, ML, Hilgert, M, Loffelholz, K, Klein, J. Glucose plus choline improve passive avoidance behaviour and increase hippocampal acetylcholine release in mice. *Neuroscience*, 2001 03, 365-71.

Kovacs, GL, de Wied, D. Peptidergic modulation of learning and memory processes. *Pharmacol. Rev*, 1994 46, 269-291.

Kozaki, H, Itakura, A, Okamura, M, Ohno, Y, Wakai, K, Mizutani, S. Maternal serum placental leucine aminopeptidase (P-LAP)/oxytocinase and preterm delivery. *Int. J. Gynaecol. Obstet*, 2001 73, 207-13.

Kramar, EA, Krishnan, R, Harding, JW, Wright, JW. Role of nitric oxide in angiotensin IV-induced increases in cerebral blood flow. *Regul. Pept*, 1998 74, 185-92.

Kramár, EA, Armstrong, DL, Ikeda, S, Wayner, MJ, Harding JW, Wright, JW. The effects of angiotensin IV analogs on long-term potentiation within the CA1 region of the hippocampus in vitro. *Brain Res*, 2001 897, 114-121.

Krebs, LT, Kramár, EA, Hanesworth, JM, Sardinia, MF, Ball, AE, Wright, JW, Harding, JW. Characterization of the binding properties and physiological action of divalinal-angiotensin IV, a putative AT4 receptor antagonist. *Regul Pept*, 1996 67, 123-130.

Krishnan, R, Hanesworth, JM, Wright, JW, Harding, JW. Structure-binding studies of the adrenal AT4 receptor: analysis of position two- and three-modified angiotensin IV analogs. *Peptides,* 1999 20, 915-20.

Laeremans, H, Demaegdt, H, De Backer, JP, Le, MT, Kersemans, V, Michotte, Y, Vauquelin, G, Vanderheyden, PML. Metal ion modulation of cystinyl aminopeptidase. *Biochem. J,* 2005 390, 352-357.

Lamirault, L, Guillou, JL, Micheau, J, Jaffard, R. Intrahippocampal injections of somatostatin dissociate acquisition from the flexible use of place responses. *Eur. J. Neurosci*, 2001 14, 567-570.

Laustsen, PG, Rasmussen, TE, Petersen, K, Pedraza-Díaz, S, Moestrup, SK, Gliemann, J, Sottrup-Jensen, L, Kristensen, T. The complete amino acid sequence of human placental oxytocinase. *Biochem. Biophys. Acta,* 1997 1352, 1–7.

Laustsen, PG, Vang, S, Kristensen, T. Mutational analysis of the active site of human insulin-regulated aminopeptidase. *Eur. J. Biochem*, 2001 268, 98-104.

Lee, J, Chai, S, Mendelsohn, FA, Morris, MJ, Allen, AM. Potentiation of cholinergic transmission in the rat hippocampus by angiotensin IV and LVV-hemorphin-7. *Neuropharmacology,* 2001 40, 618-623.

Lee, J, Mustafa, T, McDowall, SG, Mendelsohn, FA, Brennan, M, Lew, RA, Albiston, AL, Chai, SY. Structure-activity study of LVV-hemorphin-7: angiotensin AT4 receptor ligand and inhibitor of insulin-regulated aminopeptidase. *J. Pharmacol. Exp. Ther*, 2003 305, 205-11.

Lee, J, Albiston, AL, Allen, AM, Mendelsohn FA, Ping, SE, Barett, GL, Murphy M, Morris, MJ, McDowall, SG, Chai, SY. Effect of I.C.V. injection of AT4 receptor ligands, NLE1-angiotensin IV and LVV-hemorphin 7, on spatial learning in rats. *Neuroscience,* 2004 124, 341-349.

Lew, RA, Mustafa, T, Ye, S, McDowall, SG, Chai, SY, Albiston, AL. Angiotensin AT_4 ligands are potent, competitive inhibitors of insulin regulated aminopeptidase (IRAP). *J. of Neurochem*, 2003 86, 344-350.

Li, YD, Block, ER, Patel, JM. Activation of multiple signaling modules is critical in angiotensin IV-induced lung endothelial cell proliferation. *Am. J. Physiol. Lung. Cell. Mol. Physiol*, 2002 238, L707-716.

Lim, SN, Bonzelius, F, Low, SH, Wille, H, Weimbs, T, Herman, GA. Identification of Discrete Classes of Endosome-derived Small Vesicles as a Major Cellular Pool for Recycling Membrane Proteins. *Mol. Biol. Cell*, 2001 12, 981-995.

Lowther, WT and Matthews, BW. Metalloaminopeptidases: common functional themes in disparate structural surroundings. *Chem. Rev*, 2002 102, 4581-608.

Lu, J, Zhang, J, Block, ER, Patel, JM. Angiotensin IV enhances phosphorylation of 4EBP1 by multiple signaling events in lung endothelial cells. *Mol. Cell. Biochem*, 2005 275, 181-188.

Lukaszuk, A, Demaegdt, H, Szemenyei, E, Tóth, G, Tymecka, D, Misicka, A, Karoyan, V, Vanderheyden, P, Vauquelin, G, Tourwé, D. □-homo-amino acid scan of Angiotensin IV. *Br J Pharm*, 2008. In publication.

Lochard, N, Thibault, G, Silversides, DW, Touyz, RM, Reudelhuber, TL. Chronic production of angiotensin IV in the brain leads to hypertension that is reversible with an angiotensin II AT1 receptor antagonist. *Circ. Res*, 2004 94, 1451-1457.

Matsumoto, H, Nagasaka, T, Hattori, A, Rogi, T, Tsuruoka, N, Mizutani, S, Tsujimoto, M. Expression of placental leucine aminopeptidase/oxytocinase in neuronal cells and its action on neuronal peptides. *Eur. J. Biochem,* 2001 268, 3259-66.

McNay, EC, Fries, TM, Gold, PE. Decreases in rat extracellular hippocampal glucose concentration associated with cognitive demand during a spatial task. *Proc. Natl. Acad. Sci USA,* 2000 97, 2881-2885.

Miller-Wing, AV, Hanesworth, JM, Sardinia, MF, Hall, KL, Wright, JW, Speth, RC, Grove, KL, Harding, JW. Central angiotensin IV binding sites: distribution and specificity in guinea pig brain. *J. Pharmacol. Exp. Ther*, 1993 266, 1718-1726.

Mizutani, S, Safwat, MA, Goto, K, Tsujimoto, M, Nakazato, H, Itakura, A, Mizuno, M, Kurauchi, O, Kikkawa, F, Tomoda, Y. Initiating and responsible enzyme of arginine vasopressin degradation in human placenta and pregnancy serum. *Regul. Pept,* 1995 59, 371-378.

Mizutani, S. Physiological roles of placental proteases in feto-placental homeostasis. *Nagoya J. Med. Sci,* 1998 61, 85-95.

Moeller, I, Chai, SY, Oldfield, BJ, McKinley, MJ, Casley, D, Mendelsohn, FA. Localization of angiotensin IV binding sites to motor and sensory neurons in the sheep spinal cord and hindbrain. *Brain Res*, 1995 701, 301-6.

Møeller, I, Small, DH, Reed, G, Harding, JW, Mendelsohn, FA, Chai, SY. Angiotensin IV inhibits neurite outgrowth in cultured embryonic chicken sympathetic neurones. *Brain Res*, 1996a 726, 61-66.

Moeller, I, Paxinos, G, Mendelsohn, FA, Aldred, GP, Casley, D, Chai, SY. Distribution of AT4 receptors in the Macaca fascicularis brain. *Brain Res,* 1996b 712, 307-324.

Møeller, I, Lew, RA, Mendelsohn, FA, Smith, AI, Brennan, ME, Tetaz, TJ, Chai, SY. The globin fragment LVV-hemorphin-7 is an endogenous ligand for the AT4 receptor in the brain. *J. Neurochem*, 1997 68, 2530-2537.

Moeller, I, Albiston, AL, Lew, RA, Mendelsohn, FA, Chai, SY. A globin fragment, LVV-hemorphin-7, induces [3H]thymidine incorporation in a neuronal cell line via the AT4 receptor. *J. Neurochem*, 1999 73, 301-308.

Mustafa, T, Hyung, JL, Chai, SY, Albiston, AL, McDowall, SG, Mendelsohn, FAO. Bioactive angiotensin peptides: focus on angiotensin IV. *JRAAS*, 2001a 2, 205-210.

Mustafa, T, Chai, SY, Mendelsohn, FA, Møeller, I, Albiston, AL. Characterization of the AT(4) receptor in a human neuroblastoma cell line (SK-N-MC). *J. Neurochem*, 2001b 76, 1679-1687.

Nakamura, H, Itakuara, A, Okamura, M, Ito, M, Iwase, A, Nakanishi, Y, Okada, M, Nagasaka, T, Mizutani, S. Oxytocin stimulates the translocation of oxytocinase of human vascular endothelial cells via activation of oxytocin receptors. *Endocrinology*, 2000 141, 4481-4485.

Navarette Santos, A, Langner, J, Herrmann, M, Riemann, D. Aminopeptidase N/CD13 is directly linked to signal transduction pathways in monocytes. *Cell Immunol*, 2000a 201, 22-32.

Navarrete Santos, A, Langner, J, Riemann, D. Enzymatic activity is not a precondition for the intracellular calcium increase mediated by mAbs specific for aminopeptidase N/CD13. *Adv. Exp. Med. Biol,* 2000b 477, 43-47.

Patel, JM, Martens, JR, Li, YD, Gelband, CH, Raizada, MK, Block ER. Angiotensin IV receptor-mediated activation of lung endothelial NOS is associated with vasorelaxation. *Am. J. Physiol,* 1998 275, 1061-1068.

Pawlikowski, M, Gruszka, A, Mucha, S, Melen-Mucha, G. Angiotensins II and IV stimulate the rat adrenocortical cell proliferation acting via different receptors. *Endocr. Regul,* 2001 35, 139-42.

Pawlikowski, M, Meleń-Mucha, G, Mucha, S. The involvement of the renin-angiotensin system in the regulation of cell proliferation in the rat endometrium. *Cell. Mol. Life Sci,* 1999 55, 506-510.

Pederson, ES, Harding, JW, Wright, JW. Attenuation of scopolamine-induced spatial learning impairments by an angiotensin analog. *Regul. Pept,* 1998 30, 97-103.

Riederer, P, Wuketich, S. Time course of nigrostriatal degeneration in parkinson's disease. A detailed study of influential factors in human brain amine analysis., *J. Neural. Transm,* 1976 38, 277-301.

Roberts, KA, Krebs, LT, Kramár, EA, Shaffer, MJ, Harding, JW, Wright, JW. Autoradiographic identification of brain angiotensin IV binding sites and differential c-Fos expression following intracerebroventricular injection of angiotensin II and IV in rats. *Brain. Res,* 1995 682, 13-21.

Rogi, T, Tsujimoto, M, Nakazato, H, Mizutani, S, Tomoda, Y. Human placental leucine aminopeptidase/oxytocinase. A new member of type II membrane spanning zinc metallopeptidase family. *J. Biol. Chem,* 1996 271, 56–61.

Ross, SA, Herbst, JJ, Keller, SR, Lienhard, GE. Trafficking kinetics of the insulin-regulated membrane aminopeptidase in 3T3-L1 adipocytes. *Biochem. Biophys. Res. Commun,* 1997 239, 247-251.

Ross, SA, Scott, HM, Morris, NJ, Leung, WY, Mao, F, Lienhard, GE, Keller, SR. Characterization of the insulin-regulated membrane aminopeptidase in 3T3–L1 adipocytes. *J. Biol. Chem,* 1996 271, 3328-3332.

Saavedra, JM. Emerging features of brain angiotensin receptors. *Regul. Pept*, 1999 85, 31-45.

Sankar, R, Thamotharan, S, Shin, D, Moley, KH, Devaskar, SU. Insulin-responsive glucose transporters-GLUT8 and GLUT4 are expressed in the developing mammalian brain. *Brain Res. Mol. Brain Res,* 2002 107, 157-165.

Sardinia, MF, Hanesworth, JM, Krebs, LT, Harding, JW. AT4 receptor binding characteristics: D-amino acid- and glycine-substituted peptides. *Peptides*, 1993 14, 949-54.

Sardinia, MF, Hanesworth, JM, Krishnan, F, Harding, JW. AT4 receptor structure-binding relationship: N-terminal-modified angiotensin IV analogues. *Peptides*, 1994 15, 1399-1406.

Schettini, G, Florio, T, Magri, G, Grimaldi, M, Meucci, O, Landolfi, E, Marino, A. Somatostatin and SMS 201-995 reverse the impairment of cognitive functions induced by cysteamine depletion of brain somatostatin. *Eur. J. Pharmacol*, 1988 151, 399-407.

Sebret, A, Lena, I, Crete, D, Matsui, T, Roques, BP, Dauge, V. Rat hippocampal neurons are critically involved in physiological improvement of memory processes induced by cholecystokinin-B receptor stimulation. *J. Neurosci,* 1999 19, 7230-7237.

Shibasaki, Y, Asano, T, Lin, JL, Tsukuda, K, Katagiri, H, Ishihara, H, Yazaki, Y, Oka, Y. Two glucose transporter isoforms are sorted differentially and are expressed in distinct cellular compartments. *Biochem. J,* 1992 281, 829-34.

Shipp, MA, Look TA. Hematopoietic differentiation antigens that are membrane-associated enzymes: cutting is the key? *Blood,* 1993 82, 1052–1070.

Stragier, B, Sarre, S, Vanderheyden, PML, Vauquelin, G, Fournie-Zalouski, MC, Ebinger, G, Michotte, Y. Metabolism of angiotensin II is required for its in vivo effect on dopamine release in the striatum of the rat. *J. Neurochem,* 2004 90, 1251-1257.

Stragier, B, Clinckers, R, Meurs, A, De Bundel, D, Sarre, S, Ebinger, G, Michotte, Y, Smolders, I. Involvement of the somatostatin 2 receptor in the anticonvulsant effect of angiotensin IV against pilocarpine-induced limbic seizures in rats. *J. Neurochem,* 2006 98, 1100-1113.

Stragier, B, Demaegdt, H, De Bundel, D, Smolders, I, Sarre, S, Vauquelin, G, Ebinger, G, Michotte, Y, Vanderheyden, PML. Involvement of insulin-regulated aminopeptidase and/or aminopeptidase N in the angiotensin IV-induced effect on dopamine release in the striatum of the rat. *Brain Res*, 2007 1131 97-105.

Stragier, B, De Bundel, D, Sarre, S, Smolders, I, Vauquelin, G, Dupont, A, Michotte, Y, Vanderheyden, P. Involvement of insulin-regulated aminopeptidase in the effects of the renin-angiotensin fragment angiotensin IV: a review. *Heart Fail Rev,* 2007 [Epub ahead of print].

Subtil, A, Lampson, MA, Keller, SR, McGraw, TE. Characterization of the insulin-regulated endocytic recycling mechanism in 3T3-L1 adipocytes using a novel reporter molecule. *J. Biol. Chem*, 2000 275, 4787-95.

Swanson, GN, Hanesworth, JM, Sardinia, MF, Coleman, JKM, Write, JW, Hall, KL, Hall, KL, Miller Wing, AV, Stobb, JW, Cook, VI, Harding, EC, Harding, JW. Discovery of a distinct binding site for angiotensin 11(3-8), a putative angiotensin IV receptor. *Regul. Pept,* 1992 40, 409-419.

Tchekalarova, J, Kambourova, T, Georgiev, V. Interaction between angiotensin IV and adenosine A(l) receptor related drugs in passive avoidance conditioning in rats. *Behav. Brain. Res,* 2001 123, 113-116.

Tchekalarova, J, Georgiev, V. Angiotensin peptides modulatory system: how is it implicated in the control of seizure susceptibility? *Life Sci,* 2005a 76, 955-970.

Tchekalarova, J, Sotiriou, E, Georgiev, V, Kostopoulos, G, Angelatou, F. Up-regulation of adenosine A1 receptor binding in pentylenetetrazol kindling in mice: effects of angiotensin IV. *Brain Res,* 2005b 1032, 94-103.

Thoidis, G, Kandror, KV. A Glut4-vesicle marker protein, insulin-responsive aminopeptidase, is localized in a novel vesicular compartment in PC12 cells. *Traffic,* 2001 2, 577-587.

Thomas, W, Mendelsohn, FA. Angiotensin receptors: form and function and distribution. *Int. J. Biochem. Cell Biol*, 2003 35, 774-779.

Tomizawa, K, Iga, N, Lu, Y-F, Moriwaki, A, Matsushita, M, Li, S-T, Miyamoto, O, Itano, T, Matsui, H. Oxytocin improves long-lasting spatial memory during motherhood through MAP kinase cascade. *Nat. Neurosci*, 2003 6, 384-390.

Tsujimoto, M, Mizutani, S, Adachi, H, Kimura, M, Nakazato, H, Tomoda Y. Identification of human placental leucine aminopeptidase as oxytocinase. *Arch. Biochem. Biophys,* 1992 292, 388–392.

Vanderheyden, PML, Demaegdt, H, Swales, J, Lenaerts, P-J, Laeremans, H, De Backer, J-P, Vogel, LK, Vauquelin, G. Synergistic inhibition of the enzymatic activity of aminopeptidase N by divalent metal ion chelators. *Fund Clin. Pharmacol,* 2006 20, 613-619.

van Rodijnen, WF, van Lambalgen, TA, van Wijhe, MH, Tangelder, GJ, Ter Wee, PM. Renal microvascular actions of angiotensin II fragments. *Am. J. Physiol. Renal. Physiol,* 2002 283, F86-92.

Vauqelin, G, Michotte, Y, Smolders, S, Sarre, S, Ebinger, G, Dupont, A, Vanderheyden PML. Cellular targets for angiotensin II fragments: pharmacological and molecular evidence. *J. Renin. Angiotensin. Aldosterone Syst,* 2002 3, 195-204.

Vezzani, A, Hoyer, D. Brain somatostatin: a candidate inhibitory role in seizures and epileptogenesis. *Eur. J. Neurosci*, 1999 11, 3767-3776.

Vinh, A, Widdop, RE, Drummond, GR, Gaspari, TA, Chronic angiotensin IV treatment reverses endothelial dysfunction in ApoE-deficient mice. *Cardiovasc Res*, 2007.

Virgo, L, Humphries, C, Ortimer, A, Barnes, T, Hirsh, S, de Belleroche, J. Cholecystokinin messenger RNA deficit in frontal and temporal cerebral cortex in schizophrenia. *Biol. Psychiatry,* 1995 37, 694-701.

von Bohlen and Halbach, O. Angiotensin IV in the central nervous system. *Cell Tissue Res*, 2003 1, 1-9.

Wayner, MJAD, Phelix, CF, Wright, JW, Harding, JW. Angiotensin IV enhances LTP in rat dentate gyrus in vivo. *Peptides*, 2001 22, 1403-1414.

Wallis, MG, Lankford, MF, Keller, SR. Vasopressin is a physiological substrate for the insulin-regulated aminopeptidase IRAP. *Am J Physiol Endocrinol Metab*, 2007 293, E1092-102.

Waters, SB, D'Auria, M, Martin, SS, Nguyen, C, Kozma, LM, Luskey, KL. The amino terminus of insulin-responsive aminopeptidase causes Glut4 translocation in 3T3-L1 adipocytes. *J. Biol. Chem*, 1997 272, 23323-23327.

Wright, JW, Miller-Wing, AV, Shaffer, MJ, Higginson, C, Wright, DE, Hanesworth, JM, Harding, JW. Angiotensin II(3-8) (ANG IV) hippocampal binding: potential role in the facilitation of memory. *Brain. Res. Bull*, 1993 32, 497-502.

Wright, JW, Krebs, LT, Stobb, JW, Harding, JW. The angiotensin IV system: Functional implications. *Front Neuroendocrinol,* 1995 16, 23-52.

Wright, JW, Clemens, JA, Panetta, JA, Smalstig, EB, Stubley-Weatherly, LA, Kramár, EA, Pederson, ES, Mungall, BH, Harding, JW. Effects of LY231617 and angiotensin IV on ischemia-induced deficits in circular water maze and passive avoidance performance in rats. *Brain Res*, 1996a 717, 1-11.

Wright, JW, Bechtolt, AJ, Chambers, SL, Harding, JW. Angiotensin III and IV activation of the brain AT1 receptor subtype in cardiovascular function. *Peptides,* 1996b 17, 1365-1371.

Wright, JW, Stubley, L, Pedersen, ES, Kramar, EA, Hanesworth, JM, Harding, JW. Contributions to the brain angiotensin IV-AT4 receptor subtype system to spatial learning. *J. Neurosci,* 1999 19, 3952-3961.

Wright, JW, Harding, JW. The brain angiotensin system and extracellular matrix molecules in neural plasticity, learning, and memory. *Prog. Neurobiol,* 2004 72, 263-93.

Zhang, JH, Hanesworth, JM, Sardinia, M, Alt, JA, Wright, JW, Harding, JW. Structural analysis of angiotensin IV receptor (AT4) from selected bovine tissues. *J. Pharmacol. Exp. Ther,* 1999 289, 1075-1083.

Yamahara, N, Nomura, S, Suzuki, T, Itakura, A, Ito, M, Okamoto, T, Tsujimoto, M, Nakazato, H, Mizutani, S. Placental leucine aminopeptidase/oxytocinase in maternal serum and placenta during normal pregnancy. *Life Sci,* 2000 66, 1401-1410.

Yang, R, Smolders, I, De Bundel, D, Fouyn, R, Demaegdt, H, Michotte, Y, Vanderheyden, P, Dupont, AG. Brain and peripheral AT1 receptors mediate renal vasoconstrictor and blood pressure responses to Angiotensin IV in the rat. Submitted to *J Hypertension,* 2008. In publication

Yeh, TY, Sbodio, JI, Tsun, ZY, Luo, B, Chi, NW. Insulin-stimulated exocytosis of GLUT4 is enhanced by IRAP and its partner tankyrase. *Biochem. J,* 2007 402, 279-290.

Zhang, JH, Stobb, JW, Hanesworth, JM, Sardinia, MF, Harding, JW. Characterization and purification of the bovine adrenal angiotensin IV receptor (AT4) using [125I]benzoylphenyl-alanine-angiotensin IV as a specific photolabel. *J. Pharmacol. Exp. Ther,* 1998 287, 416-424.

In: Angiotensin Research Progress
Editors: H. Miura and Y. Sasaki

ISBN: 978-1-60456-537-9

Chapter V

Regulatory Effects of Angiotensin II on Cellular Function of Renal Proximal Tubule

Yu Jin Lee and Ho Jae Han[1]
Department of Veterinary Physiology, Biotherapy Human Resources Center (BK21),
College of Veterinary Medicine, Chonnam National University,
Gwangju, 500-757, Korea.

Abstract

Over the last two decades, it has become increasingly obvious that angiotensin II (ANG II), the most important effector peptide of the renin-angiotensin system (RAS), may play a central role not only in the etiology of hypertension, but also in the pathophysiology of cardiovascular and renal diseases in humans. ANG II is an essential hormone that exerts pleiotropic effects in the kidneys, including sodium-retaining, growth-promoting and pro-inflammatory properties, in addition to its physiological role in maintaining body salt and fluid balance and blood pressure homeostasis. There are two major types of angiotensin II receptor: type 1 (AT_1) receptors mediate the majority of the physiological actions of ANG II, while type 2 (AT_2) receptors have opposite function to that of AT_1 receptor. Although AT_1 receptor is coupled to G protein and has no intrinsic tyrosine kinase activity, recent *in vitro* evidence suggests that AT_1 receptor activation leads to activation of not only phospholipase C (PLC), but also a variety of intracellular signal transduction cascades, as well as transactivation of epidermal growth factor (EGF) or platelet-derived growth factor (PDGF) receptor. Specifically, ANG II-mediated AT_1 receptor activation regulates sodium transport, glucose metabolism, cell growth, or various bioactive substances (vasoactive hormones, growth factors, cytokines, etc.), and activates multiple intracellular signaling cascades in renal proximal tubule cells (PTCs). In our previous studies, ANG II had a biphasic effect on Na^+ transport in PTCs; picomolar concentrations of ANG II stimulated Na^+ transport, whereas micromolar concentrations of ANG II inhibited it. We also demonstrated that ANG II inhibited

[1] Tel: 82-62-530-2831, Fax: 82-62-530-2809, Email: hjhan@chonnam.ac.kr

Na^+/glucose cotransport (SGLT)-mediated glucose uptake. Regulation of Na^+ transport or SGLT by ANG II is commonly mediated by protein kinase C (PKC)-mitogen activated protein kinase (MAPK)-cytosolic phospholipase A_2 ($cPLA_2$) signaling cascades through the AT_1 receptor in PTCs. These actions are thought to participate in the pathophysiology of renal disease such as hypertension or diabetic nephropathy. Therefore, this review focuses on the regulatory effect of ANG II on cellular function, its physiological and pathophysiological role in PTCs, and the major ANG II-induced cell signaling cascades. Improved understanding of the regulatory effect of ANG II on renal proximal tubule cells is necessary to develop improved therapeutic strategies.

Keywords: Angiotensin II (ANG II), renal proximal tubule cells, Na^+/glucose cotransporter, Na^+ uptake, cell growth

Introduction

The renin-angiotensin system (RAS) plays a vital role in regulating the physiological processes of renal disease in humans [1]. Angiotensin II (ANG II), the most important effector of RAS, has a major role in brain [2], heart [3], adrenal glands [4], vasculature [5-7], embryo development [8, 9], and kidneys [10], and has both beneficial and pathological effects. ANG II has powerful sodium-retaining, growth-promoting and pro-inflammatory properties, in addition to a physiological role in maintaining body salt and fluid balance and blood pressure homeostasis [11]. There is now increasing evidence that ANG II may also function as an intracrine peptide, which can be synthesized and then exert biological effects within a cell [12-14]. At the cellular level, ANG II regulates cell growth, apoptosis, differentiation, cell migration, and extracellular matrix deposition. It also stimulates production of pro-inflammatory cytokines and other growth factors [e.g., platelet-derived growth factor (PDGF)] [15] and vasoconstrictors (e.g., endothelin-1) [16]. Previous cardiovascular studies showed that acute stimulation with ANG II regulates salt/water homeostasis and vasoconstriction, modulating blood pressure, while chronic stimulation promotes hyperplasia and hypertrophy of vascular smooth muscle cells [17, 18]. Our previous studies with primary cultured renal proximal tubule cells (PTCs) have shown cell proliferation [19-24], glucose uptake [25-33], and other physiological regulatory effects [34-42] in response to many growth factors and hormones. In addition, ANG II modulates Na^+/glucose cotransporter (SGLT) in renal proximal tubule cells [43] and stimulates DNA synthesis in mouse embryonic stem cells [44]. Accumulating *in vivo* evidence also supports the possibility that ANG II may directly cause renal diseases, independent of its blood pressure-elevating effect [45, 46]. The renal proximal tubule can synthesize and secrete ANG II in the lumen, and luminal ANG II levels are 100- to 1,000-fold higher than that found in plasma [47]. There are at least two seven-transmembrane G-protein-coupled receptors (GPCRs) known to mediate ANG II function, the ANG II type-1 (AT_1) receptor and the ANG II type-2 (AT_2) receptor [48, 49]. The signaling events elicited in renal cells by ANG II are divided into separate linear pathways, where intracellular targets of the phospholipase A_2 (PLA_2) pathway are distinct from those of the phospholipase C (PLC)/protein kinase C (PKC) and tyrosine kinase (TK)/mitogen-activated protein kinase (MAPK)-dependent pathways [50-51]. The functional roles of the receptors appear to be quite different, with the majority of evidence pointing to the AT_1 receptor as key for behavioral responses to ANG II. Moreover, ANG II transactivates growth factor receptors [e.g., PDGFR, epidermal growth factor receptor (EGFR), and insulin-like growth factor receptor (IGFR)] [52]. Thus, it is suggested that the multiple actions of ANG II are mediated via specific, highly complex intracellular signaling pathways that are stimulated following initial binding of the peptide to its specific receptors [53, 54]. Given its diverse range of functions in physiology, it becomes imperative to understand the biological effects of ANG II within cells and to investigate mechanisms of ANG II-induced signal transduction. This review focuses on the regulatory effect of ANG II on cellular function, its physiological and pathophysiological roles in renal proximal tubule cells, and the major ANG II-induced cell signaling cascades.

I. Physiological Actions of ANG II

A. The Renin-Angiotensin System (RAS)

The mechanisms controlling the formation and degradation of ANG II are important in determining its final physiological effect. ANG II, an octapeptide hormone, is the active component of the RAS. The RAS is an endocrine, paracrine, autocrine, and intracrine system and thus it has been difficult to delineate the quantitative contributions of systemically delivered versus locally formed ANG peptides to the levels found in any given tissue [10]. In this regard, the kidneys are unique in terms of tissue concentrations of ANG II, which are much greater than can be explained by the concentrations delivered by arterial blood flow [55]. There is substantial evidence that the selective intrarenal overproduction of ANG II stimulates endogenous intrarenal angiotensinogen (AGT) mRNA and protein expression [56]. ANG II regulates blood pressure, plasma volume (via aldosterone-regulated sodium excretion), sympathetic nervous activity, and thirst responses. In the RAS, circulating renal-derived renin cleaves hepatic-derived AGT at the N-terminus to form the decapeptide, ANG I, which is converted by the dipeptidyl carboxypeptidase, angiotensin-converting enzyme (ACE), in the lungs, to the active ANG II [57-60]. A recently identified carboxypeptidase, ACE2, cleaves one amino acid from either ANG I or ANG II [61, 62], decreasing ANG II levels and increasing the metabolite ANG 1-7, which displays vasodilator, natriuretic, diuretic, and antiproliferative actions, the opposite effects caused by ANG II [63, 64]. Thus, the balance between ACE and ACE2 is an important factor controlling ANG II levels [65]. ANG II is degraded by aminopeptidases to ANG III and ANG IV. Formation of ANG II can be reduced by agents that inhibit ACE [e.g., the ACE inhibitors (ACE-Is), such as captopril and enalapril]. The tissue-specific effect of increased ANG II levels and enhanced RAS activity depends on the cellular expression and activation of AT_1 receptor, a critical receptor in renal pathophysiology (figure 1). Moreover, all of the components necessary to generate intrarenal ANG II are present along the nephron [66].

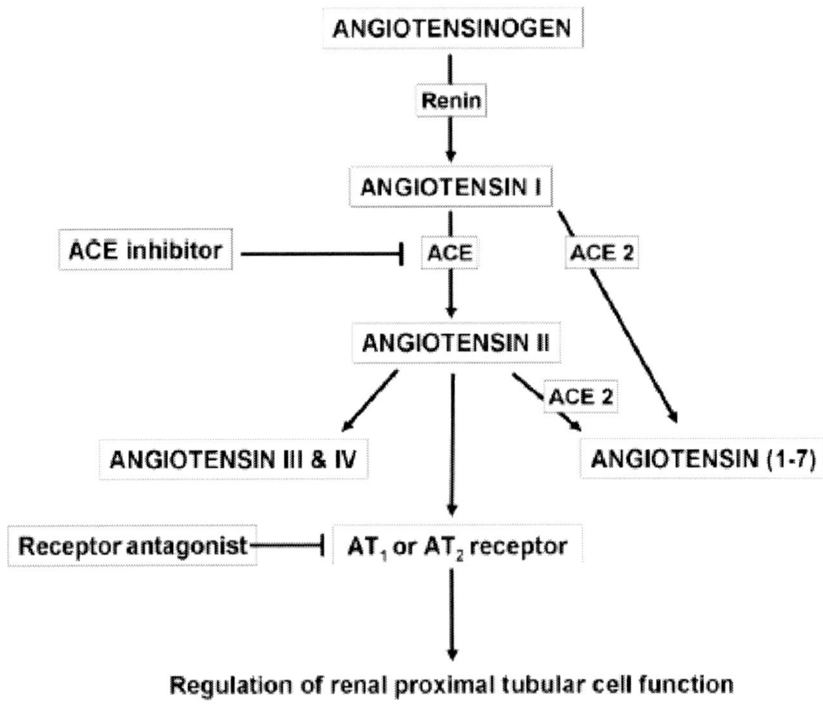

Figure 1. Biosynthesis of angiotensin II. ACE, angiotensin-converting enzyme; AT_1 or AT_2, angiotensin II type 1 or type 2 receptor.

B. ANG II Receptors

ANG II interacts with at least two classes of cell surface receptors on target tissues; these have been designated AT_1 and AT_2. These receptors are typical polypeptide proteins with seven transmembrane domains that couple to G proteins [67]. The majority of the physiological actions of ANG II, including elevation of blood pressure, vasoconstriction, increase in cardiac contractility, stimulation of aldosterone secretion from the adrenal gland, facilitation of catecholamine release from nerve endings, renal sodium and water absorption, and growth stimulation, are mediated by the AT_1 receptor [68, 69]. Expression of AT_1 receptors has been demonstrated in a variety of tissues, including adrenal gland, brain, heart, kidney, prostate and vasculature [70-74]. Because AT_1 serves as a control point for regulating the ultimate effects of ANG II, it becomes necessary to understand the mechanisms that control AT_1 receptor density on the cell membrane. Acutely, increased levels of ANG II lead to an increased level of AT_1 receptor activation; however, chronic exposure to ANG II results in downregulation of its own receptor [75-77]. Numerous growth factors upregulate or downregulate AT_1 receptor expression. For example, epidermal growth factor (EGF) has been shown to downregulate AT_1 receptor expression [78, 79]. In contrast, insulin stimulates gene expression of AT_1 receptor, providing molecular evidence of an association between hyperinsulinemia and hypertension [80]. Our previous studies also showed that high glucose downregulates AT_1 receptor expression mediated by PKC-oxidative stress- transforming growth factor (TGF)-β [39] or PKC-MAPK-$cPLA_2$ [38] signaling in renal proximal tubule cells, suggesting that ANG II receptor is associated with pathophysiology of diabetic nephropathy.

Another ANG II receptor, AT_2, which is expressed at much lower levels in adult kidneys [81, 82] and might possess counter-regulatory actions against the AT_1 receptor, has also been described and can be blocked by selective AT_2 antagonists such as PD123319. AT_2 receptors mediate a wide variety of actions, including vasodilatation, inhibition of cell growth, cell differentiation, and apoptosis [83]. Although freely expressed during fetal development, AT_2 receptor expression is reduced at birth, thereby allowing the AT_1 receptor to predominate. Several pathological states, including ischemic heart disease, dilated cardiomyopathy, and atrial fibrillation, are associated with increased AT_2 receptor expression. It was recently reported that AT_2 receptors cause hypertrophy in cardiomyocytes, independent of ANG II, and do not block AT_1 receptor-mediated hypertrophy [84]. Indeed, some effects of ANG II antagonists might involve inhibition of ANG II at the AT_1 receptor plus stimulation of the unblocked AT_2 receptor. However, the exact role and the extent to which AT_2 receptor plays a role in pathology is unclear as various studies have produced conflicting results.

C. Receptor Transactivation Induced by ANG II

Since 1996, when Axel Ullrich's group discovered a pathway showing how G-protein coupled receptors such as the AT_1-receptor could cause phosphorylation and activation of the EGFR [85], the EGFR transactivation pathway induced by ANG II has been well documented in many studies [17, 86]. In addition to EGFR, ANG II can induce platelet-derived growth factor receptor (PDGFR) or insulin-like growth factor-1 receptor (IGF-1R) transactivation, which may also contribute to the regulation of cellular function [87-89].

The mechanism of EGFR transactivation and its pathological significance are currently one of the major topics of signal transduction research, and recently several interesting findings have been reported on possible components involved in EGFR transactivation. First, EGFR transactivation by ANG II appears to require second messengers, such as elevation of intracellular Ca^{2+} [90-93], activation of PKC, and generation of ROS [94-99]. Second, cytosolic nonreceptor tyrosine kinases, such as Src and proline-rich tyrosine kinase 2 (Pyk2), may be involved in EGFR transactivation [100, 101]. Third, a metalloprotease, such as a disintegrin and metalloprotease (ADAM) or matrix metalloprotease (MMP) causing release of heparin-binding EGF (HB-EGF), is required for EGFR transactivation by ANG II. Recently, many studies identified ADAM17 as mediating EGFR transactivation by ANG II in ACHN tumor cells [102], COS7 cells [103], or vascular smooth muscle cells (VSMCs) [104]. Similar to EGF stimulation, ANG II rapidly transactivates the EGFR leading to activation of the Ras/ERK cascade. Thus, the subsequent EGFR transactivation plays critical roles in cellular functional regulation induced by ANG II.

The protein kinase activity of PDGFR is involved in ANG II-dependent hypertensive renal damage [105]. PDGFR consists of five extracellular IgG-like domains and two intracellular kinase domains. Similar to EGFR, the PDGFR has an intrinsic tyrosine kinase activity. Both PDGF and ANG II activate several common pathways (such as MAPK and PLC) that are implicated in cell growth regulation [106-109]. ANG II-induced PDGFR transactivation may be redox-sensitive and independent of intracellular Ca^{2+} increase [110]. In addition, PDGFR mediates ANG II-induced p44/42 MAPKs phosphorylation and astrocyte growth and may exhibit synergism to regulate the effects of the peptide in cells [88]. Interestingly, it has been recently reported that ANG II induces phosphorylation of ectodomain-truncated PDGFR-β at Tyr751 and Tyr1021 and recruits the p85 subunit of PI3K without the enhancement of PDGFR kinase activity [111].

The significance of IGF-1R transactivation by ANG II has been less clear in renal physiology. It was reported that ANG II induced Src-dependent IGF-1R transactivation in VSMCs [112, 113]. Touyz et al. [114] suggested that ANG II induces ROS production partially through IGF-1R transactivation, leading to p38 MAPK and ERK5 activation in VSMCs. Obviously, further research is necessary to fully determine the significance of IGF-1R transactivation by ANG II in renal pathophysiology. Therefore, it will be important to elucidate the precise molecular mechanisms by which ANG II induces EGFR, PDGFR, or IGF-1R transactivation in renal physiology.

II. Cellular Physiology of ANG II

A. Effect of ANG II on Regulation of Na^+ Transport

Seventy to 80% of urinary Na^+ is reabsorbed in proximal tubule and this mechanism is controlled by various hormones including ANG II. The RAS plays a critical role in the regulation of renal Na^+ and water metabolism through a variety of physiological pathways. Indeed, several reports have associated hypertension with altered Na^+ transport. These reports have described inhibition of active Na^+ transport [115] and increased passive membrane permeability for Na^+ [116] in essential hypertension. In particular, ANG II is widely recognized as an important regulator of renal sodium excretion [117, 118]. Recently, studies have shown that increased abundance of apical Na^+-H^+ exchange (NHE3) in proximal tubule brush border may contribute to enhanced renal Na^+ reabsorption in response to long-term ANG II treatment [119-121]. In the proximal tubule, there is a well-characterized biphasic effect of ANG II on transport [122, 123]. Burns and colleagues [124] have shown that low concentrations of ANG II (< 10^{-9} M) stimulate transcellular sodium and bicarbonate reabsorption via activation of apical Na^+-H^+ exchange, basolateral Na^+/HCO^{3-} cotransport, and basolateral Na^+-K^+-ATPase, and via insertion of H^+-ATPase into the apical membrane. These transport processes are associated with endocytosis of AT_1 receptors, coupling to inhibitory G proteins and G_q-like protein, and subsequent reduction of cAMP levels and protein kinase C-ξ (PKC-ξ) activation. High concentrations of ANG II (> 10^{-9} M) reduce sodium and bicarbonate reabsorption via inhibition of apical Na^+/HCO^{3-} cotransport, and basolateral Na^+-K^+-ATPase. This inhibition is also associated with AT_1 receptor-dependent endocytosis and coupling to G_q-like proteins, with subsequent elevation of intracellular calcium levels and activation of PLA_2 and cytochrome P450 epoxygenase [124, 125].

In cultured proximal tubular cells, ANG II also has a biphasic effect on Na^+ transport; picomolar concentrations of ANG II stimulate Na^+ transport, whereas micromolar concentrations of ANG II inhibit it [41]. In this study, 5,6 epoxy-eicosatrienoic acids (5,6-

EET), a metabolic product of PLA_2-derived arachidonic acid (AA, metabolized via a cytochrome P450 pathway), appears to mediate the ANG II-induced inhibition of Na^+ uptake. In addition, a PLC/PKC signaling pathway is also involved in mediating the effect of ANG II. The initial activation of PLA_2 by ANG II involves the interaction of ANG II with the AT_1 receptor and occurs in a calcium-dependent manner. Thus, a calcium-dependent PLA_2 appears to be involved in mediating the inhibitory effect of ANG II on Na^+ uptake (figure 2). Similarly, Robey et al. [126] reported that the stimulatory effect of ANG II (10^{-10} M) on Na^+/HCO_3^- cotransport activity was dependent on activation of the Src family of tyrosine kinases and the classic MAPK pathway. ANG II stimulated Src as well as the classic MAPK module via binding to AT_1 receptors. These results suggest a central role for these effectors in the coordinated regulation of epithelial transport by ANG II. Moreover, the same group showed that interaction of the nonreceptor tyrosine kinase, Pyk2, with Src was required to mediate ANG II-stimulated Na^+/HCO^{3-} cotransport activity. This suggests a potential coupling of increased cell growth responses to enhanced sodium transport, a mechanism that may be adaptive in surviving nephrons after renal injury [127]. As shown above, AT_1 receptor may regulate sodium transport in renal proximal tubule though diverse intracellular signaling pathways. Therefore, ANG II may represent an important protective mechanism for maintaining extracellular fluid volume during states of volume contraction. Similarly, activation of ANG II in pathophysiological conditions might be expected to lead to elevation of blood pressure in the absence of other adaptive changes to increase urinary sodium excretion.

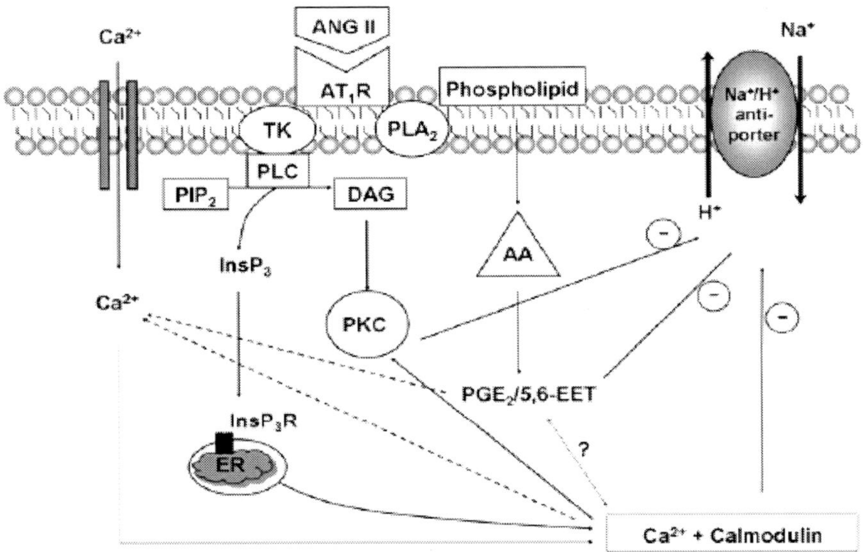

Figure 2. The hypothetical model of Na^+ uptake by high-dose ANG II in renal proximal tubule cells. ANG II binding to AT_1 receptor leads to activation of PLA_2, production of AA metabolites, and subsequent inhibition of Na^+ uptake. ANG II also activates PLC to generate IP_3 and DAG. In turn, IP_3 induces the release of Ca^{2+} from intracellular storage pool and DAG activates protein kinase C, which inhibits Na^+ uptake. AT_1R, angiotensin II receptor type 1; TK, tyrosine kinase; PLC, phospholipase C; PLA_2, phospholipase A_2; PIP_2, phosphatidylinositol-4,5-biphosphate; AA, arachidonic acid; $InsP_3$, 1,4,5-inositol-triphosphate; $InsP_3R$, 1,4,5-inositol-triphosphate receptor; DAG, diacylglycerol; PKC, protein kinase C; ER, endoplasmic reticulum; 5,6-EET, 5,6 epoxy-eicosatrienoic acids. Solid line denotes the proposed pathway and the dotted line is the suspected pathway.

B. Effect of ANG II on Glucose Metabolism

Glucose is a major fuel and important metabolic substrate in cell growth, differentiation, and viability in kidney, vascular smooth muscle, and embryo. Both dietary glucose and glucose synthesized within the body must be transported from the circulation into target cells. Glucose can be transported across membranes by two different mechanisms: a Na^+-coupled active carrier system (SGLT1-6) and a growing family of structurally related Na^+-independent glucose transporter glycoproteins (GLUT 1-12, HMIT) [128]. ANG II is a multifunctional hormone that influences the regulation of glucose metabolism. There are many reports showing that one of the biological functions of ANG II is to regulate glucose uptake and metabolism and thus maintain or enable anabolic pathways required for cell proliferation and survival [129, 130]. Interestingly, studies have reported that ANG II caused both increased and decreased glucose uptake in different types of cells and experiments. For example, studies of VSMCs have shown that ANG II inhibits insulin-induced glucose uptake via MMP or p44/42 MAPKs pathway [129, 131, 132]. For that reason, the increased glucose uptake caused by ACE inhibitor, which blocks ANG II production, may contribute to the treatment of ischemic heart disease [133, 134]. Inhibition of AT_1 receptor-induced increase of glucose uptake provides fresh insight into improved therapeutic approaches to treat type 2 diabetes, hypertension, and metabolic syndrome [135, 136]. However, Low et al. [137] and Quinn and McCumbee [138] indicated that ANG II stimulates glucose uptake into VSMCs. Consistent with this, Han et al. [139] demonstrated that ANG II increases 2-deoxyglucose uptake by PKC activation via AT_1 receptor in mouse embryonic stem cells. These results suggest that ANG II can help cells meet the increased metabolic demands of the growth processes that it initiates by concurrently stimulating glucose transport.

Glucose reabsorption in kidney has important roles in controlling the plasma glucose levels and has received increasing attention as a therapeutic target in patients with diabetes mellitus [140]. Streptozotocin-induced diabetic animals showed an increase in intrarenal angiotensinogen mRNA and interstitial levels of ANG II [141-143]. Treatment with an ACEI prevented increases in intrarenal ANG II and reversed manifestations of diabetic nephropathy [134]. In our studies of renal physiology, we demonstrated that ANG II inhibited SGLT-mediated glucose uptake in renal proximal tubule cells (PTCs) [32]. Our previous reports suggest that ANG II is involved in the high glucose-induced inhibition of α-MG uptake through an increase in ANG II by the high glucose condition, which is a model of diabetic nephropathy [144, 145]. Consistent with this, Kawano et al. [146] showed that ANG II inhibited α-MG uptake into LLC-PK_1 cells caused by inactivation of protein kinase A (PKA) and decrease of PI3K activity. However, Bautista and colleagues [147] demonstrated that ANG II increased expression of SGLT in renovascular hypertension. They thought that increased Na^+ reabsorption through the SGLT may participate in the development of hypertension and that increased glucose epithelial uptake may play an important role in the adaptation of glucose transport to hypertension. Consistent with our present and previous results [41], Burns and colleague [148] reported that the binding of ANG II to the brush-border membrane of rabbit renal PTCs was inhibited by AT_1 antagonist but not by AT_2 antagonist, which suggests that the effects of ANG II are mediated by AT_1 receptor in rabbit renal PTCs. Binding of ANG II to the AT_1 receptor stimulates PLC/PKC, PLA_2, TK, and

MAPKs, and its events are early signaling processes that occur within minutes [51]. Han et al. [32] demonstrated that ANG II inhibited α-MG uptake via TK-PKC-MAPK-cytosolic (c) PLA_2 signaling cascades in PTCs. They previously reported that ANG II had a biphasic effect on Na^+ transport such that micromolar concentrations of ANG II inhibited it [41]. However, there is no biphasic effect of ANG II on SGLT, indicating that ANG II differentially regulates α-MG uptake, unlike Na^+ transport. As shown in Figure 3, ANG II activates AT_1 receptor and stimulates PKC, which triggers p44/42 MAPKs activation. Subsequently, p44/42 MAPKs activation induces the release of AA, which is metabolized to 5,6-EET or PGE_2 by cyclooxygenase and cytochrome P450. These molecules may induce a decrease in SGLT protein expression, which is involved in the inhibition of SGLT activity in renal PTCs. Thus, regulation of SGLTs by increased ANG II may play a pivotal role in the pathogenesis of renal diseases, such as diabetic nephropathy, which might be treated with ACE inhibitors and AT_1-receptor antagonists [11, 149].

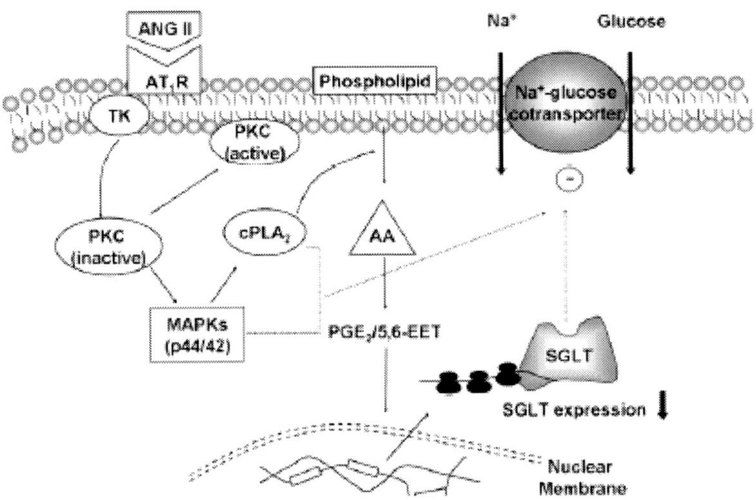

Figure 3. The hypothetical model of inhibition of Na^+/glucose cotransporters (SGLTs, α-MG uptake) by ANG II in renal proximal tubule cells. ANG II binding to AT_1 receptor stimulates TK or PKC, which induces p44/42 MAPKs activation. Subsequently, p44/42 MAPKs activation induces the release of $[^3H]$-AA, which may be metabolized to cyclooxygenase and cytochrome P-450 metabolites such as PGE_2 and 5,6-EET. These molecules may induce a decrease in SGLTs expression, which is involved in inhibition of α-MG uptake. AT_1R, angiotensin II receptor type 1; TK, tyrosine kinase; PKC, protein kinase C; p44/42 MAPKs, p44/42 mitogen activated protein kinases; $cPLA_2$, cytosolic phospholipase A_2; AA, arachidonic acid; PGE_2, prostaglandin E_2; 5,6-EET, 5,6 epoxy-eicosatrienoic acids. Solid line denotes the proposed pathway and the dotted line is the suspected pathway.

C. Effect of ANG II on Cell Growth

ANG II has emerged as an important growth factor that controls growth by cellular proliferation (DNA replication) and hypertrophy (increase in cell size, cell protein, and mRNA content without DNA replication) and/or by inducing apoptosis [150, 151]. ANG II-mediated cell growth depends on the cell type, the subtype of ANG II receptor, and the presence of other growth factors and cytokines. Although ANG II is an important growth modulator of blood vessels and renal organogenesis during development, such ANG II-mediated modulation of growth processes may underlie various pathophysiological processes such as atherosclerosis, vascular and cardiac remodeling, and progression of chronic renal disease [50, 51, 152]. A large body of experimental data demonstrates that ANG II stimulates proliferation of mesangial cells [153-155] and VSMCs [156-159]. Many studies have demonstrated the signaling pathway modulating the growth-promoting effects of ANG II. Recently, Zhuo et al. [160] showed that AT_1 receptor-mediated endocytosis of extracellular ANG II may be required for Ang II-induced NF-κB activation and subsequent cell proliferation in PTCs. ANG II-stimulated proliferation is also mediated by AT_1 receptor-dependent Ca^{2+}/PKC and EGF receptor-dependent PI3K/Akt/mTOR/p70S6K1 signaling pathways in mouse ES cells [44]. In addition, the proliferative response by AT_1 receptor activation can be mediated by the ROS/EGFR/c-Jun NH_2-terminal kinase (JNK) pathway [153, 154]. Diep et al. [161] also demonstrated that ANG II infusion into rats significantly increased [^{3}H] thymidine incorporation into mesenteric arteries. In that study, cell cycle regulatory proteins (cyclin D4 and CDK4) were strongly expressed and expression of $p21^{Cip1}$ and $p27^{kip1}$ was suppressed after ANG II infusion.

Extensive studies have indicated that AT_1 is the predominant receptor. However, the AT_2 receptor is mainly expressed in fetal tissues and neonatal kidney [162, 163, 164], suggesting that AT_2 receptor might play an important role in renal organogenesis. AT_2 receptor expression in the kidney decreases after birth [162, 163] and is expressed at low levels in the normal adult rat kidney [81, 164, 165]. Cao et al. [166] have reported that AT_2 receptor is expressed in adult rat kidney proximal tubular cells and has a role in mediating ANG II-induced proliferation and apoptosis in proximal tubule cells. Zhang et al. [167] recently demonstrated that, after renal hypoxia, ANG II stimulates cell proliferation via AT_2 receptor and the Janus kinase 2 (JAK2)/signal transducers and activators of transcription (STAT) signaling pathway, which may be important in renal repair following injury.

Hypertrophy is defined as an increase in cell size, RNA content, and protein synthesis, without concomitant changes in cellular DNA content. ANG II induces hypertrophy, but not proliferation, in VSMCs [168, 169]. Like VSMCs, hypertrophy of cultured renal proximal tubule cells (PTCs) was induced by ANG II [86, 170, 171]. Indeed, ANG II-induced ROS generation plays a pivotal role in several pathophysiological situations via activation of p44/42 MAPKs and $p27^{kip1}$ pathways, leading to renal growth regulation and remodeling after injury [172]. Moreover, hypertrophy of different renal cells occurs early in diabetic nephropathy. Similar to observations obtained with ANG II, growing mesangial cells in high glucose leads to $p27^{kip1}$-dependent cell cycle arrest and hypertrophy [173, 174]. Consistent with that study, Park and colleagues [175] reported that high glucose inhibited PTC proliferation, suggesting that renal proximal tubule enlargement in diabetes results from cellular hypertrophy, not hyperplasia. These responses may, in fact, be mediated by ANG II because high glucose stimulates the generation of ANG II in kidney cells [175-177]. In addition, ANG II stimulates the expression of connective tissue growth factor and increases the total protein content as well as the size of cultured proximal tubular cells [56, 178, 179]. Taken together, depending on the target cell, the ANG II receptor and the signaling pathway, ANG II may induce either proliferation or hypertrophy. Therefore, it is necessary to understand that ANG II modulates cell growth in renal diseases and to develop improved therapeutic strategies.

D. Pro-Inflammatory Effect of ANG II

ANG II is a pro-inflammatory cytokine and is involved in the pathogenesis of renal diseases [180-182]. By promoting proliferation, inflammation and fibrosis, ANG II contributes to chronic diseases, such as hypertension, atherosclerosis, cardiac hypertrophy and renal injury [183]. Since the early observation in 1970 that ANG II infusion leads to infiltration of immunocompetent cells in the kidney [184], many studies have clearly demonstrated that ANG II causes the adhesion of circulating cells to endothelial and mesangial cells and the migration of inflammatory cells in the kidney. In the inflammatory process, AT_1 regulates several pro-inflammatory genes, including cytokines (interleukin-6 [IL-6]), chemokines (monocyte chemoattractant protein 1 [MCP-1]), and adhesion molecules (vascular cell adhesion molecule 1 [VCAM-1]) [183], while others, such as the chemokine RANTES, are regulated by AT_2 [185]. Some evidence suggests that AT_2 participates in the inflammatory response in renal and vascular tissues [183, 185, 186]. NF-κB is an important transcription factor in inflammatory diseases, and activation of NF-κB by ANG II stimulates transcription of many sodium transporters, cytokines and chemokines, including angiotensinogen, MCP-1, TGFβ, and RANTES [187-190]. In kidney damage, increased tissue NF-κB activity and upregulation of pro-inflammatory parameters have been described, and those processes were diminished by pharmacological blockade of RAS [191, 192]. In addition, *in vivo* and *in vitro* studies have shown that ANG II activates NF-κB in the kidney; however, different receptors are involved depending on the cell type. In glomerular mesangial and vascular smooth muscle cells, NF-κB activation occurs via both AT_1 and AT_2 [193, 194], whereas in tubuloepithelial cells [195] and in PTCs [196], it occurs only through AT_1 and in glomerular endothelial cells via AT_2 [189]. Taken together and consistent with the study of Esteban et al. [197], renal NF-κB activation blockade, ACE inhibition, or combined therapy with AT_1 plus AT_2 antagonists could abolish the inflammatory process in the kidney, supporting the importance of the regulation of this transcription factor during renal damage.

Conclusion

During the last few years, the potential role of intracellular ANG II has been studied by many groups using various tissues and kidney cells. Our current understanding of intracellular ANG II and its signaling in proximal tubule cells has demonstrated that ANG II is an essential hormone that exerts pleiotropic effects in kidneys including regulation of sodium transport, glucose metabolism, cell growth, and pro-inflammatory action. The effects of ANG II are mediated primarily by the AT_1 receptor, and the AT_2 receptor may oppose the AT_1 receptor. In the classical view, ANG II binds a cell surface receptor (AT_1, AT_2), resulting in receptor phosphorylation and activation of a downstream signal transduction pathway and intracellular responses. However, Zhuo and Li [11] recently explained that binding of ANG II to its cell surface AT_1 receptor also activates endocytotic (or internalization) processes that promote trafficking of both the effector and the receptor into intracellular compartments. In addition, there are conflicting data concerning the role of AT_2 receptor in cardiovascular and renal disease. Although the molecular, cellular, and physiological functions of AT_2 receptor have not been clearly identified, AT_2 receptor exerts antiproliferative effects on neonatal rat cardiac myocytes [198], fibroblasts [199], and rat coronary endothelial cells [200]. All of these findings support the beneficial role of AT_2 receptor in cardiovascular and renal disease [201].

ANG II activates multiple signaling pathways in renal proximal tubule cells. These major intracellular signaling cascades do not function independently and are actively engaged in cross-talk. The regulation of Na^+ transport, Na^+/glucose cotransport, cell hypertrophy, and pro-inflammatory properties in proximal tubule cells by ANG II play a pivotal role in pathogenesis of hypertension, diabetic nephropathy, and renal inflammation. Therefore, modulating these cell signaling cascades by treatment with ACE inhibitors, AT_1-receptor antagonists, or AT_2 receptor agonists are possible therapeutic strategies for renal disease. Further studies are necessary to advance our knowledge of the role of ANG II in the pathogenesis of renal disease and to develop therapeutic strategies.

Acknowledgements

This work was supported by a grant (code # 20070401034006) from BioGreen 21 Program, Rural Development Administration, Republic of Korea. The authors acknowledge a graduate fellowship provided by the Ministry of Education and Human Resources Development through the Brain Korea 21 project, Korea.

References

[1] Wolf G, Ziyadeh FN. (1997). The role of angiotensin II in diabetic nephropathy: emphasis on nonhemodynamic mechanisms. *Am. J. Kidney Dis, 29*, 153-163.

[2] Baltatu O, Silva JA Jr, Ganten D, Bader M. (2000). The brain renin-angiotensin system modulates angiotensin II-induced hypertension and cardiac hypertrophy. *Hypertension, 35*, 409–412.

[3] Dell'Italia LJ, Meng QC, Balcells E, Wei CC, Palmer R, Hageman GR, Durand J, Hankes GH, Oparil S. (1997). Compartmentalization of angiotensin II generation in the dog heart Evidence for independent mechanisms in intravascular and interstitial spaces. *J. Clin. Invest, 100*, 253–258.

[4] Mazzocchi G, Malendowicz LK, Markowska A, Albertin G, Nussdorfer GG. (2000). Role of adrenal renin-angiotensin system in the control of aldosterone secretion in sodium-restricted rats. *Am. J. Physiol. Endocrinol. Metab, 278,* E1027–E1030.

[5] Danser AH, Admiraal PJ, Derkx FH, Schalekamp MA. (1998). Angiotensin I-to-II conversion in the human renal vascular bed. *J. Hypertens*, 16, 2051–2056.

[6] Fukui T, Ishizaka N, Rajagopalan S, Laursen JB, Capers Qt Taylor WR, Harrison DG, de Leon H, Wilcox JN, Griendling KK. (1997). p22phox mRNA expression and NADPH oxidase activity are increased in aortas from hypertensive rats. *Circ. Res, 80,* 45–51.

[7] Griendling KK, Minieri CA, Ollerenshaw JD, Alexander RW. (1994). Angiotensin II stimulates NADH and NADPH oxidase activity in cultured vascular smooth muscle cells. *Circ. Res, 74,* 1141–1148.

[8] Tebbs C, Pratten MK, Broughton Pipkin F. (1995). Angiotensin II is a growth factor in the peri-implantation rat embryo. *J. Anat, 195,* 75-86.

[9] Leung PS. (2004). The peptide hormone angiotensin II: its new functions in tissues and organs. *Curr. Protein. Pept. Sci, 5,* 267-273.

[10] Navar LG, Harrison-Bernard LM, Nishiyama A, Kobori H. (2002). Regulation of intrarenal angiotensin II in hypertension. *Hypertension, 39,* 316–322.

[11] Zhuo JL, Li XC. (2007). Novel roles of intracrine angiotensin II and signalling mechanisms in kidney cells. *J. Renin. Angiotensin. Aldosterone. Syst, 8,* 23-33.

[12] Re RN. (2000). On the biological actions of intracellular angiotensin. *Hypertension, 35,* 1189-1190.

[13] Cook JL, Zhang Z, Re RN. (2001). In vitro evidence for an intracellular site of angiotensin action. *Circ. Res, 89,* 1138-1146.

[14] Lavoie JL, Liu X, Bianco RA, Beltz TG, Johnson AK, Sigmund CD. (2006). Evidence supporting a functional role for intracellular renin in the brain. *Hypertension, 47,* 461-466.

[15] Cook JL, Giardina JF, Zhang Z, Re RN. (2002). Intracellular angiotensin II increases the long isoform of PDGF mRNA in rat hepatoma cells. *J. Mol. Cell. Cardiol, 34,* 1525-1537.

[16] An SJ, Boyd R, Zhu M, Chapman A, Pimentel DR, Wang HD. (2007). NADPH oxidase mediates angiotensin II-induced endothelin-1 expression in vascular adventitial fibroblasts. *Cardiovasc. Res, 75,* 702-709.

[17] Mehta PK, Griendling KK. (2007). Angiotensin II cell signaling: physiological and pathological effects in the cardiovascular system. *Am. J. Physiol. Cell. Physiol, 292,* C82-C97.

[18] Geisterfer AA, Peach MJ, Owens GK. (1988). Angiotensin II induces hypertrophy, not hyperplasia, of cultured rat aortic smooth muscle cells. *Circ. Res, 62,* 749-756.

[19] Han HJ, Lim MJ, Lee YJ, Lee JH, Yang IS, Taub M. (2007). Uric acid inhibits renal proximal tubule cell proliferation via at least two signaling pathways involving PKC, MAPK, $cPLA_2$, and NF-κB. *Am. J. Physiol. Renal. Physiol, 292,* F373-F381.

[20] Lee YJ, Lee JH, Han HJ. (2006). Extracellular adenosine triphosphate protects oxidative stress-induced increase of p21(WAF1/Cip1) and p27(Kip1) expression in primary cultured renal proximal tubule cells: role of PI3K and Akt signaling. *J. Cell Physiol, 209,* 802-810.

[21] Lee YJ, Han HJ. (2006). Role of ATP in DNA synthesis of renal proximal tubule cells: involvement of calcium, MAPKs, and CDKs. *Am. J. Physiol. Renal. Physiol, 291,* F98-F106.

[22] Lee YJ, Lee JH, Han HJ. (2004). Ethanol-inhibited [^{3}H] thymidine incorporation via protein kinase C-p44/42 mitogen-activated protein kinase/phospholipase A_2 signal pathway in renal proximal tubule cells. *Alcohol. Clin. Exp. Res, 28,* 1172-1179.

[23] Han HJ, Lim MJ, Lee YJ. (2004). Oxalate inhibits renal proximal tubule cell proliferation via oxidative stress, p38 MAPK/JNK, and $cPLA_2$ signaling pathways. *Am. J. Physiol. Cell Physiol, 287,* C1058-C1066.

[24] Han HJ, Sigurdson WJ, Nickerson PA, Taub M. (2004). Both mitogen activated protein kinase and the mammalian target of rapamycin modulate the development of functional renal proximal tubules in matrigel. *J. Cell Sci, 117,* 1821-1833.

[25] Lee YJ, Heo JS, Suh HN, Lee MY, Han HJ. (2007). Interleukin-6 stimulates alpha-MG uptake in renal proximal tubule cells: involvement of STAT3, PI3K/Akt, MAPKs, and NF-κB. *Am. J. Physiol. Renal. Physiol, 293,* F1036-F1046.

[26] Lee YJ, Park SH, Han HJ. (2005). ATP stimulates Na^+-glucose cotransporter activity via cAMP and p38 MAPK in renal proximal tubule cells. *Am. J. Physiol. Cell Physiol, 289,* C1268-C1276.

[27] Han HJ, Lee YJ, Park SH, Lee JH, Taub M. (2005). High glucose-induced oxidative stress inhibits Na^+/glucose cotransporter activity in renal proximal tubule cells. *Am. J. Physiol. Renal. Physiol, 288,* F988-F996.

[28] Han HJ, Lim MJ, Lee YJ, Kim EJ, Jeon YJ, Lee JH. (2005). Effects of TCDD and estradiol-17β on the proliferation and Na^+/glucose cotransporter in renal proximal tubule cells. *Toxicol In Vitro, 19,* 21-30.

[29] Han HJ, Oh YJ, Lee YJ. (2005). Effect of albumin on ^{14}C-α-Methyl-D-Glucopyranoside uptake in primary cultured renal proximal tubule cells: involvement of PLC, MAPK, and NF-κB. *J. Cell Physiol, 202,* 246-254.

[30] Han HJ, Lee YJ, Park JY, Kim EJ, Lee JH, Taub ML. (2005). Effect of EGF on H2O2-induced inhibition of α-MG uptake in renal proximal tubule cells: involvement of MAPK and AA release. *J. Cell Physiol, 203,* 217-225.

[31] Kim EJ, Lee YJ, Lee JH, Han HJ. (2004). Effect of epinephrine on α-methyl-D-glucopyranoside uptake in renal proximal tubule cells. *Cell Physiol. Biochem, 14,* 395-406.

[32] Han HJ, Park SH, Lee YJ. (2004). Signaling cascade of ANG II-induced inhibition of α-MG uptake in renal proximal tubule cells. *Am. J. Physiol. Renal. Physiol, 286,* F634-F642.

[33] Han HJ, Jeon YJ, Lee YJ. (2003). Involvement of NF-κB in high glucose-induced alteration of α-methyl-D-glucopyranoside (α-MG) uptake in renal proximal tubule cells. *Cell Physiol. Biochem, 13,* 375-384.

[34] Han JY, Heo JS, Lee YJ, Lee JH, Taub M, Han HJ. (2007). Dopamine stimulates $^{45}Ca^{2+}$ uptake through cAMP, PLC/PKC, and MAPKs in renal proximal tubule cells. *J. Cell Physiol, 211,* 486-494.

[35] Lee YJ, Park SH, Jeung TO, Kim KW, Lee JH, Han HJ. (2005). Effect of adenosine triphosphate on phosphate uptake in renal proximal tubule cells: involvement of PKC and p38 MAPK. *J. Cell Physiol, 205,* 68-76.

[36] Lee YJ, Han HJ. (2005). Effect of adenosine triphosphate in renal ischemic injury: involvement of NF-κB. *J. Cell Physiol, 204,* 792-799.

[37] Park SH, Lee YJ, Lim MJ, Kim EJ, Lee JH, Han HJ. (2004). High glucose inhibits fructose uptake in renal proximal tubule cells: involvement of cAMP, PLC/PKC, p44/42 MAPK, and $cPLA_2$. *J. Cell Physiol, 200*, 407-416.
[38] Park SH, Woo CH, Kim JH, Lee JH, Yang IS, Park KM, Han HJ. (2002). High glucose down-regulates angiotensin II binding via the PKC-MAPK-$cPLA_2$ signal cascade in renal proximal tubule cells. *Kidney Int, 61*, 913-925.
[39] Park SH, Han HJ. (2002). The mechanism of angiotensin II binding downregulation by high glucose in primary renal proximal tubule cells. *Am. J. Physiol. Renal. Physiol, 282*, F228-F237.
[40] Park SH, Shin SS, Han HJ. (2001). High glucose levels alter angiotensin II-induced Ca^{2+} uptake via PKC and cAMP pathways in renal proximal tubular cells. *Kidney Blood Press. Res, 24*, 84-91.
[41] Han HJ, Park SH, Koh HJ, Taub M. (2000). Mechanism of regulation of Na^+ transport by angiotensin II in primary renal cells. *Kidney Int, 57*, 2457-2467.
[42] Han HJ, Koh HJ, Park SH. (1999). A signaling pathway for stimulation of Na^+ uptake induced by angiotensin II in primary cultured rabbit renal proximal tubule cells. *J. Vet. Med. Sci, 61*, 135-141.
[43] Lee YJ, Lee YJ, Han HJ. (2007). Regulatory mechanisms of Na^+/glucose cotransporters in renal proximal tubule cells. *Kidney Int. Suppl, 106*, S27-S35.
[44] Han HJ, Han JY, Heo JS, Lee SH, Lee MY, Kim YH. (2007). ANG II-stimulated DNA synthesis is mediated by ANG II receptor-dependent Ca^{2+}/PKC as well as EGF receptor-dependent PI3K/Akt/mTOR/p70S6K1 signal pathways in mouse embryonic stem cells. *J. Cell Physiol, 211*, 618-629.
[45] Carey RM, Siragy HM. (2003). The intrarenal renin-angiotensin system and diabetic nephropathy. *Trends Endocrinol. Metab, 14*, 274-281.
[46] Ortiz PA, Garvin JL. (2001). Intrarenal transport and vasoactive substances in hypertension. *Hypertension, 38*, 621-624.
[47] Wang CT, Navar LG, Mitchell KD. (2003). Proximal tubular fluid angiotensin II levels in angiotensin II-induced hypertensive rats. *J. Hypertens, 21*, 353-360.
[48] Murphy TJ, Alexander RW, Griendling KK, Runge MS, Bernstein KE. (1991). Isolation of a cDNA encoding the vascular type-1 angiotensin II receptor. *Nature, 351*, 233-236.
[49] Kambayashi Y, Bardhan S, Takahashi K, Tsuzuki S, Inui H, Hamakubo T, Inagami T. (1993). Molecular cloning of a novel angiotensin II receptor isoform involved in phosphotyrosine phosphatase inhibition. *J. Biol. Chem, 268*, 24543-24546.
[50] Kim S, Iwao H. (2000). Molecular and cellular mechanisms of angiotensin II-mediated cardiovascular and renal diseases. *Pharmacol. Rev, 52*, 11-34.
[51] Touyz RM, Schiffrin EL. (2000). Signal transduction mechanisms mediating the physiological and pathophysiological actions of angiotensin II in vascular smooth muscle cells. *Pharmacol. Rev, 52*, 639-672.
[52] Higuchi S, Ohtsu H, Suzuki H, Shirai H, Frank GD, Eguchi S. (2007). Angiotensin II signal transduction through the AT_1 receptor: novel insights into mechanisms and pathophysiology. *Clin. Sci. (Lond), 112*, 417-428.

[53] Touyz RM, Berry C. (2002). Recent advances in angiotensin II signaling. *Braz. J. Med. Biol. Res, 35*, 1001-1015.

[54] Daniels D, Yee DK, Fluharty SJ. (2007). Angiotensin II receptor signalling. *Exp. Physiol, 92*, 523-527.

[55] Ingert C, Grima M, Coquard C, Barthelmebs M, Imbs JL. (2002). Contribution of angiotensin II internalization to intrarenal angiotensin II levels in rats. *Am. J. Physiol. Renal. Physiol, 283,* F1003–F1010.

[56] Kobori H, Ozawa Y, Satou R, Katsurada A, Miyata K, Ohashi N, Hase N, Suzaki Y, Sigmund CD, Navar LG. (2007). Kidney-specific enhancement of ANG II stimulates endogenous intrarenal angiotensinogen in gene-targeted mice. *Am. J. Physiol. Renal. Physiol, 293*, F938-F945.

[57] Skeggs LT, Lentz KE, Gould AB, Hochstrasser H, Kahn Jr. (1967). Biochemistry and kinetics of the renin-angiotensin system. *Fed. Proc, 26,* 42-47.

[58] Dorer FE, Kahn, Jr, Lentz KE, Levine M, Skeggs LT. (1972). Purification and properties of angiotensin-converting enzyme from hog lung. *Circ. Res, 31,* 356-366.

[59] Phillips MI, Speakman EA, Kimura B. (1993). Levels of angiotensin and molecular biology of the tissue renin angiotensin systems. *Regul. Pept, 43,* 1-20.

[60] Inagami T. (1998). A memorial to Robert Tiegerstedt. The centennial of renin discovery. *Hypertension, 32,* 953-957.

[61] Crackower MA, Sarao R, Oudit GY, Yagil C, Kozieradzki I, Scanga SE, Oliveira-dos-Santos AJ, da Costa J, Zhang L, Pei Y, Scholey J, Ferrario CM, Manoukian AS, Chappell MC, Backx PH, Yagil Y, Penninger JM. (2002). Angiotensin-converting enzyme 2 is an essential regulator of heart function. *Nature, 417*, 822-828.

[62] Li N, Zimpelmann J, Cheng K, Wilkins JA, Burns KD. (2005). The role of angiotensin converting enzyme 2 in the generation of angiotensin 1-7 by rat proximal tubules. *Am. J. Physiol. Renal. Physiol, 288*, F353-F362.

[63] Su Z, Zimpelmann J, Burns KD. (2006). Angiotensin-(1-7) inhibits angiotensin II-stimulated phosphorylation of MAP kinases in proximal tubular cells. *Kidney Int, 69*, 2212-2218.

[64] Ferrario CM. (2003). Contribution of angiotensin-(1-7) to cardiovascular physiology and pathology. *Curr. Hypertens. Rep, 5*, 129-134.

[65] Danilczyk U, Penninger JM. (2006). Angiotensin-converting enzyme II in the heart and the kidney. *Circ. Res, 98*, 463-471.

[66] Kobori H, Ozawa Y, Suzaki Y, Prieto-Carrasquero MC, Nishiyama A, Shoji T, Cohen EP, Navar LG. (2006). Young Scholars Award Lecture: Intratubular angiotensinogen in hypertension and kidney diseases. *Am. J. Hypertens, 19*, 541-550.

[67] de Gasparo M, Catt KJ, Inagami T, Wright JW, Unger T. (2000). International union of pharmacology. XXIII. The angiotensin II receptors. *Pharmacol. Rev, 52*, 415-472.

[68] Silverstein RL, Fenves AZ, Ram CV. (2004). ARBs and target organ protection. Exploring benefits beyond their antihypertensive effects. *Postgrad. Med, 116*, 31-38.

[69] Wagenaar LJ, Voors AA, Buikema H, van Gilst WH. (1999). Angiotensin receptors in the cardiovascular system. *Can. J. Cardiol, 18*, 1331-1339.

[70] Allen AM, Yamada H, Mendelsohn FA. (1990). In vitro autoradiographic localization of binding to angiotensin receptors in the rat heart. *Int. J. Cardiol, 28*, 25-33.

[71] Montiel M, Barker S, Vinson GP, Jiménez E. (1993). Angiotensin II receptor isoforms in the rat adrenal gland: studies with the selective subtype antagonists DuP 753 and CGP42112A. *J. Mol. Endocrinol, 11*, 69-75.

[72] Saavedra JM, Viswanathan M, Shigematsu K. (1993). Localization of angiotensin AT1 receptors in the rat heart conduction system. *Eur. J. Pharmacol, 235*, 301-303.

[73] Song K, Allen AM, Paxinos G, Mendelsohn FA. (1992). Mapping of angiotensin II receptor subtype heterogeneity in rat brain. *J. Comp. Neurol, 316*, 467-484.

[74] Zhuo J, Alcorn D, Allen AM, Mendelsohn FA. (1992). High resolution localization of angiotensin II receptors in rat renal medulla. *Kidney Int, 42*, 1372-1380.

[75] Griendling KK, Delafontaine P, Rittenhouse SE, Gimbrone MA Jr, Alexander RW. (1987). Correlation of receptor sequestration with sustained diacylglycerol accumulation in angiotensin II-stimulated cultured vascular smooth muscle cells. *J. Biol. Chem, 262*, 14555-14562.

[76] Lassegue B, Alexander RW, Nickenig G, Clark M, Murphy TJ, Griendling KK. (1995). Angiotensin II down-regulates the vascular smooth muscle AT_1 receptor by transcriptional and post-transcriptional mechanisms: evidence for homologous and heterologous regulation. *Mol. Pharmacol, 48*, 601-609.

[77] Touyz RM, He G, Deng LY, Schiffrin EL. (1999). Role of extracellular signal-regulated kinases in angiotensin II-stimulated contraction of smooth muscle cells from human resistance arteries. *Circulation, 99*, 392-399.

[78] Guo DF, Inagami T. (1994). Epidermal growth factor-enhanced human angiotensin II type 1 receptor. *Hypertension, 23*, 1032-1035.

[79] Ullian ME, Raymond JR, Willingham MC, Paul RV. (1997). Regulation of vascular angiotensin II receptors by EGF. *Am. J. Physiol, 273*, C1241-C1249.

[80] Nickenig G, Baumer AT, Temur Y, Kebben D, Jockenhovel F, Bohm M. Statin-sensitive dysregulated AT_1 receptor function and density in hypercholesterolemic men. *Circulation, 100*, 2131-2134.

[81] Wang ZQ, Millatt LJ, Heiderstadt NT, Siragy HM, Johns RA, Carey RM. (1999). Differential regulation of renal angiotensin subtype AT_1A and AT_2 receptor protein in rats with angiotensin-dependent hypertension. *Hypertension, 33*, 96-101.

[82] Miyata N, Park F, Li XF, Cowley AW Jr. (1999). Distribution of angiotensin AT_1 and AT_2 receptor subtypes in the rat kidney. *Am. J. Physiol, 277*, F437-F446.

[83] Ferrario CM. (2006). Role of angiotensin II in cardiovascular disease therapeutic implications of more than a century of research. *J. Renin. Angiotensin. Aldosterone. Syst, 7*, 3-14.

[84] D'Amore A, Black MJ, Thomas WG. (2005). The angiotensin II type 2 receptor causes constitutive growth of cardiomyocytes and does not antagonize angiotensin II type 1 receptor-mediated hypertrophy. *Hypertension, 46*, 1347-1354.

[85] Daub H, Weiss FU, Wallasch C, Ullrich A. (1996). Role of transactivation of the EGF receptor in signalling by G-protein-coupled receptors. *Nature, 379*, 557-560.

[86] Wolf G. (2005). "As time goes by": angiotensin II-mediated transactivation of the EGF receptor comes of age. *Nephrol. Dial. Transplant, 20*, 2050-2053.

[87] Escano CS Jr, Keever LB, Gutweiler AA, Andresen BT (2008). Angiotensin II activates extracellular signal-regulated kinase independently of receptor tyrosine

kinases in renal smooth muscle cells: implications for blood pressure regulation. *J. Pharmacol. Exp. Ther, 324*, 34-42.

[88] Clark MA, Gonzalez N. (2007). Angiotensin II stimulates rat astrocyte mitogen-activated protein kinase activity and growth through EGF and PDGF receptor transactivation. *Regul. Pept, 144*, 115-122.

[89] Cruzado MC, Risler NR, Miatello RM, Yao G, Schiffrin EL, Touyz RM. (2005). Vascular smooth muscle cell NAD(P)H oxidase activity during the development of hypertension: Effect of angiotensin II and role of insulinlike growth factor-1 receptor transactivation. *Am. J. Hypertens, 18*, 81-87.

[90] Eguchi S, Numaguchi K, Iwasaki H, Matsumoto T, Yamakawa T, Utsunomiya H, Motley ED, Kawakatsu H, Owada KM, Hirata Y, Marumo F, Inagami T. (1998). Calcium-dependent epidermal growth factor receptor transactivation mediates the angiotensin II-induced mitogen-activated protein kinase activation in vascular smooth muscle cells. *J. Biol. Chem, 273*, 8890-8896.

[91] Murasawa S, Mori Y, Nozawa Y, Gotoh N, Shibuya M, Masaki H, Maruyama K, Tsutsumi Y, Moriguchi Y, Shibazaki Y, Tanaka Y, Iwasaka T, Inada M, Matsubara H. (1998). Angiotensin II type 1 receptor-induced extracellular signal-regulated protein kinase activation is mediated by Ca^{2+}/calmodulin-dependent transactivation of epidermal growth factor receptor. *Circ. Res, 82*, 1338-1348.

[92] Wang D, Yu X, Cohen RA, Brecher P. (2000). Distinct effects of N-acetylcysteine and nitric oxide on angiotensin II-induced epidermal growth factor receptor phosphorylation and intracellular Ca^{2+} levels. *J. Biol. Chem, 275*, 12223-12230.

[93] Che Q, Carmines PK. (2002). Angiotensin II triggers EGFR tyrosine kinase-dependent Ca^{2+} influx in afferent arterioles. *Hypertension, 40*, 700-706.

[94] Seshiah PN, Weber DS, Rocic P, Valppu L, Taniyama Y, Griendling KK. (2002). Angiotensin II stimulation of NAD(P)H oxidase activity: upstream mediators. *Circ. Res, 91*, 406-413.

[95] Ushio-Fukai M, Griendling KK, Becker PL, Hilenski L, Halleran S, Alexander RW. (2001). Epidermal growth factor receptor transactivation by angiotensin II requires reactive oxygen species in vascular smooth muscle cells. *Arterioscler. Thromb. Vasc. Biol, 21*, 489-495.

[96] Bokemeyer D, Schmitz U, Kramer HJ. (2000). Angiotensin II-induced growth of vascular smooth muscle cells requires an Src-dependent activation of the epidermal growth factor receptor. *Kidney Int, 58*, 549-558.

[97] Ushio-Fukai M, Zuo L, Ikeda S, Tojo T, Patrushev NA, Alexander RW. (2005). cAbl tyrosine kinase mediates reactive oxygen species- and caveolin-dependent AT_1 receptor signaling in vascular smooth muscle: role in vascular hypertrophy. *Circ. Res, 97*, 829-836.

[98] Frank GD, Eguchi S. (2003). Activation of tyrosine kinases by reactive oxygen species in vascular smooth muscle cells: significance and involvement of EGF receptor transactivation by angiotensin II. *Antioxid. Redox. Signal, 5,* 771-780.

[99] Ohtsu H, Frank GD, Utsunomiya H, Eguchi S. (2005). Redox-dependent protein kinase regulation by angiotensin II: mechanistic insights and its pathophysiology. *Antioxid. Redox. Signal, 7*, 1315-1326.

[100] Clark MA, Gonzalez N. (2007). Src and Pyk2 mediate angiotensin II effects in cultured rat astrocytes. *Regul .Pept, 143*, 47-55.
[101] Andreev J, Galisteo ML, Kranenburg O, Logan SK, Chiu ES, Okigaki M, Cary LA, Moolenaar WH, Schlessinger J. (2001). Src and Pyk2 mediate G-protein-coupled receptor activation of epidermal growth factor receptor (EGFR) but are not required for coupling to the mitogen-activated protein (MAP) kinase signaling cascade. *J. Biol. Chem, 276,* 20130–20135.
[102] Schafer, B., Marg, B., Gschwind, A. and Ullrich, A. (2004). Distinct ADAM metalloproteinases regulate G proteincoupled receptor-induced cell proliferation and survival. *J. Biol. Chem, 279*, 47929–47938.
[103] Mifune M, Ohtsu H, Suzuki H, Nakashima H, Brailoiu E, Dun NJ, Frank GD, Inagami T, Higashiyama S, Thomas WG, Eckhart AD, Dempsey PJ, Eguchi S. (2005). G protein coupling and second messenger generation are indispensable for metalloprotease-dependent, heparin-binding epidermal growth factor shedding through angiotensin II type-1 receptor. *J. Biol Chem, 280,* 26592-26599.
[104] Ohtsu H, Dempsey PJ, Frank GD, Brailoiu E, Higuchi S, Suzuki H, Nakashima H, Eguchi K, Eguchi S. (2006). ADAM17 mediates epidermal growth factor receptor transactivation and vascular smooth muscle cell hypertrophy induced by angiotensin II. *Arterioscler. Thromb. Vasc. Biol, 26*, e133-e137.
[105] de Borst MH, Diks SH, Bolbrinker J, Schellings MW, van Dalen MB, Peppelenbosch MP, Kreutz R, Pinto YM, Navis G, van Goor H. (2007). Profiling of the renal kinome: a novel tool to identify protein kinases involved in angiotensin II-dependent hypertensive renal damage. *Am. J. Physiol. Renal. Physiol, 293*, F428-F437.
[106] Suzuki H, Eguchi S. (2006). Growth factor receptor transactivation in mediating end organ damage by angiotensin II. *Hypertension, 47*, 339-340.
[107] Schellings MW, Baumann M, van Leeuwen RE, Duisters RF, Janssen SH, Schroen B, Peutz-Kootstra CJ, Heymans S, Pinto YM. (2006). Imatinib attenuates end-organ damage in hypertensive homozygous TGR(mRen2)27 rats. *Hypertension, 47*, 467-474.
[108] Mondorf UF, Geiger H, Herrero M, Zeuzem S, Piiper A. (2000). Involvement of the platelet-derived growth factor receptor in angiotensin II-induced activation of extracellular regulated kinases 1 and 2 in human mesangial cells. *FEBS Lett, 472*, 129-132.
[109] Ishizaka N, Matsuzaki G, Saito K, Noiri E, Mori I, Nagai R. (2006). Expression and localization of PDGF-B, PDGF-D, and PDGF receptor in the kidney of angiotensin II-infused rat. *Lab. Invest, 86*, 1285-1292.
[110] Heeneman S, Haendeler J, Saito Y, Ishida M, Berk BC. (2000). Angiotensin II induces transactivation of two different populations of the platelet-derived growth factor beta receptor. Key role for the p66 adaptor protein Shc. *J. Biol. Chem, 275*, 15926-15932.
[111] Gao BB, Hansen H, Chen HC, Feener EP. (2006). Angiotensin II stimulates phosphorylation of an ectodomain-truncated platelet-derived growth factor receptor-beta and its binding to class IA PI3K in vascular smooth muscle cells. *Biochem. J, 397*, 337-344.
[112] Zahradka P, Litchie B, Storie B, Helwer G. (2004). Transactivation of the insulin-like growth factor-I receptor by angiotensin II mediates downstream signaling from the

angiotensin II type 1 receptor to phosphatidylinositol 3-kinase. *Endocrinology, 145*, 2978-2987.

[113] Azar ZM, Mehdi MZ, Srivastava AK. (2007). Insulin-like growth factor type-1 receptor transactivation in vasoactive peptide and oxidant-induced signaling pathways in vascular smooth muscle cells. *Can. J. Physiol. Pharmacol, 85*, 105-111.

[114] Touyz RM, Cruzado M, Tabet F, Yao G, Salomon S, Schiffrin EL. (2003). Redox-dependent MAP kinase signaling by Ang II in vascular smooth muscle cells: role of receptor tyrosine kinase transactivation. *Can. J. Physiol. Pharmacol, 81*, 159-167.

[115] Hilton PJ. (1991). Na^+ transport in hypertension. *Diabetes Care, 14*, 233-239.

[116] Blaustein MP. (1984). Sodium transport and hypertension. Where are we going? *Hypertension, 6*, 445-453.

[117] Li L, Wang YP, Capparelli AW, Jo OD, Yanagawa N. (1994). Effect of luminal angiotensin II on proximal tubule fluid transport: role of apical phospholipase A_2. *Am. J. Physiol, 266*, F202-F209.

[118] Burns KD, Homma T, Harris RC. (1993). The intrarenal renin-angiotensin system. *Semin. Nephrol, 13*, 13-30.

[119] Kwon TH, Nielsen J, Kim YH, Knepper MA, Frøkiaer J, Nielsen S. (2003). Regulation of sodium transporters in the thick ascending limb of rat kidney: response to angiotensin II. *Am. J. Physiol. Renal. Physiol, 285*, F152-F165.

[120] Xu L, Dixit MP, Nullmeyer KD, Xu H, Kiela PR, Lynch RM, Ghishan FK. (2006). Regulation of Na^+/H^+ exchanger-NHE3 by angiotensin-II in OKP cells. *Biochim. Biophys. Acta, 1758*, 519-526.

[121] Dixit MP, Xu L, Xu H, Bai L, Collins JF, Ghishan FK. (2004). Effect of angiotensin-II on renal Na^+/H^+ exchanger-NHE3 and NHE2. *Biochim. Biophys. Acta, 1664*, 38-44.

[122] Houillier P, Chambrey R, Achard JM, Froissart M, Poggioli J, Paillard M. (1996). Signaling pathways in the biphasic effect of angiotensin II on apical Na^+/H^+ antiport activity in proximal tubule. *Kidney Int, 50*, 1496-1505.

[123] Musa-Aziz R, Oliveira-Souza M, Mello-Aires M. (2005). Signaling pathways in the biphasic effect of ANG II on Na^+/H^+ exchanger in T84 cells. *J. Membr. Biol, 205*, 49-60.

[124] Burns KD, Li N. (2003). The role of angiotensin II-stimulated renal tubular transport in hypertension. *Curr. Hypertens. Rep, 5*, 165-171.

[125] Kennedy CR, Burns KD. (2001). Angiotensin II as a mediator of renal tubular transport. *Contrib. Nephrol, 135*, 47-62.

[126] Robey RB, Ruiz OS, Espiritu DJ, Ibanez VC, Kear FT, Noboa OA, Bernardo AA, Arruda JA. (2002). Angiotensin II stimulation of renal epithelial cell Na/HCO_3 cotransport activity: a central role for Src family kinase/classic MAPK pathway coupling. *J. Membr. Biol, 187*, 135-145.

[127] Espiritu DJ, Bernardo AA, Robey RB, Arruda JA. (2002). A central role for Pyk2-Src interaction in coupling diverse stimuli to increased epithelial NBC activity. *Am. J. Physiol. Renal. Physiol, 283,* F663-F670.

[128] Wood IS, Trayhurn P. (2003). Glucose transporters (GLUT and SGLT): expanded families of sugar transport proteins. *Br. J. Nutr, 89*, 3-9.

[129] Nazari H, Takahashi A, Harada N, Mawatari K, Nakano M, Kishi K, Ebina Y, Nakaya Y. (2007). Angiotensin II inhibits insulin-induced actin stress fiber formation and glucose uptake via ERK1/2. *J. Med. Invest, 54*, 19-27.

[130] Barron JT, Sasse MF, Nair A. (2004). Effect of angiotensin II on energetics, glucose metabolism and cytosolic NADH/NAD and NADPH/NADP redox in vascular smooth muscle. *Mol. Cell Biochem, 262,* 91-99.

[131] Izawa Y, Yoshizumi M, Fujita Y, Ali N, Kanematsu Y, Ishizawa K, Tsuchiya K, Obata T, Ebina Y, Tomita S, Tamaki T. (2005). ERK1/2 activation by angiotensin II inhibits insulin-induced glucose uptake in vascular smooth muscle cells. *Exp. Cell Res, 308*, 291-299.

[132] Masori M, Hamamoto A, Mawatari K, Harada N, Takahasi A, Nakaya Y. (2007). Angiotensin II Decreases Glucose Uptake by Downregulation of GLUT1 in the Cell Membrane of the Vascular Smooth Muscle Cell Line A10. *J. Cardiovasc. Pharmacol, 50,* 267-273.

[133] Randsbaek F, Lund S, Kimose HH, Grove E, Botker HE, Schmitz O, Nielsen TT. (2003). Angiotensin II inhibition increases cellular glucose transport during reperfusion but not ischemia in pig hearts. *Scand. Cardiovasc. J., 37*, 205-210.

[134] Sharma AM, Engeli S. (2006). The role of renin-angiotensin system blockade in the management of hypertension associated with the cardiometabolic syndrome. *J. Cardiometab. Syndr, 1*, 29-35.

[135] Mathur G, Noronha B, Rodrigues E, Davis G. (2007). The role of angiotensin II type 1 receptor blockers in the prevention and management of diabetes mellitus. *Diabetes. Obes. Metab, 9*, 617-629.

[136] Fujimoto M, Masuzaki H, Tanaka T, Yasue S, Tomita T, Okazawa K, Fujikura J, Chusho H, Ebihara K, Hayashi T, Hosoda K, Nakao K. (2004). An angiotensin II AT_1 receptor antagonist, telmisartan augments glucose uptake and GLUT4 protein expression in 3T3-L1 adipocytes. *FEBS Lett, 576*, 492-497.

[137] Low BC, Ross IK, Grigor MR. (1992). Angiotensin II stimulates glucose transport activity in cultured vascular smooth muscle cells. *J. Biol. Chem, 267*, 20740-20745.

[138] Quinn LA, McCumbee WD. (1998). Regulation of glucose transport by angiotensin II and glucose in cultured vascular smooth muscle cells. *J. Cell Physiol, 177*, 94-102.

[139] Han HJ, Heo JS, Lee YJ. (2005). ANG II increases 2-deoxyglucose uptake in mouse embryonic stem cells. *Life Sci, 77*, 1916-1933.

[140] Ehrenkranz JR, Lewis NG, Kahn CR, Roth J. (2005). Phlorizin: a review. *Diabetes Metab. Res. Rev, 21*, 31-38.

[141] Zimpelmann J, Kumar D, Levine DZ, Wehbi G, Imig JD, Navar LG, Burns KD. (2000). Early diabetes mellitus stimulates proximal tubule renin mRNA expression in the rat. *Kidney Int, 58*, 2320-2330.

[142] Siragy HM, Awad A, Abadir P, Webb R. (2003). The angiotensin II type 1 receptor mediates renal interstitial content of tumor necrosis factor-alpha in diabetic rats. *Endocrinology, 144*, 2229-2233.

[143] Singh R, Singh AK, Leehey DJ. (2005). A novel mechanism for angiotensin II formation in streptozotocindiabetic rat glomeruli. *Am. J. Physiol, 288*, F1183-F1190.

[144] Han HJ, Choi HJ, Park SH. (2000). High glucose-induced inhibition of α-methyl-D-glucopyranoside uptake is mediated by protein kinase C-dependent activation of arachidonic acid release in primary cultured rabbit renal proximal tubule cells. *J. Cell Physiol, 183*, 355-363.

[145] Han HJ, Choi HJ, Park SH. (2000). High glucose inhibits glucose uptake in renal proximal tubule cells by oxidative stress and protein kinase C. *Kidney Int, 57*, 918-926.

[146] Kawano K, Ikari A, Nakano M, Suketa Y. (2002). Phosphatidylinositol 3-kinase mediates inhibitory effect of angiotensin II on sodium/glucose cotransporter in renal epithelial cells. *Life Sci, 71*, 1-13.

[147] Bautista R, Manning R, Martinez F, Avila-Casado Mdel C, Soto V, Medina A, Escalante B. (2004). Angiotensin II-dependent increased expression of Na^+-glucose cotransporter in hypertension. *Am. J. Physiol. Renal. Physiol, 286,* F127-F133.

[148] Burns KD, Inagami T, Harris RC. (1993). Cloning of a rabbit kidney cortex AT_1 angiotensin II receptor that is present in proximal tubule epithelium. *Am. J. Physiol, 264*, F645-F654.

[149] Donaire JA, Ruilope LM. (2007). Angiotensin receptor blockade in diabetic renal disease - focus on candesartan. *Diabetes Res. Clin. Pract, 76*, S22-S30.

[150] deBlois D, Tea BS, Than VD, Tremblay J, Hamet P. (1997). Smooth muscle apoptosis during vascular regression in spontaneously hypertensive rats. *Hypertension, 29*, 340-349.

[151] Diep QN, Li JS, Schiffrin EL. (1999). In vivo study of AT_1 and AT_2 angiotensin receptors in apoptosis in rat blood vessels. *Hypertension, 34*, 617-624.

[152] Wolf G, Wenzel UO. (2004). Angiotensin II and cell cycle regulation. *Hypertension, 43*, 693-698.

[153] Ding G, Zhang A, Huang S, Pan X, Chen R, Yang T. (2007). Angiotensin II induces c-Jun NH_2-terminal kinase activation and proliferation of human mesangial cells via redox-sensitive transactivation of the EGF receptor. *Am. J. Physiol. Renal. Physiol, 293,* F1889-F1897.

[154] Zhang A, Ding G, Huang S, Wu Y, Pan X, Guan X, Chen R, Yang T. (2005). c-Jun NH_2-terminal kinase mediation of angiotensin II-induced proliferation of human mesangial cells. *Am. J. Physiol. Renal. Physiol, 288*, F1118-F1124.

[155] Wolf G, Haberstroh U, Neilson EG. (1992). Angiotensin II stimulates the proliferation and biosynthesis of type I collagen in cultured murine mesangial cells. *Am. J. Pathol, 140*, 95-107.

[156] Wang S, Liu Y, Fan F, Yan J, Wang X, Chen J. (2007). Inhibitory effects of emodin on the proliferation of cultured rat vascular smooth muscle cell-induced by angiotensin II. *Phytother. Res. 2007 Sep 20.*

[157] Kuma S, Oki E, Onohara T, Komori K, Maehara Y. (2007). Angiotensin II-induced growth of vascular smooth muscle cells is associated with modulation of cell surface area and platelet-derived growth factor receptor expression. *Clin. Exp. Pharmacol. Physiol, 34,* 153-160.

[158] Zhou CH, Qian ZY, Zheng SG, Xiang M. (2006). ERK1/2 pathway is involved in the inhibitory effect of crocetin on angiotensin II-induced vascular smooth muscle cell proliferation. *Eur. J. Pharmacol, 535*, 61-68.

[159] Hafizi S, Chester AH, Allen SP, Morgan K, Yacoub MH. (1998). Growth response of human coronary smooth muscle cells to angiotensin II and influence of angiotensin AT_1 receptor blockade. *Coron. Artery. Dis, 9*, 167-175.

[160] Zhuo JL, Carretero OA, Li XC. (2006). Effects of AT_1 receptor-mediated endocytosis of extracellular Ang II on activation of nuclear factor-κB in proximal tubule cells. *Ann. N. Y. Acad. Sci, 1091*, 336-345.

[161] Diep QN, El Mabrouk M, Touyz RM, Schiffrin EL. (2001). Expression of cell cycle proteins in blood vessels of angiotensin II-infused rats: role of AT_1 receptors. *Hypertension, 37*, 604-608.

[162] Grady EF, Sechi LA, Griffin CA, Schambelan M, Kalinyak JE. (1991). Expression of AT_2 receptors in the developing rat fetus. *J. Clin. Invest, 88*, 921-933.

[163] Ozono R, Wang ZQ, Moore AF, Inagami T, Siragy HM, Carey RM. (1997). Expression of the subtype 2 angiotensin (AT_2) receptor protein in rat kidney. *Hypertension, 30*, 1238-1246.

[164] Shanmugam S, Llorens-Cortes C, Clauser E, Corvol P, Gasc JM. (1995). Expression of angiotensin II AT_2 receptor mRNA during development of rat kidney and adrenal gland. *Am. J. Physiol, 268*, F922-F930.

[165] Wang ZQ, Millatt LJ, Heiderstadt NT, Siragy HM, Johns RA, Carey RM. (1999). Differential regulation of renal angiotensin subtype AT_1A and AT_2 receptor protein in rats with angiotensin-dependent hypertension. *Hypertension, 33*, 96-101.

[166] Cao Z, Kelly DJ, Cox A, Casley D, Forbes JM, Martinello P, Dean R, Gilbert RE, Cooper ME. (2000). Angiotensin type 2 receptor is expressed in the adult rat kidney and promotes cellular proliferation and apoptosis. *Kidney Int, 58*, 2437-2451.

[167] Zhang SL, Guo J, Moini B, Ingelfinger JR. (2004). Angiotensin II stimulates Pax-2 in rat kidney proximal tubular cells: impact on proliferation and apoptosis. *Kidney Int, 66*, 2181-2192.

[168] Geisterfer AA, Peach MJ, Owens GK. (1988). Angiotensin II induces hypertrophy, not hyperplasia, of cultured rat aortic smooth muscle cells. *Circ. Res, 62*, 749-756.

[169] Bunkenburg B, van Amelsvoort T, Rogg H, Wood JM. (1992). Receptor-mediated effects of angiotensin II on growth of vascular smooth muscle cells from spontaneously hypertensive rats. *Hypertension, 20*, 746-754.

[170] Wolf G, Jablonski K, Schroeder R, Reinking R, Shankland SJ, Stahl RA. (2003). Angiotensin II-induced hypertrophy of proximal tubular cells requires $p27^{Kip1}$. *Kidney Int, 64*, 71-81.

[171] Hannken T, Schroeder R, Zahner G, Stahl RA, Wolf G. (2000). Reactive oxygen species stimulate p44/42 mitogen-activated protein kinase and induce p27(Kip1): role in angiotensin II-mediated hypertrophy of proximal tubular cells. *J. Am. Soc. Nephrol, 11*, 1387-1397.

[172] Sachse A, Wolf G. (2007). Angiotensin II-induced reactive oxygen species and the kidney. *J. Am. Soc. Nephrol, 18*, 2439-2446.

[173] Wolf G, Wenzel U, Hannken T, Stahl RA. (2001). Angiotensin II induces p27(Kip1) expression in renal tubules in vivo: role of reactive oxygen species. *J. Mol. Med, 79*, 382-389.

[174] Wolf G, Schroeder R, Zahner G, Stahl RA, Shankland SJ. (2001). High glucose-induced hypertrophy of mesangial cells requires p27(Kip1), an inhibitor of cyclin-dependent kinases. *Am. J. Pathol, 158*, 1091-1100.

[175] Park SH, Choi HJ, Lee JH, Woo CH, Kim JH, Han HJ. (2001). High glucose inhibits renal proximal tubule cell proliferation and involves PKC, oxidative stress, and TGF-beta 1. *Kidney Int, 59*, 1695-1705.

[176] Hsieh TJ, Fustier P, Zhang SL, Filep JG, Tang SS, Ingelfinger JR, Fantus IG, Hamet P, Chan JS. (2003). High glucose stimulates angiotensinogen gene expression and cell hypertrophy via activation of the hexosamine biosynthesis pathway in rat kidney proximal tubular cells. *Endocrinology, 144*, 4338-4349.

[177] Vidotti DB, Casarini DE, Cristovam PC, Leite CA, Schor N, Boim MA. (2004). High glucose concentration stimulates intracellular renin activity and angiotensin II generation in rat mesangial cells. *Am. J. Physiol. Renal. Physiol, 286*, F1039-F1045.

[178] Liu BC, Chen L, Sun J, Huang HQ, Ma KL, Liu H, Liu DG, Zhang XL. (2006). Connective tissue growth factormediated angiotensin II-induced hypertrophy of proximal tubular cells. *Nephron. Exp. Nephrol, 103*, e16-e26.

[179] Chen J, Chen JK, Neilson EG, Harris RC. (2006). Role of EGF receptor activation in angiotensin II-induced renal epithelial cell hypertrophy. *J. Am. Soc. Nephrol, 17*, 1615-1623.

[180] Wolf G, Ziyadeh FN, Zahner G, Stahl RA. (1995). Angiotensin II-stimulated expression of transforming growth factor beta in renal proximal tubular cells: attenuation after stable transfection with the c-mas oncogene. *Kidney Int, 48*, 1818-1827.

[181] Ruiz-Ortega M, Esteban V, Rupérez M, Sánchez-López E, Rodríguez-Vita J, Carvajal G, Egido J. (2006). Renal and vascular hypertension-induced inflammation: role of angiotensin II. *Curr. Opin. Nephrol. Hypertens, 15*, 159-166.

[182] Ruiz-Ortega M, Lorenzo O, Suzuki Y, Rupérez M, Egido J. (2001). Proinflammatory actions of angiotensins. *Curr. Opin. Nephrol. Hypertens, 10*, 321-329.

[183] Ruiz-Ortega M, Ruperez M, Esteban V, Egido J. (2003). Molecular mechanisms of angiotensin II-induced vascular injury. *Curr. Hypertens. Rep, 5*, 73-79.

[184] Olsen F. (1970). The secondary and tertiary inflammatory cellular reaction in arterioles damaged by acute angiotensin-hypertension in rats. *Acta Pathol. Microbiol. Scand, 78*, 451-457.

[185] Wolf G, Ziyadeh FN, Thaiss F, Tomaszewski J, Caron RJ, Wenzel U, Zahner G, Helmchen U, Stahl RA. (1997). Angiotensin II stimulates expression of the chemokine RANTES in rat glomerular endothelial cells. Role of the angiotensin type II receptor. *J. Clin. Invest, 100,* 1047-1058.

[186] Akishita M, Horiuchi M, Yamada H, Zhang L, Shirakami G, Tamura K, Ouchi Y, Dzau VJ. (2000). Inflammation influences vascular remodeling through AT_2 receptor expression and signaling. *Physiol. Genomics, 24,* 13–20.

[187] Ruiz-Ortega M, Ruperez M, Lorenzo O, Esteban V, Blanco J, Mezzano S, Egido J. (2002). Angiotensin II regulates the synthesis of proinflammatory cytokines and chemokines in the kidney. *Kidney Int. Suppl, 82*, 12-22.

[188] Muller DN, Dechend R, Mervaala EM, Park JK, Schmidt F, Fiebeler A, Theuer J, Breu V, Ganten D, Haller H, Luft FC. (2000). NF-κB inhibition ameliorates angiotensin II-induced inflammatory damage in rats. *Hypertension, 35*, 193-201.

[189] Wolf G, Wenzel U, Burns KD, Harris RC, Stahl RA, Thaiss F. (2002). Angiotensin II activates nuclear transcription factor-κB through AT_1 and AT_2 receptors. *Kidney Int, 61*, 1986-1995.

[190] Zhuo JL. (2004). Monocyte chemoattractant protein-1: a key mediator of angiotensin II-induced target organ damage in hypertensive heart disease? *J. Hypertens, 22*, 451-454.

[191] Ruiz-Ortega M, Bustos C, Hernández-Presa MA, Lorenzo O, Plaza JJ, Egido J. (1998). Angiotensin II participates in mononuclear cell recruitment in the kidney through nuclear factor-κB activation and monocyte chemoattractant protein-1 gene expression. *J. Immunol., 161*, 430-439.

[192] Morrissey JJ, Klahr S. (1997). Enalapril decreases nuclear factor B activation in the kidney with ureteral obstruction. *Kidney Int, 52*, 926-933.

[193] Lorenzo O, Ruiz-Ortega M, Suzuki Y, Rupérez M, Esteban V, Sugaya T, Egido J. (2002). Angiotensin III activates nuclear transcription factor B in cultured mesangial cells mainly via AT_2 receptors. Studies in AT_1 receptor-knockout mice. *J. Am. Soc. Nephrol, 13*, 1162-1171.

[194] 194. Ruiz-Ortega M, Lorenzo O, Ruperez M, König S, Wittig B, Egido J. (2000). Role of AT_1 and AT_2 receptors in angiotensin II-induced nuclear transcription factor B activation in vascular smooth muscle cells. *Circ. Res, 86*, 1266-1272.

[195] Ruiz-Ortega M, Lorenzo O, Ruperez M, Blanco J, Egido J. (2001). Systemic Infusion of Angiotensin II into normal rats activates nuclear factor k-B and AP-1 in the kidney. Role of AT_1 and AT_2 receptors. *Am. J. Pathol, 158*, 1743-1756.

[196] Zhuo JL, Carretero OA, Li XC. (2006). Effects of AT_1 receptor-mediated endocytosis of extracellular Ang II on activation of nuclear factor-κB in proximal tubule cells. *Ann. N. Y. Acad. Sci, 1091*, 336-345.

[197] Esteban V, Lorenzo O, Rupérez M, Suzuki Y, Mezzano S, Blanco J, Kretzler M, Sugaya T, Egido J, Ruiz-Ortega M. (2004). Angiotensin II, via AT_1 and AT_2 receptors and NF-κB pathway, regulates the inflammatory response in unilateral ureteral obstruction. *J. Am. Soc. Nephrol, 15*, 1514-1529.

[198] Booz GW, Baker KM. (1996). Role of type 1 and type 2 angiotensin receptors in angiotensin II-induced cardiomyocyte hypertrophy. *Hypertension, 28*, 635-640.

[199] Ohkubo N, Matsubara H, Nozawa Y, Mori Y, Murasawa S, Kijima K, Maruyama K, Masaki H, Tsutumi Y, Shibazaki Y, Iwasaka T, Inada M. (1997). Angiotensin type 2 receptors are reexpressed by cardiac fibroblasts from failing myopathic hamster hearts and inhibit cell growth and fibrillar collagen metabolism. *Circulation, 96*, 3954-3962.

[200] Stoll M, Steckelings UM, Paul M, Bottari SP, Metzger R, Unger T. (1995). The angiotensin AT_2-receptor mediates inhibition of cell proliferation in coronary endothelial cells. *J. Clin. Invest, 95*, 651-657.

[201] Matsubara H. (1998). Pathophysiological role of angiotensin II type 2 receptor in cardiovascular and renal diseases. *Circ. Res, 83*, 1182-1191.

In: Angiotensin Research Progress
Editors: H. Miura and Y. Sasaki
ISBN: 978-1-60456-537-9

Chapter VI

Advances in the Physiology and Pharmacology of Angiotensin

Christian Höcht*[*1, 2], Marcos Mayer[1, 2] and Carlos A. Taira[1, 2]
[1] Cátedra de Farmacología
[2] Instituto de Fisiopatología y Bioquímica Clínica
Facultad de Farmacia y Bioquímica, Universidad de Buenos Aires, Junín 956, (C1113AAD) Buenos Aires, Argentina.

1. Abstract

In the last decade, new biologically active components of the renin angiotensin system (RAS) were found. Angiotensin-(1-7) (ANG-(1-7)), a metabolite of angiotensin I and angiotensin II (ANG II), is considered the most pleiotropic component of RAS. ANG-(1-7) lacks the pressor, dipsogenic and aldosterone secretrory actions elicited by ANG II, and, conversely to the octapeptide, systemic administration of ANG-(1-7) induces natriuresis, diuresis and vasodilation. However, this heptapeptide increases vasopressin release, prostaglandin synthesis and facilitates peripheral noradrenergic neurotransmission in the same manner as ANG II.

In addition, activity of ANG-(1-7) is enhanced in subjects with increased activity of the RAS, acting as a counteregulatory modulator of the harmful effects induced by ANG II.

Considering the physiological actions of ANG-(1-7) and its role in the pathophysiology of cardiovascular disease, an increased interest exists in the development of therapeutic agents able to enhance ANG-(1-7) endogenous levels or activation of ANG-(1-7) receptors, such as the Mas receptor. Several reports have described that angiotensin converting enzyme (ACE) inhibitors, such as captopril, increase endogenous levels of ANG-(1-7). Considering physiological actions of ANG-(1-7), enhancement of the heptapeptide levels by ACE inhibition may participate in the beneficial effects of these therapeutic agents. In addition, non-peptidic ANG-(1-7)

[*] Phone: +(54-11)-4964-8265; Fax: +(54-11)-4508-3645; e-mail: chocht@ffyb.uba.ar

agonists are under development and could have a role in the treatment of cardiovascular diseases.

The aim of the present chapter is to review the biochemical and physiological actions of ANG-(1-7), the therapeutic strategies designed to enhanced ANG-(1-7) activity and the possible role of ANG-(1-7) in the treatment of cardiovascular diseases.

2. Introduction

According to its classical definition, the renin-angiotensin system (RAS) is a peptidergic system with endocrine characteristics. Its main substrate, angiotensinogen, is released from the liver and cleaved in the circulation by the enzyme renin (that is secreted from the juxtaglomerular apparatus of the kidney) to form the decapeptide angiotensin (ANG) I. ANG I is then activated to the octapeptide ANG II by angiotensin converting enzyme (ACE), which is predominantly expressed in high concentrations on the surface of endothelial cells in the pulmonary circulation (Paul 2006).

This view of the RAS had to be expanded by recent findings that increased the complexity of the system, such as the existence of different ANG receptors (AT1, AT2, AT4), signal transduction pathways and alternative metabolic cascades; and the discovery of additional truncated peptides such as Ang-(2-8) (ANG III), Ang-(3-8) (ANG IV), and ANG-(1-7) (Paul 2006).

As described by Ferrario (2006) the study of ANG-(1-7) began with the fortuitous observation that the addition of ANG I to canine brain stem homogenates resulted in the appearance of a peak in the high-performance liquid chromatography chromatogram that corresponded to the NH2-terminal truncated peptide ANG-(1-7). Afterwards, the finding that it induced the release of vasopressin from neurohypophysial explants in a similar manner than ANG II (Schiavone 1988), finished with the previous consensus that angiotensin peptides with a truncated COOH-terminal had no biological activity.

These earlier experiments were followed by many that explored the different and opposed effects of ANG-(1-7) with respect to ANG II, and lead to the concept that the heptapeptide acts as a counterregulatory factor of ANG II, and changed our view of the RAS.

Taking into account the importance of ANG-(1-7) in normal physiology and pathophysiology, the aim of the present chapter is to review its main mechanisms of action, physiological functions and the pharmacological perspectives regarding ANG-(1-7) modulation.

3. Angiotensin-(1-7) Metabolism

Three known enzymes can regulate the formation of ANG-(1-7) from ANG I (Trask 2007). They include neprilysin 24.11 (NEP), thimet oligopeptidase 24.15 (TOP), and prolyl oligopeptidase 21.26 (POP) (Trask 2007). As evidenced by many researchers (Shaltout 2007; Yamamoto 1992) NEP can metabolize ANG I directly into ANG-(1-7), and ANG II into the inactive peptides ANG-(1-4) and ANG-(3-4). In a similar manner, TOP (Chappell 2000) and

POP (Yamamoto 1992) are able to generate ANG-(1-7) from the substrate ANG I. Moreover, POP can also cleave ANG II into ANG-(1-7) (Trask 2007).

An important advance in the interpretation of ANG-(1-7) synthesis and metabolism occurred in 2000, when Donoguet et al. (2000) cloned ACE2. This enzyme, which shares 42% sequence identity to the catalytic domain of ACE (Trask 2007) but is insensitive to ACE inhibitors, mediates the conversion of ANG I into ANG-(1-9), as well as the conversion of ANG II into ANG-(1-7). In this manner, it regulates in a double fashion the production of ANG-(1-7): indirectly through the cleavage of ANG I into ANG-(1-9) (which can be metabolized into ANG-(1-7) by ACE), and directly by means of the conversion of ANG II into ANG-(1-7). However, this direct mechanism seems to be most kinetically favored, considering that ACE 2 shows a higher substrate preference for ANG II than for ANG I (Vickers et al. 2002).

ANG-(1-7) has a short half-life (about 10 seconds in plasma) because of its peptidic nature that allows it to be readily acted upon by peptidases (Trask 2007). On the other hand, ACE also contributes to its metabolism, leading to the formation of the inactive metabolite ANG-(1-5) (Chappell 1998a; Deddish et al. 1998; Allred 2000). According to Allred et al. (2000), the lung is the primary site for the metabolism of ANG-(1-7) by ACE and, after the conversion of ANG-(1-7) to ANG- (1-5) by ACE, the pentapeptide is further metabolized to ANG-(3-5) by amino- or dipeptidyl aminopeptidases. Alternative pathways might occur in different tissues, such as renal tissue where NEP can hydrolyze the 4-tyrosine,5-isoleucine (Tyr4-Ile5) bond of ANG-(1-7) to form ANG-(1-4) (Allred 2000).

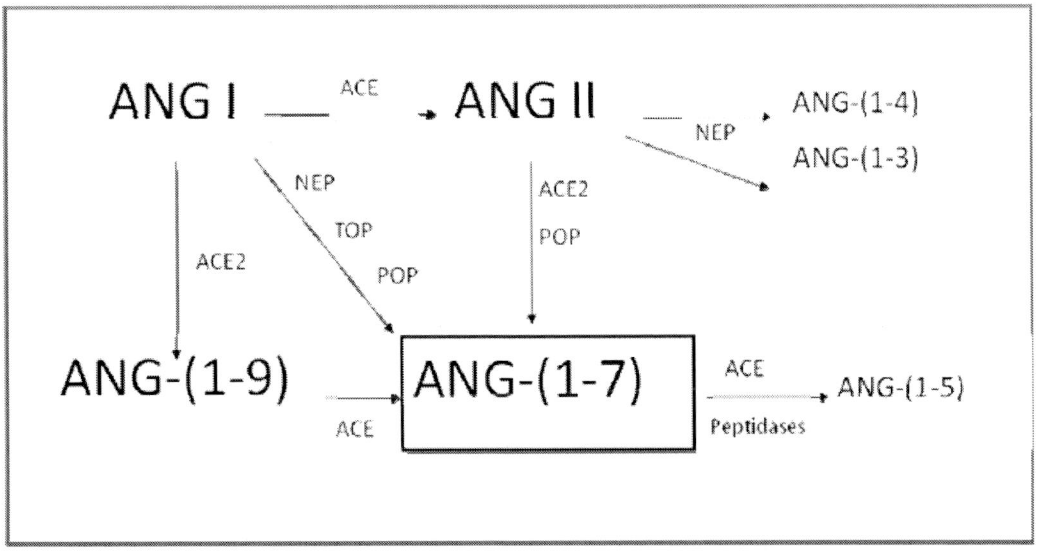

Figure 1. Angiotensin-(1-7) synthesis and metabolism. ANG I: Angiotensin I; ANG II: Angiotensin II; ANG-(1-4): Angiotensin-(1-4); ANG-(1-3): Angiotensin-(1-3); ANG-(1-9): Angiotensin-(1-9); ANG-(1-7): Angiotensin-(1-7); ACE: Angiotensin Converting Enzyme; ACE2: Angiotensin Converting Enzyme 2; NEP: neprilysin 24.11; TOP: Thimet oligopeptidase 24.15; POP: Prolyl oligopeptidase 21.26.

4. Pharmacodynamics

Recently, the receptor-mediated mechanisms underlying the actions of ANG-(1-7) were identified (Santos 2007). Particularly, after the discovery that the orphan G protein-coupled receptor Mas was a functional receptor for ANG-(1-7) (Santos et al. 2003), many advances were carried out in this area. This G-protein-coupled receptor mediates the current known actions of ANG-(1-7), as most of them can be blocked by its specific antagonist, D-Ala7-ANG-(1-7) (Trask 2007). At present time, the main signaling pathways involved in ANG-(1-7) responses by means of Mas receptor stimulation include PI3K-dependent AKT phosphorilation (Sampaio 2006), which leads to nitric oxide synthase phosphorilation and nitric oxide synthesis; inhibition of MAP kinase phosphorilation (Tallant 2005); phospholipase A2 activity stimulation leading to release of arachidonic acid for prostanoid production (Ferrario 2005); and prostacyclin-mediated production of cAMP and activation of cAMP-dependent protein kinase.

In a recent study, Sampaio et al. (2007) determined *in vitro* that ANG-(1-7), through Mas receptor activation, stimulates eNOS activation and NO production through Akt-dependent pathways .

There is increasing evidence to suggest an interaction between ANG-(1-7) and other vasoactive peptides. The mitogen-activated protein kinase (MAPK) cascade is an important intracellular mediator of responses related to cell growth and differentiation, survival, and cell death. ANG II is a well-known activator of this signaling pathway, and ANG II is also capable of inducing the synthesis of the profibrotic cytokine transforming growth factor-β1 (TGF- β1), associated with subsequent alterations in cell growth and matrix production in different tissues. In rat proximal tubular cells, it has been reported that ANG-(1–7) inhibits ANG II-mediated MAPK phosphorylation, and partly suppresses ANG II-stimulated increases in TGF- β1 (Su 2006), acting as a counterregulatory factor to ANG II. Interestingly, many of this negative effects on MAPK and TGF- β1 are not observed in the absence of ANG II, indicating that ANG-(1-7) depends on previous AT-1 (and possibly AT-2) receptor activation in order to exert some of its antiproliferative actions (Su 2006).

ANG-(1–7) has also been shown to exert antiproliferative effects through the release of prostaglandins, subsequent production of cAMP, and activation of cAMP dependent protein kinase A (Magaldi 2003; Tallant 2003). On the other hand, Gianni et al (2007) have recently postulated that ANG-(1–7) could be a positive physiological contributor to the actions of insulin in heart, considering that ANG-(1–7) stimulated cardiac Akt phosphorylation, meanwhile acute in vivo insulin-induced cardiac Akt phosphorylation was inhibited by ANG II. Interestingly, coadministration of insulin with an equimolar mixture of ANG II and ANG-(1–7) reversed this inhibitory effect. These studies clearly demonstrate the opposite effects of ANG II and ANG-(1-7) at subcellular level.

Another interaction was described between ANG-(1-7) and bradykinin, which would participate in the potentiation of bradykinin response after ACE inhibitors therapy. Although the exact mechanisms involved are not completely understood, it has been proposed that ACE inhibition, non-hydrolytic interaction with ACE favoring a crosstalk between ACE and BK-B2 receptor and MAS receptor-mediated changes in the coupling and/or signaling of BK (Santos 2005a) could be some of the mechanisms involved.

Although most of ANG-(1-7) effects seem to be mediated by MAS receptor´s activation, it has been described that this heptapeptide is able to interact with AT1 and AT2 receptors, as well as with another yet unknown receptor (Outdot et al. 2005). In the case of AT1 and AT2 receptors, the low affinity of ANG-(1-7) for them makes difficult to conceive that ANG-(1-7) acts as a physiological agonist in normal conditions. However, experimental data confirm the ability of ANG-(1-7) to interact with AT1 and AT2 receptors when it is administered in high pharmacological doses, acting as a partial agonist.

On the other hand, the possible existence of another ANG-(1-7) receptor has been suggested, although there is not enough data to support this theory at present time (Silva et al. 2007).

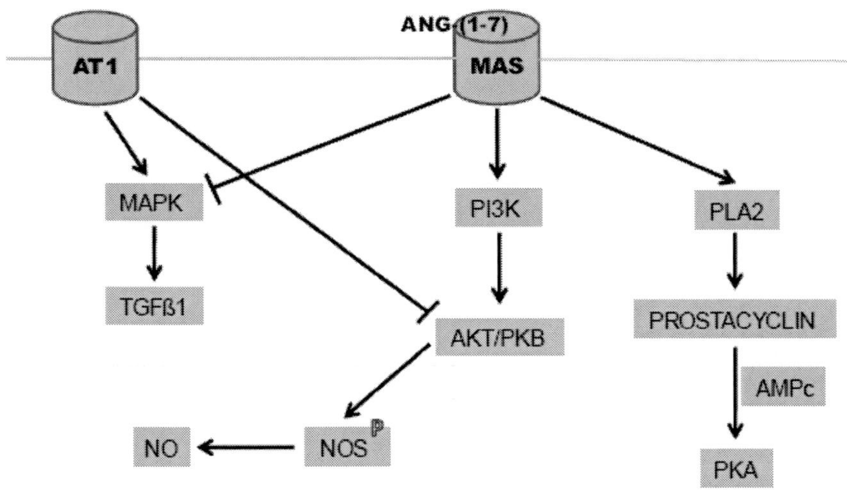

Figure 2. Angiotensin-(1-7) pharmacodynamics. AT1: AT1 receptor; MAS: MAS receptor; PLA2: Phospholipase A2; PKA: Protein kinase A; PI3K: Phosphatidil inositol 3 kinase; MAPK: MAP kinase; TGFß: Transforming growth factor ß; NOS: Nitric oxide synthase; NO: Nitric oxyde.

5. Pathophysiological Actions of Angiotensin-(1-7)

ANG-(1–7) is considered the most pleiotropic component of RAS. In addition, in many tissues ANG-(1-7) exerts counterregulatory actions with regard to ANG II limiting the deleterious effect of the octapeptide. Although most researches focused on the evaluation of the cardiovascular effects of the heptapeptide, ANG-(1-7) also regulates hematopoyesis, wound healing and exerts antineoplastic actions.

Another important aspect is the fact that the cardiovascular effects of ANG-(1-7) are enhanced in experimental animals with cardiovascular diseases, suggesting a critical role of the hormone in the development and maintenance of experimental hypertension and heart failure among others.

Although cardiovascular effects of ANG-(1-7) have been extensively studied in both basic research and clinical setting, the exact mechanism of action involved in these physiological actions remains partially elusive. As previously mentioned, it is well known the existence of a specific receptor for ANG-(1-7), called Mas receptor. However, blockade of AT1 and AT2 receptors also prevented most actions of the heptapeptide, suggesting a crosstalk between the Mas and AT1 /AT2 receptors. In addition, increase of nitric oxyde (NO), bradykinin and prostacyclin (PGI2) production is also involved in the mechanism of action of ANG-(1-7) effects.

In this section, we reviewed current knowledge of the pathophysiological effects of ANG-(1-7) in both basic and clinical research.

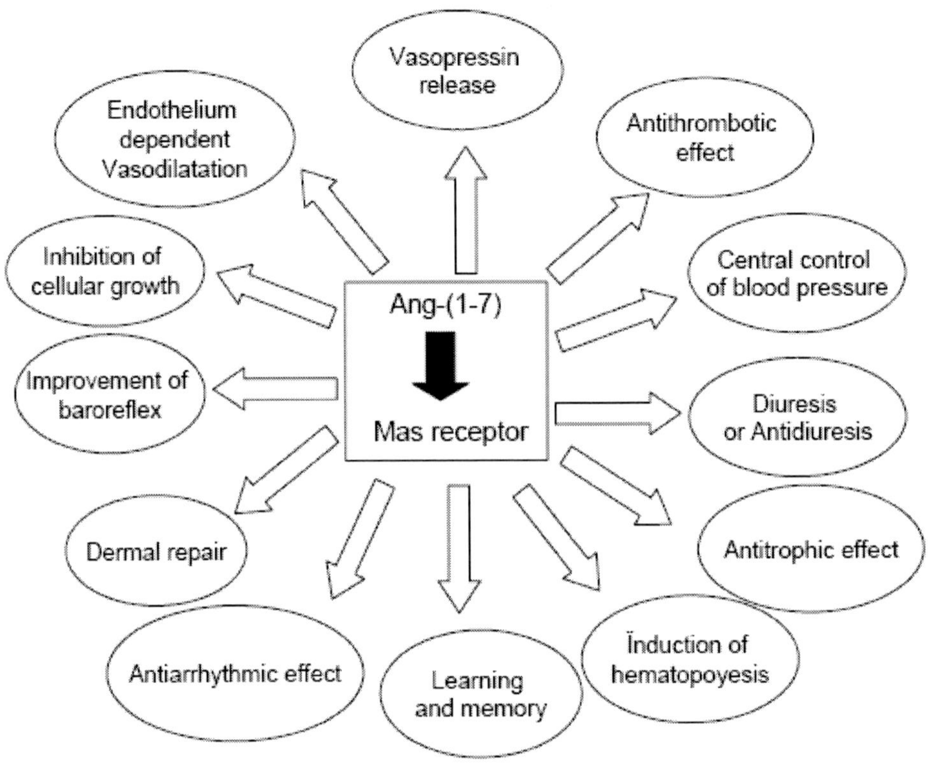

Figure 3. Summary of the pathophysiological actions of angiotensin-(1-7).

5.1. Effects of Angiotensin-(1-7) on Hemostasis

Considering that ANG-(1-7) stimulates the release of NO and PGI2, it is expected that this hormone could exert an antithrombotic action. For instance, Kucharewicz et al (2002) have found that ANG-(1-7) significantly reduced the thrombus weight in 2K-1C hypertensive rats but not in normotensive control animals by means of activation of Mas receptors. Moreover, the antithrombotic effect of ANG-(1-7) was also blocked by an AT_1 antagonist, suggesting a crosstalk between Mas and AT_1 receptors. The antithrombotic effect of ANG-(1-7) was also prevented by the coadministration of NG-nitro-L-arginine methyl ester (L-NAME) and indomethacin, suggesting the involvement of NO and prostaglandins in ANG-(1-7) thrombus formation inhibition (Kucharewicz et al 2002).

In another work, Kucharewicz et al. (2000) have also demonstrated that ANG-(1-7) exerts an antithrombotic effect by means of reducing platelet adhesion to fibrillar collagen.

5.2. Hemodynamic Effects of Angiotensin-(1-7)

The effect of ANG-(1-7) on blood flow distribution in different tissues, cardiac output, and total peripheral resistance were investigated in anaesthetized normotensive rats by means of fluorescent microspheres (Sampaio et al 2003). Whilst ANG-(1–7) infusion significantly increased blood flow to the kidney, mesentery, brain, and skin, the heptapeptide did not change perfusion of the spleen, muscle, lungs, and adrenals. Although ANG-(1-7) produced a significant increase in cardiac index (30%) and a decrease in total peripheral resistance, blood pressure and heart rate were not altered by the heptapetide administration (Sampaio et al 2003). Interestingly, application of a 100-fold higher dose of ANG-(1-7) exerted oppositive effects on tissue blood flow distribution with regard to the lower dose, possibly as a consequence of AT_1 receptor activation by ANG-(1-7).

5.3. Effects of Angiotensin-(1-7) on Vascular Tone

The vascular actions of ANG-(1-7) have been studied in different vascular beds and animal species. In addition, alteration of ANG-(1-7) vascular effects in pathological situations have been also evaluated. In most studies, ANG-(1-7) exerts a vasodilatadory action, although the mechanism of action is highly dependent of the vascular bed and animal specie. In this way, it has been demonstrated a vasodilatory effect of the heptapeptide in several vascular beds, including canine and porcine coronary arteries, canine middle cerebral artery, porcine piglet pial arterioles, feline systemic vasculature, rabbit renal afferent arterioles, rat aortic rings, neovasculature and mesenteric microvessels of normotensive and hypertensive rats (for review see Santos et al. 2005a).

Vascular effects of ANG-(1-7) were also evaluated in normotensive and hypertensive patients showing conflicting results. Sasaki et al. (2001) found that ANG-(1-7) induced relaxation in forearm circulation of both normotensive and hypertensive subjects through a mechanism independent of NO, considering that a NO synthase inhibitor did not affect the blood flow response to ANG-(1-7). In addition, the vasodilatadory effect of the heptapeptide was not altered in hypertensive patients (Sasaki et al. 2001). Conversely, Davie and Murray (1999) and Ueda et al (2000) did not find a direct vasodilatadory effect of ANG-(1-7) in patients with heart failure treated with ACE inhibitors and young volunteers, respectively. Taking together, vascular actions of ANG-(1-7) are highly dependent on the experimental conditions of the study.

Other authors show a biphasic effect of ANG-(1-7) in rat aorta (Haulica et al. 2003). Whilst low concentrations of the heptapeptide induced a endothelium dependent vasorelaxation, high doses (10^{-6}–10^{-5} M) exerted vasoconstriction through an AT_1 receptor dependent mechanism (Haulica et al. 2003).

Regarding the mechanism of ANG-(1-7) vasodilatadory action, several reports show that the vascular effect of ANG-(1-7) is endothelium dependent (Santos et al. 2005a). In these cells, the heptapeptide increases NO synthesis, prostaglandins or endothelium-derived relaxing factor (for review see Paul et al. 2006). In addition, other authors also found that a complex interaction between Mas receptor and bradykinin B_2 and angiotensin AT_2 receptors could be involved in vascular actions of ANG-(1-7). In this way, ANG-(1-7)-induced vasodilatation of canine and coronary arteries was prevented by the specific bradykinin B_2 receptor antagonist Hoe 140 (Brosnihan et al. 1996; Gorelik et al. 1998).

In addition, several studies characterized the receptor involved in the vasorelaxation of ANG-(1-7). Neves et al (2003) have demonstrated that ANG-(1-7)-induced dilatation of mesenteric arteries could be blocked by the application of both a Mas receptor and an AT_1 receptor antagonist.

More recently, a new receptor for ANG-(1-7) has been proposed to be involved in the vascular action of the peptide in rat aorta. In this work, the authors found that the vasodilator effect of ANG-(1-7) is mediated by activation of ANG-(1-7) receptors sensitive to D-Pro7-ANG-(1-7), but not to A-779, suggesting the existence of a different ANG-(1-7) receptor subtype (Silva et al. 2007).

5.4. Renal Effects of Angiotensin-(1-7)

Ang-(1–7) also regulates cardiovascular function by acting at the kidneys, producing complex renal effects (Paul et al. 2006). In this way, ANG-(1-7) can regulate fluid homeostasis through different mechanisms, including stimulation of vasopressin release and by acting in different nephron segments with variable effects on sodium and water excretion. Several studies have demonstrated a diuretic and natriuretic effect of ANG-(1-7) in different experimental conditions, including sodium replete anaesthetized rats with denervated kidneys (Handa 2000; Handa et al. 1996), several in vitro preparations (Handa et al. 1996; Dellipizzi et al 1994; Inagami 1999), and in anaesthetized animals (Chappell et al. 1998b; Ferrario et al. 1998a). ANG-(1-7) diuretic effect is related to the action of the peptide at proximal tubules (Chappell et al. 1998b) and Henle's loops (Vallon et al. 1997).

Conversely, through an action at the distal nephron, ANG-(1-7) induces opposite effects generating an antidiuretic effect in water-loaded rats (Baracho et al. 1998; Santos et al. 1996; Simões e Silva et al. 1997). This effect is prevented by Mas receptor antagonist but not by blockade of the vasopressin V_2 receptors (Baracho et al. 1998; Santos et al. 1996). In addition, the heptapeptide potently increases water conductivity in the inner medullary collecting duct (Santos et al. 1996) through a mechanism that involves an interaction between Mas receptor and vasopressin V_2 receptor (Magaldi et al. 2003).

Taking together, ANG-(1-7) exerts opposite effects on water and sodium excretion in different segments of the nephron. However, the antidiuretic effect of the heptapeptide seems to predominate, considering that application of a Mas antagonist, A-779, stimulates renal excretion of sodium and water.

Studies of vascular effects of ANG-(1-7) in renal vasculature have demonstrated that the peptide does not exert vasodilatation in in-vitro and in-vivo experiments (van der Wouden et al. 2006). In addition, although ANG-(1-7) attenuates vasoconstriction induced by ANG II in-vitro in isolated small renal arteries and the isolated perfused kidney, the heptapeptide has no effects on afferent and efferent arteriolar responses to ANG II in anaesthetized rats. A possible explanation to these conflicting results is the fact that ANG-(1-7) levels in in-vivo experiments are considerably lower than the concentrations used in the in-vitro setting (van der Wouden et al. 2006).

5.5. Effects of ANG-(1-7) at the Central Nervous System

At the central nervous system, ANG-(1-7) is mainly generated in central nuclei related to blood pressure regulation, such as medullary areas and the hypothalamus (Santos et al 2000; Gironacci 2005; Ferrario et al 1997). In normotensive rats, intracerebroventricular administration of the peptide does not modify blood pressure or heart rate, although it facilitates baroreceptor reflex control of heart rate (Campagnole-Santos et al. 1992). Conversely, ANG-(1-7) induces a depressor response in situations involving overactivity of ANG II. For instance, cerebroventricular infusion of ANG-(1-7) caused a significant decrease in blood pressure and injections of a specific ANG-(1–7) monoclonal antibody increased blood pressure and heart rate in [mRen-2]27 transgenic hypertensive (Tg+) rats but not in normotensive rats (Dobruch et al. 2003; Moriguchi et al. 1995). In previous studies from our laboratory (Höcht et al. 2006; Höcht et al. 2007), we found that ANG-(1-7) blocked the pressor response to intrahypothalamic administration of ANG II in spontaneously hypertensive and sinoaortic denervated rats withouth affecting the hypothalamic effects of ANG II in control normotensive animals.

Although the mechanisms by which ANG-(1-7) elicited a central hypotensive effect were not fully investigated, Alzamora et al. (2002) have shown that the hypotensive effect of ANG-(1–7) when microinjected at the caudal ventrolateral medulla involves a NO-related mechanism. Release of excitatory neurotransmitters may also participate in the hypotensive response of the heptapeptide (Wang et al. 2005). In addition, we have previously shown that ANG-(1-7) acts as a sympathoinhibitory neuromodulator since it decreases norepinephrine release in hypothalamus isolated from normotensive (Gironacci et al. 2000) and SHR rats (Gironacci et al. 2004). Diz and Pirro (1992) have shown that ANG-(1–7) facilitated a potassium evoked release of substance P in isolated rat hypothalamus.

Central effects of the peptide are also dose-dependent. Whilst the intrahypothalamic administration of a low dose of ANG-(1-7) (5 pmol) in sinoaortic denervated rats induced a reduction of blood pressure, a higher dose (50 pmol) was withouth effect (Höcht et al. 2007). In addition, we demonstrated that the higher dose of ANG-(1-7) caused a reduction in MAP in the presence of the AT_1 antagonist, suggesting that ANG-(1–7) at high doses does not only bind to Mas receptors but also to AT_1 receptors (Höcht et al. 2007). At the central nervous system, Diz and Ferrario (1996) have demonstrated that ANG-(1-7) is able to compete for almost all of the specific ANG II binding in the rostral but not the caudal region of the nuclei of the dorsal medulla, with a potency 30–40 fold less than ANG II.

In addition, ANG-(1-7) may also affect central regulation of blood pressure by changing vasopressin release. It was demonstrated that ANG-(1-7) stimulates release of vasopressin with a same potency that ANG II (Ferrario et al. 1998a). However, conversely to ANG II, the heptapeptide does not induce a dipsogenic effect after intracerebroventricular administration (Santos et al 2005b).

ANG-(1-7) also affects neuron excitability. The heptapeptide, as well as ANG II, increases the excitability of neurons within the vagal-solitary complex (Ferrario et al. 1990), paraventricular neurons (Felix et al. 1991) and neurons in the CA1 region of the hippocampus (Hellner et al. 2005).

Although the effects of administration of ANG-(1-7) on behavior have not been evaluated, the fact that Mas-deficient mice display higher levels of anxiety suggests that ANG-(1-7) may play a role in emotional behavior (Walther et al. 1998).

In addition, ANG-(1-7) regulates neurotransmission of dopaminergic, noradrenergic and γ-aminobutyric acid (GABA) neurons. The heptapeptide does not only inhibit the release of noradrenaline (Gironacci et al. 2000) but also the ANG II-stimulated noradrenaline outflow in hypothalamic nuclei from hypertensive rats (Gironacci et al. 2004). Inhibition of noradrenaline release induced by ANG-(1-7) involved both Mas and AT_2 receptors and is mediated through a bradykinin / NO mechanism. Conversely, ANG-(1-7) reduced release of dopamine and GABA in the striatum of normotensive rats (Stragier et al. 2005). Whilst release of dopamine is induced by conversion of ANG-(1-7) to Ang-(3-7) and is not blocked by either AT_1, AT_2 or Mas receptor blockade, ANG-(1-7) enhances GABA release by activating Mas receptors and by increasing NO striatal levels (Stragier et al. 2005). On the contrary, the peptide does not modify striatal glutamate release (Stragier et al. 2005).

5.6. Effects of Angiotensin-(1-7) on the Heart

Several lines of evidence demonstrate a cardioprotective role of ANG-(1-7) and these cardiac actions of the heptapeptide oppose to those exerted by ANG II. In this way, ANG-(1-7) induces antiarrhythmogenic, antifibrotic, antihypertrofic effects and vasodilatation in heart vasculature. In addition, ANG-(1-7) endogenous levels are elevated in cardiovascular diseases, including heart failure and acute myocardial infarction.

Coronary artery vasodilatation induced by ANG-(1-7) is dependent of bradykinin receptors and the release of NO (Porsti et al. 1994; Gorelik et al. 1998; Brosnihan et al. 1996). In other experiments, it was also demonstrated that the vasodilatory effect of the heptapeptide in perfused mouse hearts involved an interaction of Mas with AT_1 and AT_2 receptors inducing the release of prostacyclin and NO (Castro el al. 2005). Conversely, other authors found a reduction of coronary flow after application of higher concentrations of the peptide (Neves et al. 1997; Kumagai et al. 1990). Taking together, although most evidences demonstrated a vasodilatory effect of ANG-(1-7) on coronary vessels, higher concentrations of the peptide could exert opposite effects possibly by means of AT_1 receptors activation.

Activation of the RAS generates malignant re-entrant arrhythmias through impairment of cell communication and impulse propagation. In addition, ANG II decreases conduction velocity and enhances the dispersion of action potential duration in the failing heart inducing cardiac arrhythmias. Considering the counterregulatory actions of ANG-(1-7) with regards to the octapeptide, an antiarrhythmogenic effect of ANG-(1-7) has been postulated. In a pioneer report, Ferreira et al. (2001) have demonstrated that ANG-(1-7) perfusion reduced the incidence and duration of reperfusion arrhythmias after local ischemia induced by coronary ligation. The antiarrhythmic effect of low concentrations of ANG-(1-7) is mediated by Mas receptor activation and release of endogenous prostaglandins. Conversely to other beneficial effects of the heptapeptide, neither bradykinins nor NO participate in the antiarrhythmogenic effect (Ferreira et al. 2001).

De Mello et al. (2007) also demonstrated an antiarrhythmic effect of ANG-(1-7) in the failing heart of hamsters. The heptapeptide induces hyperpolarization of heart cells and enhances conduction velocity in cardiomyophatic hamsters (De Mello et al. 2007). In addition, ANG-(1-7) increases cardiac refractoriness, partially, by activation of Mas receptor. However, the authors also found that high concentrations of ANG-(1-7) induced proarrhythmic effects by increasing duration of the action potential and consequently generating early-afterdepolarisations (De Mello et al. 2007). Taking together, although activation of Mas receptor exerts a cardioprotective effect in the failing heart, excessive generation of the heptapeptide in the heart could be harmful by means of increasing the incidence of proarrhythmia.

ANG-(1-7) also exerts cardioprotective effects through inhibition of cardiac hypertrophy and remodeling induced by different adverse stimuli. Myocyte hypertrophy and fibrosis is a pathological process that can lead to reduced cardiac function and increased wall stiffness with diastolic dysfunction (Schaper 1998).

Different authors found that an increase in cardiac ANG-(1-7) levels, either by means of overexpression of ACE2 or chronic infusion of ANG-(1-7), prevents cardiac hypertrophy and fibrosis in chronically ANG II infused rats (Huentelman et al. 2005), SHR (Yuan et al. 2006) and DOCA-salt rat model (Grobe et al. 2006). It is also important to mention that antiremodeling effects of the heptapeptide are induced by activation of cardiac Mas receptor and are independent of effects on blood pressure or ANG II receptors in the cardiac tissue (Grobe et al. 2007).

ANG-(1-7) also prevents cardiac fibrosis, considering that the heptapeptide reduces collagen synthesis and decreases mRNA expression of growth factors in adult rat cardiac fibroblasts (Iwata et al. 2005). In addition, pretreatment with ANG-(1-7) also inhibits ANG II-induced increases in collagen synthesis and in mRNA expression of growth factors (Iwata et al. 2005). Effects of ANG-(1-7) on cardiac fibrosis was mediated by means of Mas receptor activation, considering that application of A-779 but not AT_1 or AT_2 reduced the affinity of the heptapeptide to adult rat cardiac fibroblasts (Iwata et al. 2005).

5.7. Inhibition of Angiogenesis by Angiotensin-(1-7)

Considering that ANG-(1-7) acts as an inhibitory growth factor, an anti-angiogenic action of the peptide has been described. ANG-(1-7) exerts an opposite effect on angiogenesis with regard to ANG II, reducing angiogenesis and fibrovascular tissue infiltration in the sponge (Machado et al 2000). In addition, antiangiogenic action of ANG-(1-7) in a mouse sponge model of angiogenesis was prevented by the application of a Mas receptor antagonist and an inhibitor of NO synthase (Machado et al. 2001).

5.8. Antiproliferative Actions

ANG-(1-7) induces antitrophic effects on vascular smooth muscle cells of normotensive rats, possibly by means of increased release of prostaglandins, PGI_2 and PGE_2 (Tallant et al. 1999). It was demonstrated that ANG-(1-7) reduced in a dose dependent manner the incorporation of [^{3}H]thymidine in vascular smooth muscle cells obtained from aorta, a biomarker of cellular proliferation, induced by different stimuli, including fetal bovine serum, platelet-derived growth factor (PDGF) and ANG II (Freeman et al 1996). Moreover, antiproliferative effect of ANG-(1-7) was similar to those previously reported for atrial natriuretic factor (Hutchinson et al 1997).

In addition, it has been found that chronic administration of ANG-(1-7) reduced neointimal formation after vascular injury withouth affecting medial area of the injured or the contralateral uninjured artery. It is also important to mention that these effects occurred in the absence of changes in blood pressure (Strawn et al 1999).

Regarding the receptor involved in the antiproliferative actions of ANG-(1-7), reduced of thymidine incorporation induced by ANG-(1-7) was not affected by pretreatment with AT_1 or AT_2 antagonists, but was prevented by application of [D-Ala^7]-ANG-(1-7), a Mas receptor blocker (Tallant et al 1999).

5.9. Antineoplastic Effects of Angiotensin-(1-7)

Taking into account the antiproliferative effects of ANG-(1-7), the ability of the peptide to inhibit human lung cell growth has been evaluated. Gallagher and Tallant (2004) have shown that the heptapeptide reduced DNA synthesis and cell proliferation of cultured human lung cancer cells. The molecular mechanism of inhibition of cell growth by ANG-(1-7) involves a reduction in cellular proliferation and induction of apoptosis. The antineoplastic effect of ANG-(1-7) was prevented by application of a Mas receptor antagonist but was not affected by AT_1 / AT_2 receptors blockers. It is also important to mention that the antiproliferative response to ANG-(1-7) was achieved with concentrations of the heptapeptide in the sub-nanomolar range, similar to endogenous ANG-(1-7) levels after treatment of rats with an ACE inhibitor, suggesting a potential antineoplastic activity of this therapeutic group.

In addition, ANG-(1-7) reduces adenocarcinoma growth in vivo possibly through the reduction in the cyclooxygenase-2/proinflammatory pathway, and this observation could explain the reduced incidence of certain types of cancer induced by ACE inhibitor therapy (Menon et al. 2007).

5.10. Effect of ANG-(1-7) on Hematopoyesis

Recently, it was demonstrated that the biological active peptides of the RAS, such as ANG II, are potent regulators of tissue regeneration and can act as hematopoietic factors

(Mrug et al. 1997; Nakao et al. 1967). However, adverse effects of ANG II on blood pressure regulation limits the clinical application of the octapeptide for stimulation of hematopoyesis.

More recently, it has been demonstrated that ANG-(1-7) increases the number of hematopoietic progenitors in the bone marrow of myelosuppressed mice (Rodgers et al. 2002). Another study has shown that administration of ANG-(1-7) led to a slight increase in the number of circulating leukocytes and platelets in a 5-fluorouracil myelosuppressed mice (Ellefson et al. 2004; Rodgers et al. 2003). Moreover, a synergistic effect on the number of circulating leukocytes was observed when ANG-(1-7) was given in combination with a subclinical dose of Neupogen, suggesting that these hematopoietic agents act at different sites within the hematopoietic cascade and that combining these two agents may be beneficial in the treatment of hematopoietic disorders (Ellefson et al. 2004).

The beneficial effect of ANG-(1-7) at a dose of 100 µg/kg per day, in attenuating multilineage cytopenias following chemotherapy was studied in Phase I/II dose escalation study in patients with newly diagnosed breast cancer. At this dose, ANG-(1-7) reduced the frequency of thrombocytopenia, anemia, and lymphopenia as compared to filgrastim. In addition, use of ANG-(1-7) was safe and well tolerated with no dose-limiting toxicity, suggesting a possible therapeutic application of the heptapeptide for reducing multilineage cytopenias induced by myelosuppressive chemotherapy.

5.11. Effects of ANG-(1-7) on Tissue Repair

Preclinical experiments have demonstrated a high efficacy of ANG-(1-7) in accelerating dermal healing in different experimental models, including rat and diabetic mouse full-thickness excisional wounds, rat random flap, and guinea pig partial thickness thermal injury (Rodgers et al. 2001). Moreover, efficacy of ANG-(1-7) on wound repair seems to be superior than platelet-derived growth factor, the active ingredient in the only approved growth factor formulation available for wound repair, considering that ANG-(1-7) exerts a beneficial effect on keratinocyte proliferation that is absent when platelet-derived growth factor is used (Rodgers et al. 2001).

In another report by the same authors, the beneficial effects of NorLeu^3-ANG-(1-7) in wound healing was also demonstrated (Rodgers et al. 2003). Moreover, the healing responses to NorLeu^3-ANG-(1-7) was superior to the naturally occurring angiotensin peptide Ang (1–7) and to platelet-derived growth factor. Wound repair induced by NorLeu^3-ANG-(1-7) is a consequence of Mas receptor activation, considering that the action of this peptide was blocked by A-779. This study suggests a possible clinical application of NorLeu^3-ANG-(1-7) for accelerating wound repair and reducing scar formation (Rodgers et al. 2003).

5.12. Effects of ANG-(1-7) on Circadian Rhythm

Recently, the effect of Mas receptor activation on circadian rhythm of blood pressure and heart rate has been investigated in spontaneously hypertensive rats (Wessel et al. 2007). It was shown that chronic administration of a non-peptidic Mas receptor agonist, AVE0991, have detrimental effects on circadian blood pressure variability by reducing the increase of blood pressure that normally occurs during night-time in SHR (Wessel et al. 2007).

6. Therapeutic Perspectives of Angiotensin-(1-7)

As comment above, ANG-(1-7), at physiological or pharmacological concentrations, regulates activity of several systems, including cardiovascular, renal, central nervous, hematopoietic systems among others. In addition, ANG-(1-7) exerts antiproliferative actions, that could be of interest in several cardiovascular diseases, including heart failure and atherosclerosis.

The potential therapeutic benefit of ANG-(1-7) is also supported by the fact that endogenous ANG-(1-7) levels or the effect of the heptapeptide are modified in pathological conditions.

Most preclinical studies demonstrated a role of ANG-(1-7) in blood pressure regulation of different models of hypertension, including spontaneously hypertensive rats (Benter et al. 1995; Widdop et al. 1999), spontaneously hypertensive rats treated with L-NAME, a NO synthase inhibitor (Benter et al. 2006), salt depleted hypertensive rats (Iyer et al. 2000) and sinoaortic denervated rats (Höcht et al. 2007).

In the clinical setting, a reduced urinary excretion of ANG-(1-7) was found in essential hypertensive patients with regard to normotensive control subjects (Ferrario et al. 1998b). Moreover, urinary concentrations of ANG-(1-7) correlated inversely with systolic and diastolic blood pressure, providing a potential marker for the characterization of forms of essential hypertension associated with reduced production or activity of vasodilatory hormones.

In addition, beneficial effect of ANG-(1-7) in hypertension is not only related to blood pressure lowering effect but also to reduction in target organ damage. Grobe et al. (2006) have found that chronic infusion of ANG-(1-7) prevented collagen deposition in the DOCA-salt model of hypertension withouth affecting blood pressure and cardiac hypertrophy. In addition, the heptapeptide is also able to prevent cardiac remodeling induced by ANG II. This effect of ANG-(1-7) is not mediated through modulation of blood pressure or altered cardiac angiotensin receptor populations (Grobe et al. 2006). ANG-(1–7) also improves renal, cardiac, and vascular function in spontaneously hypertensive rats treated with L-NAME and these effects were similar to many of the beneficial effects of RAS blockade with an ACE inhibitor (Benter et al. 2006).

ANG-(1-7) seems also to play a protective role during acute myocardial infarction. Several authors have demonstrated that chronic infusion of the heptapeptide reduces arrhythmias (Ferreira et al. 2001) and attenuates development of heart failure after myocardial infarction in rats (Loot et al. 2002).

Another indirect evidence that supports a role for ANG-(1-7) in cardiovascular pathologies is the fact that ACE2-deficient mice develop cardiomyopathy associated with increased oxidative stress and inflammation, as well as MAP kinase and phosphatidylinositol 3-kinase activation. Moreover, Burrell et al. (2005) have demonstrated the existence of an increase in ACE2 after miocardial infarction, suggesting that the enzyme plays an important role in the negative modulation of the RAS after cardiac injury.

Clinical studies also indirectly show a protective role of ANG-(1-7) in heart failure. Activity of NEP and ACE2 were increased in failing human heart ventricles, suggesting that enhanced ANG-(1-7) cardiac levels could exert a counterregulatory, cardioprotective effect in heart failure (Zisman et al. 2003).

ANG-(1-7) may also have a therapeutic role in the treatment of diabetic complications. For instance, Benter et al. (2007) have demonstrated that ANG-(1-7) prevents cardiovascular dysfunction in streptozotocin-treated rats, an experimental model of diabetes, suggesting that activation of Ang-(1–7) signal transduction could be an important therapeutic strategy to reduce cardiovascular events in diabetic patients. Diabetic nephropathy could also be reduced by activation of Mas receptors. For instance, ANG-(1-7) or its receptor agonist AVE 0991 attenuates proteinuria and improves renal vascular activity in the diabetic rat but did not reverse the urinary excretion of lysozyme, a marker of tubulointerstitial damage (Benter et al. 2007).

More indirect evidences demonstrated that renal ACE2 is reduced in the proximal tubules of the streptozotocin-induced model of type 1 diabetes, and the attenuation of renal injury by ACE inhibition is associated with increased ACE2 expression (Tikellis et al. 2003). Moreover, chronic ACE2 inhibition in the diabetic db/db mice exacerbates the extent of albuminuria in a 3-fold manner (Wysocki et al. 2006).

Although endogenous levels of ANG-(1-7) and the vasodilator response to the peptide are enhanced during normal pregnancy (Brosnihan et al. 2004), there is a decrease in plasma in preeclampsia (Brosnihan et al. 2004) and gestational diabetes.

Considering the pathophysiological effects of ANG-(1-7) and the expression of this peptide in different pathological conditions, it could be suggested a wide range of therapeutic opportunities (table X).

7. Therapeutic Strategies to Enhance Endogenous ANG-(1-7) Levels or Mas Receptor Activation

Taking into account the beneficial effects of ANG-(1-7) and the involvement of the peptide in the pathophysiological state of several diseases, enhancement of endogenous ANG-(1-7) levels or Mas receptor activation could be a novel therapeutic strategy in the treatment of hypertension, heart failure, arrhythmias, diabetic complications, among other diseases.

Actually, several therapeutic groups, including ACE inhibitors, AT1 receptor antagonist and vasopeptidase inhibitors are able to increase endogenous ANG-(1-7) levels. ACE inhibitors increase endogenous levels of Ang-(1–7) by 25-fold through different mechanism (Ferrario et al. 2005). As commented above, Ang-(1–7) can be formed by the degradation of Ang I by means of neutral endopeptidase (Ferrario et al. 2005). If degradation of Ang I to ANG II is blocked by ACE inhibitors, a greater conversion of Ang I to Ang-(1–7) is expected. In addition, Ang-(1–7) is converted to Ang-(1-5) by ACE, and this pathway is the main mechanism of degradation of the heptapeptide (Santos et al. 2000). For instance, ACE inhibition increases plasma half-life of ANG-(1-7) 4- to 6-fold (Yamada et al. 1998). Several works demonstrate that enhancement of endogenous ANG-(1-7) by ACE inhibitors is involved, at least in part, in several beneficial effects of this therapeutic group, including blood pressure lowering effect, antithrombotic effect, baroreflex improvement, among others. However, van der Wouden et al. [51] have found that ANG-(1-7) plays no major role in the anti-proteinuric and blood pressure lowering effects of ACE inhibitors in an experimental model of nephrosis.

Similarly to ACE inhibitors, AT_1 receptor antagonists also enhance ANG-(1-7) endogenous levels by as much as 25-fold. For this therapeutic group, ANG-(1-7) increment could be explained by the fact that ANG II affinity for AT1 receptors is a 1000 times higher than its affinity to ACE2. Therefore, after blockade of AT1 receptors, ANG II would accumulate and converted to ANG-(1-7) by ACE2. Another explanation is the fact that AT1 receptor antagonist induces an overexpression and an increased activity of ACE2 (Ferrario et al. 2005). As described for ACE inhibitors, ANG-(1-7) also participates partially in the blood pressure lowering effect of AT1 receptor antagonists (Iyer et al. 1998).

Another therapeutic strategy to increase Mas receptor activation is the administration of an exogenous Mas receptor agonist. Therapeutic use of ANG-(1-7) is theoretically precluded considering its short plasma half-life (about 10 min) by a rapid proteolysis due to ACE activity and its poor bioavailability and absorption by oral or topical routes, considering its peptidergic nature and its high water-solubility.

However, recently promising strategies have been proposed to increase the therapeutic potential of ANG-(1-7), including the use of liposomes as sustained and controlled release systems and the use of cyclodextrins to improve peptide bioavailability by oral route (Santos et al. 2005b). For instance, Santos et al. (2005b) have demonstrated that inclusion of ANG-(1-7) in liposomes prolongs its biological activity by preventing its degradation and by the slow liberation of ANG-(1-7) from the liposome. The authors have found a sustained pressor response of ANG-(1-7)-containing liposomes after application in the rostral ventrolateral medulla (Silva-Barcellos et al. 2001). In another report, the authors found that pegylated, rigid liposomes (200 nm), containing 50 ng of ANG-(1-7), elicited a significant increase of blood pressure during 7 days, conversely to empty liposomes or non-pegylated liposomes (Silva-Barcellos et al. 2004).

The design of non-peptidic Mas receptor agonists is probably the most attractive therapeutic strategy in order to enhance the beneficial physiological actions of the heptapeptide. In 2001, Aventis Pharma designed the first nonpeptide and orally active analog of ANG-(1-7), AVE 0991 (5-Formyl-4-methoxy-2-phenyl-1-[[4-[2-(ethylaminocarbonyl-sulfonamido)-5-isobutyl-3-thienyl]-phenyl]-methyl]-imidazole). AVE 0991 has a 10-fold higher affinity for Mas receptor than ANG-(1-7) (Santos and Ferreira 2006). Although the pharmacokinetic properties of this new chemical entity have not been studied, it is well known that Ave 0991 mimics ANG-(1-7) effects in several organs (Santos and Ferreira 2006). AVE 0991 evokes similar effects with regard to ANG-(1–7) in cultured bovine aortic endothelial cells (Wiemer et al. 2002), kidneys (Pinheiro et al. 2004), and vessels (Faria-Silva et al. 2005, Lemos et al. 2005). In addition, Benter et al. (2006) showed that chronic administration of ANG-(1–7) or its synthetic analog, AVE-0991, significantly improved the ischemia-reperfusion recovery in isolated perfused spontaneously hypertensive rat hearts treated with L-NAME. More recently, Ferreira et al. (2007a) have demonstrated the ability of AVE 0991 to attenuate postischemic heart failure in a model of myocardial infarction induced by left coronary artery ligation. In addition, AVE 0991 also exerts a cardioprotective role on cardiac remodeling induced by cardiac β-adrenoceptor overstimulation (Ferreira et al. 2007b). Pharmacological actions of AVE 0991 are listed in table 1.

Table 1. Pharmacological effects of the nonpeptide Mas receptor agonist AVE 0991

Alteration of circadian rhythm of blood pressure
Antidiuresis
Attenuation of cardiac remodeling
Blood pressure lowering effect
Endothelium dependent vasodilatation
Improvement of endothelial function
Improvement of baroreflex sensitivity
Prevention of diabetic-induced cardiovascular dysfunction

8. Conclusions

Nowadays, several investigational lines have clearly demonstrated a protective role of ANG-(1-7) on the cardiovascular system. Increasing endogenous ANG-(1-7) levels limits the deleterious effect of ANG II through activation of a specific Mas receptor and by upregulating the action of several cellular mediators, including NO, prostaglandins and bradykinin, among others. In addition, Mas receptor activation seems to regulate activity of ANG II receptors, such as AT_1 and AT_2 receptors, considering that several beneficial effects of ANG-(1-7) are prevented after blocked of these receptors. In addition, it was also demonstrated a role of ANG-(1-7) in prevention of certain cancer types, treatment of chemotherapy-induced cytopenias and dermal repair.

Taking together, activation of Mas receptors, either by enhancement of endogenous ANG-(1-7) concentrations or by application of a Mas receptor agonist, could have a therapeutic role in the treatment of several diseases, including hypertension, heart failure, kidney disease, cancer, cutopenias and dermal repair.

9. References

Allred AJ, Diz DI, Ferrario CM, Chappell MC (2000) Pathways for angiotensin-(1-7) metabolism in pulmonary and renal tissues. *Am. J. Physiol. Renal. Physiol.* 279:841-850.

Alzamora AC, Santos RAS, Campagnole-Santos MJ (2002) Hypotensive effect of ANG II and Ang-(1–7) at the caudal ventrolateral medulla involves different mechanisms. *Am. J. Physiol.* 283:R1187–95.

Baracho NCV, Simões e Silva AC, Khosla MC, Santos RAS (1998) Effect of selective angiotensin antagonists on the antidiuresis produced by angiotensin-(1–7) in water-loaded rats. *Braz. J. Med. Biol. Res.* 31:1221–1227.

Benter IF, Ferrario CM, Morris M, Diz DI (1995) Antihypertensive actions of angiotensin-(1-7) in spontaneously hypertensive rats. *Am. J. Physiol.* 269:H313-H319.

Benter IF, Yousif MH, Anim JT, Cojocel C, Diz DI (2006) Angiotensin-(1–7) prevents development of severe hypertension and end-organ damage in spontaneously hypertensive rats treated with L-NAME. *Am. J. Physiol. Heart. Circ. Physiol.* 290: H684–H691.

Benter IF, Yousif MH, Cojocel C, Al-Maghrebi M, Diz DI (2007) Angiotensin-(1-7) prevents diabetes-induced cardiovascular dysfunction. *Am. J. Physiol. Heart. Circ. Physiol.* 292:H666-H672.

Brosnihan KB, Li P, Ferrario CM (1996) Angiotensin-(1-7) dilates canine coronary arteries through kinins and nitric oxide. *Hypertension* 27: 523-528.

Brosnihan KB, Neves LA, Anton L, Joyner J, Valdes G, Merrill DC (2004) Enhanced expression of ANG-(1-7) during pregnancy. *Braz. J. Med. Biol. Res.* 37:1255-1262.

Burrell LM, Risvanis J, Kubota E, Dean RG, MacDonald PS, Lu S, Tikellis C, Grant SL, Lew RA, Smith AI, Cooper ME, Johnston CI (2005) Myocardial infarction increases ACE2 expression in rat and humans. *Eur. Heart. J.* 26:369-375.

Campagnole-Santos MJ, Heringer SB, Batista EN, Khosla MC, Santos RAS (1992) Differential baroreceptor reflex modulation by centrally infused angiotensin peptides. *Am. J. Physiol.* 263:R89–94.

Castro CH, Santos RAS, Ferreira AJ, Bader M, Alenina N, Almeida AP (2005) Evidence for a functional interaction of the angiotensin-(1-7) receptor Mas with AT1 and AT2 receptors in the mouse heart. *Hypertension* 46: 937-942.

Chappell MC, Pirro NT, Sykes A, FerrarioCM (1998a) Metabolism of angiotensin-(1-7) by angiotensin-converting enzyme. *Hypertension* 31:362-367.

Chappell MC, Diz DI, Yunis C, Ferrario CM (1998b) Differential actions of angiotensin-(1–7) in the kidney. *Kidney Int. Suppl.* 68:S3–S6.

Chappell MC, Gomez MN, Pirro NT, Ferrario CM (2000) Release of angiotensin-(1–7) from the rat hindlimb: Influence of angiotensin-converting enzyme inhibition. *Hypertension* 35:348-352.

Davie AP, McMurray JJ (1999) Effect of angiotensin-(1-7) and bradykinin in patients with heart failure treated with an ACE inhibitor. *Hypertension* 34:457–460.

De Mello WC, Ferrario CM, Jessup JA (2007) Beneficial versus harmful effects of Angiotensin (1-7) on impulse propagation and cardiac arrhythmias in the failing heart. *JRAAS* 8; 74-80.

Deddish PA, Marcic B, Jackman HL, Wang HZ, Skidgel RA, Erdos EG (1998) N-domain-specific substrate and C-domain inhibitors of angiotensin-converting enzyme: Angiotensin-(1–7) and keto-ACE. *Hypertension* 31:912-917.

Dellipizzi A, Hilchley SD, McGiff JC, Bell-Quilley CP (1994) Natriuretic action of angiotensin-(1–7). *Br. J. Pharmacol.* 11:1–4.

Diz DE, Pirro NT (1992) Differential actions of angiotensin II and angiotensin-(1–7) on transmitter release. *Hypertension* 19(suppl II):II–41–II-48.

Diz DI, Ferrario CM (1996) Angiotensin receptor heterogeneity in the dorsal medulla oblongata as defined by angiotensin-(1–7). In: Raizada MK, editor. Recent advances in cellular and molecular aspects of angiotensin receptors. *New York: Plenum Press*; p. 225–35.

Dobruch J, Paczwa P, Lon S, Khosla MC, Szczepanska-Sadowska E (2003) Hypotensive function of the brain angiotensin-(1–7) in Sprague Dawley and renin transgenic rats. *J.Physiol. Pharmacol.* 54:371–81.

Donoghue M, Hsieh F, Baronas E, Godbout K, Gosselin M, Stagliano N, Donovan M, Woolf B, Robison K, Jeyaseelan R, et al. (2000) A novel angiotensin-converting enzyme-related carboxypeptidase (ACE2) converts angiotensin I to angiotensin 1-9. *Circ. Res.* 87:E1-E9.

Ellefson DD, Espinoza T, Roda N, Maldonado S, Rodgers KE (2004) Synergistic effects on co-administration of Angiotensin 1–7 and Neupogen on hematopoietic recovery in mice. *Cancer Chemother Pharmacol* 53(1):15–24.

Faria-Silva R, Duarte FV, Santos RAS (2005) Short-term angiotensin-(1–7) receptor Mas stimulation improves endothelial function in normotensive rats. *Hypertension* 46:948–952.

Felix D, Khosla MC, Barnes KL, Imboden H, Montani B, Ferrario CM (1991) Neurophysiologicla responses to angiotensin-(1-7*). Hypertension* 17: 1111-1114.

Ferrario CM, Barnes KL, Block CH, Brosnihan KB, Diz DI, Khosla MC, Santos RAS (1990) Pathways of angiotensin formation and function in the brain. *Hypertension* 15:I13-I19.

Ferrario CM, Chappell MC, Tallant EA, Brosnihan KB, Diz DI (1997) Counterregulatory actions of angiotensin-(1–7). *Hypertension* 30(Part 2):535–41.

Ferrario CM, Chappell MC, Dean RH, Iyer SN (1998a) Novel angiotensin peptides regulate blood pressure, endothelial function, and natriuresis. *J. Am. Soc. Nephrol.* 9:1716–1722.

Ferrario CM, Martell N, Yunis C, Flack JM, Chappell MC, Brosnihan KB, Dean RH, Fernandez A, Novikov SV, Pinillas C, Luque M (1998b) Characterization of angiotensin-(1-7) in the urine of normal and essential hypertensive subjects. *Am. J. Hypertens.* 11:137-146.

Ferrario CM, Jessup J, Chappell MC, Averill DB, Brosnihan KB, Tallant EA, Diz DI, Gallagher PE (2005) Effect of angiotensin-converting enzyme inhibition and angiotensin II receptor blockers on cardiac angiotensin-converting enzyme 2. *Circulation* 111:2605-2610.

Ferrario CM, Trask AJ, Jessup JÁ (2005) Advances in biochemical and functional roles of angiotensin-converting enzyme 2 and angiotensin-(1-7) in regulation of cardiovascular function. *Am. J. Physiol. Heart. Circ. Physiol.* 289:2281-2290.

Ferrario CM (2006) Angiotensin-Converting enzyme 2 and angiotensin-(1-7). An evolving story in cardiovascular regulation. *Hypertension* 47:515-521.

Ferreira AJ, Santos RA, Almeida AP (2001) Angiotensin (1-7): cardioprotective effect in myocardial ischemia/reperfusion. *Hypertension* 38:665-668.

Ferreira AJ, Jacoby BA, Araujo CA, Macedo FA, Silva GA, Almeida AP, Caliari MV, Santos RA (2007a) The nonpeptide angiotensin-(1-7) receptor Mas agonist AVE-0991 attenuates heart failure induced by myocardial infarction. *Am. J. Physiol. Heart. Circ. Physiol.* 292:H1113-1119.

Ferreira AJ, Oliveira TL, Castro MC, Almeida AP, Castro CH, Caliari MV, Gava E, Kitten GT, Santos RA (2007b) Isoproterenol-induced impairment of heart function and remodeling are attenuated by the nonpeptide angiotensin-(1-7) analogue AVE 0991. Life Sci 81:916-923.

Freeman EJ, Chisolm GM, Ferrario CM, Tallant EA (1996) Angiotensin-(1-7) inhibits vascular smooth muscle cell growth. *Hypertension* 28: 104-108.

Gallagher PE, Tallant EA (2004) Inhibition of human lung cancer cell growth by angiotensin-(1-7). *Carcinogenesis* 25:2045-2052.

Giani JF, Gironacci MM, Muñoz MC, Peña C, Turyn D, Dominici FP (2007) Angiotensin-(1 7) stimulates the phosphorylation of JAK2, IRS-1 and Akt in rat heart in vivo: role of the AT1 and Mas receptors.*Am. J. Physiol. Heart. Circ. Physiol.* 293:H1154-H1163.

Gironacci MM, Vatta M, Rodriguez-Fermepín M, Fernández BE, Peña C (2000) Angiotensin-(1–7) reduces norepinephrine release through a nitric oxide mechanism in rat hypothalamus. *Hypertension* 35:1248–52.

Gironacci MM, Valera MS, Yujnovsky I, Pena C (2004) Angiotensin-(1–7) inhibitory mechanism of norepinephrine release in hypertensive rats. *Hypertension* 44:783–7.

Gironacci MM (2005) Contribution of angiotensin-(1–7) to the pathophysiology of hypertension. *Braz. J. Hypertens* 12:147–52.

Gorelik G, Carbini LA, Scicli AG (1998) Angiotensin 1-7 induces bradykinin-mediated relaxation in porcine coronary artery. .*J Pharmacol. Exp. Ther.* 286: 403-410.

Grobe JL, Mecca AP, Mao H, Katovich MJ (2006) Chronic angiotensin-(1–7) prevents cardiac fibrosis in the DOCA-salt model of hypertension. *Am. J. Physiol. Heart. Circ. Physiol.* 290: H2417–H2423.

Grobe JL, Mecca AP, Lingis M, Shenoy V, Bolton TA, Machado JM, Speth RC, Raizada MK, Katovich MJ (2007) Prevention of angiotensin II-induced cardiac remodeling by angiotensin-(1–7). *Am. J. Physiol. Heart. Circ. Physiol.* 292:H736-H742.

Handa RK, Ferrario CM, Strandhoy JW (1996) Renal actions of angiotensin-(1–7): in vivo and in vitro studies. *Am. J. Physiol.* 270:F141–F147.

Handa RK (2000) Metabolism alters the selectivity of angiotensin-(1–7) receptor ligands for angiotensin receptors. *J. Am. Soc. Nephrol.* 11:1377–1386.

Haulica I, Bild W, Mihaila CN, Ionita T, Boisteanu CP, Neagu B (2003) Biphasic effects of angiotensin (1-7) and its interactions with angiotensin II in rat aorta. *J. Renin. Angiotensin. Aldosterone. Syst.* 4:124-128.

Hellner K, Walther T, Schubert M, Albrecht D (2005) Angiotensin-(1-7) enhances LTP in the hippocampus through G-protein-coupled receptor Mas. *Mol. Cell. Neurosci.* 29: 427-435.

Höcht C, Gironacci MM, Mayer MA, Schuman M, Bertera FM, Taira CA (2007) Involvement of angiotensin-(1–7) in the hypothalamic hypotensive effect of captopril in sinoaortic denervated rats. *Regul. Pept*, in press.

Höcht C, Opezzo JA, Gironacci MM, Pena C, Taira CA (2006) Hypothalamic cardiovascular effects of Ang-(1–7) in spontaneously hypertensive rats. *Regul. Pept.* 135:39–44.

Huentelman MJ, Grobe JL, Vazquez J, Stewart JM, Mecca AP, Katovich MJ, Ferrario CM, Raizada MK (2005) Protection from angiotensin II-induced cardiac hypertrophy and fibrosis by systemic lentiviral delivery of ACE2. *Exp. Physiol.* 90: 783–790.

Hutchinson HG, Trindade PY, Cunanan DB, Wu CF, Pratt RE (1997) Mechanisms of natriuretic-peptide-induced growth inhibition of vascular smooth muscle cells. *Cardiovasc. Res.* 35: 158-167.

Inagami T (1999) Molecular biology and signaling of angiotensin receptors: an overview. *J. Am. Soc. Nephrol.* 10:S2–S7

Iwata M, Cowling RT, Gurantz D, Moore C, Zhang S, Yuan JX, Greenberg BH (2005) Angiotensin-(1-7) binds to specific receptors on cardiac fibroblasts to initiate antifibrotic and antitrophic effects. *Am. J. Physiol. Heart. Circ. Physiol.* 289:H2356-H2363.

Iyer SN, Ferrario CM, Chappell MC (1998) Angiotensin-(1-7) contributes to the antihypertensive effects of blockade of the renin-angiotensin system.*Hypertension* 31:356-361.

Iyer SN, Averill DB, Chappell MC, Yamada K, Allred AJ, Ferrario CM (2000) Contribution of angiotensin-(1-7) to blood pressure regulation in salt-depleted hypertensive rats. *Hypertension* 36:417-422.

Kucharewicz I, Chabielska E, Pawlak D, Matys T, Rólkowski R, Buczko W (2000) The antithrombotic effect of angiotensin-(1–7) closely resembles that of losartan. *J. Renin. Angiotensin. Aldosterone. Syst. 1*: 268-272.

Kumagai H, Khosla M, Ferrario CM, Fouad-Tarazi FM (1990) Biological activity of angiotensin-(1-7) heptapeptide in the hamster heart. *Hypertension* 15: I29-I33.

Lemos VS, Silva DM, Walther T, Alenina N, Bader M, Santos RAS (2005) The endothelium-dependent vasodilator effect of the nonpeptide ANG-(1–7) mimic AVE 0991 is abolished in the aorta of Mas-knockout mice. *J. Cardiovasc. Pharmacol.* 46: 274–279.

Loot AE, Roks AJ, Henning RH, Tio RA, Suurmeijer AJ, Boomsma F, van Gilst WH (2002) Angiotensin-(1-7) attenuates the development of heart failure after myocardial infarction in rats. *Circulation* 105:1548-1550.

Machado RD, Santos RA, Andrade SP (2000) Opposing actions of angiotensins on angiogenesis. *Life Sci.* 66:67-76.

Machado RD, Santos RA, Andrade SP (2001) Mechanisms of angiotensin-(1-7)-induced inhibition of angiogenesis. *Am. J. Physiol. Regul. Integr. Comp. Physiol.* 280:R994-R1000.

Magaldi AJ, Cesar KR, de Araujo M, Simoes e Silva AC, Santos RA (2003) Angiotensin-(1-7) stimulates water transport in rat inner medullary collecting duct: evidence for involvement of vasopressin V2 receptors. *Pflugers Arch* 447:223-230.

Menon J, Sota-Pantoja DR, Callahan MF, Cline JM, Ferrario CM, Tallant EA, Gallagher PE (2007) Angiotensin-(1-7) inhibits growth of a human lung adenocarcinoma xenografts in nude mice through a reduction in cycloxygenase-2. *Cancer Res.* 67:2809–2815.

Moriguchi A, Tallant EA, Matsumura K, Reilly TM, Walton H, Ganten D, Ferrario CM (1995) Opposing actions of angiotensin-(1–7) and angiotensin II in the brain of transgenic hypertensive rats. *Hypertension* 25:1260–5.

Mrug M, Stopka T, Julian BA, Prchal JF, Prchal JT (1997) Angiotensin II stimulates proliferation of normal erythroid progenitors. *J. Clin. Invest.* 100:2310–2314.

Nakao K, Shirakura T, Azuma M, Maekawa T (1967) Studies on erythropoietic action of angiotensin II. *Blood* 29:754–760.

Neves LAA, Almeida AP, Khosla MC, Campagnole-Santos MJ, Santos RAS (1997) Effect of angiotensin-(1-7) on reperfusion arrhythmias in isolated rat hearts. *Braz. J. Med. Biol. Res.* 30: 801-809.

Neves LA, Averill DB, Ferrario CM, Chappell MC, Aschner JL, Walkup MP, Brosnihan KB (2003) Characterization of angiotensin-(1-7) receptor subtype in mesenteric arteries. *Peptides* 24:455-462.

Nogueira AI, Souza Santos RA, Simoes E Silva AC, Cabral AC, Vieira RL, Drumond TC, Machado LJ, Freire CM, Ribeiro-Oliveira A Jr. (2007) The pregnancy-induced increase of plasma angiotensin-(1-7) is blunted in gestational diabetes. *Regul. Pept.* 141:55-60.

Oudit GY, Kassiri Z, Patel MP, Chappell MC, Buany J, Backx PH, Tsushima RG, Scholey JW, Khokha R, Penninger JM (2007) Angiotensin II oxidative stress and inflammation mediate the age dependent cardiomyopathy in ACE2 null mice. *Cardiovascular Res.* 75:29–39.

Oudot A, Vergely C, Ecarnot-Laubriet A, Rochette L (2005) Pharmacological concentration of angiotensin-(1–7) activates NADPH oxidase after ischemia–reperfusion in rat heart through AT1 receptor stimulation. *Regul. Pept.* 127:101-110.

Paul M, Mehr AP, Kreutz R (2006) Physiology of local renin–angiotensin systems. *Physiol. Rev.* 86:747–803.

Pinheiro SV, Simoes e Silva AC, SampaioWO, Paula RD,Mendes EP, Bontempo ED, Pesquero JB, Walther T, Alenina N, Bader M, Bleich M, Santos RAS (2004) Nonpeptide AVE 0991 is an angiotensin-(1–7) receptor Mas agonist in the mouse kidney. *Hypertension* 44: 490–496.

Porsti I, Bara AT, Busse R, Hecker M (1994) Release of nitric oxide by angiotensin-(1-7) from pocine coronary endothelium: implications for a novel angiotensin receptor. *Br. J. Pharmacol.* 111: 652-654.

Rodgers KE, Espinoza T, Felix J, Roda N, Maldonado S, diZerega G (2003) Acceleration of healing, reduction of fibrotic scar, and normalization of tissue architecture by an angiotensin analogue, NorLeu3-A(1-7*). Plast. Reconstr. Surg.* 111:1195-1206.

Rodgers KE, Xiong S, Felix J, Roda N, Espinoza T, Maldonado S, Dizerega G (2001) Development of angiotensin(1–7) as an agent to accelerate dermal repair. *Wound Repair Regen.* 9:238-247.

Rodgers KE, Xiong S, diZerega GS (2002) Accelerated recovery from irradiation injury by angiotensin II. *Cancer Chemother. Pharmacol.* 49:403–411.

Rodgers KE, Xiong S, diZerega GS (2003) Effect of Angiotensin II and Angiotensin (1-7) on white blood cell recovery after intravenous chemotherapy. *Cancer Chemother. Pharmacol.* 51:97–106.

Rodgers KE, Oliver J, diZerega GS (2006) Phase I/II dose escalation study of angiotensin 1-7 [A(1-7)] administered before and after chemotherapy in patients with newly diagnosed breast cancer. *Cancer Chemother. Pharmacol.* 57: 559–568.

Sampaio WO, Nascimento AAS, Santos RAS (2003) Systemic and regional hemodynamic effects of angiotensin-(1–7) in rats. *Am. J. Physiol. Heart. Circ. Physiol.* 284: H1985–H1994.

Sampaio WO, Santos RAS, Faria-Silva R, da Mata Machado LT, Schiffrin EL, Touyz RM (2007) Angiotensin-(1-7) Through Receptor Mas Mediates Endothelial Nitric Oxide Synthase Activation via Akt-Dependent Pathways. *Hypertension*. 49:1-8.

Santos RAS, Simões e Silva AC, Magaldi AJ, Khosla MC, Cesar KR, Passaglio KT, Baracho NCV (1996) Evidence for a physiological role of angiotensin-(1–7) in the control of hydroelectrolyte balance. *Hypertension* 27:875–884

Santos RAS, Campagnole-Santos MJ, Andrade SP (2000) Angiotensin-(1–7): an update. *Regul. Pept.* 91:45–62.

Santos RA, Simoes e Silva AC, Maric C (2003) Angiotensin-(1-7) is an endogenous ligand for the G protein-coupled receptor Mas. Proc Natl Acad Sci USA 100:8258-8263.

Santos RAS, Ferreira AJ, Pinheiro SV, Sampaio WO, Touyz R, Campagnole-Santos MJ (2005a) Angiotensin-(1–7) and its receptor as a potential targets for new cardiovascular drugs. Expert Opin Investig Drugs 14:1019–1031.

Santos RAS, Frézard F, Ferreira AJ (2005b) Angiotensin-(1-7): Blood, Heart, and Blood Vessels. Curr Med Chem Cardiovasc Hematol Agents 3:383-391

Santos RA, Ferreira AJ (2006) Pharmacological effects of AVE 0991, a nonpeptide angiotensin-(1-7) receptor agonist. Cardiovasc *Drug. Rev* .24:239-246.

Santos RA, Ferreira AJ (2007) Angiotensin-(1-7) and the renin-angiotensin system. Curr Opin Nephrol Hypertens 16:122-128.

Sasaki S, Higashi Y, Nakagawa K, Matsuura H, Kajiyama G, Oshima T (2001) Effects of angiotensin-(1-7) on dorearm circulation in normotensive subjects and patients with essential hypertension. *Hypertension* 38:90-94.

Schaper J (1998) Effect of multiple ischemic events on human myocardium: an ultrastructural study. *Eur. Heart. J.* 9: 141–149.

Schiavone MT, Santos RA, Brosnihan KB, Khosla MC, Ferrario CM (1988) Release of vasopressin form the rat hypothalamo-neurohypophysiil system by angiotensin-(1-7) heptapeptide. *Proc. Natl. Acad. Sci. U.S.A.* 85:4095-4098.

Shaltout HA, Westwood BM, Averill DB, Ferrario CM, Figueroa JP, Diz DI, Rose JC, Chappell MC (2007) Angiotensin metabolism in renal proximal tubules, urine, and serum of sheep: Evidence for ACE2-dependent processing of angiotensin II. *Am. J. Physiol. Renal. Physiol.* 292:F82-F91.

Silva DM, Vianna HR, Cortes SF, Campagnole-Santos MJ, Santos RA, Lemos VS (2007) Evidence for a new angiotensin-(1-7) receptor subtype in the aorta of Sprague-Dawley rats. *Peptides* 28:702-707.

Silva-Barcellos NM, Frezard F, Caligiorne S, Santos RA (2001) Long-lasting cardiovascular effects of liposome-entrapped angiotensin-(1-7) at the rostral ventrolateral medulla. *Hypertension*38:1266-1271.

Silva-Barcellos NM, Caligiorne S, dos Santos RA, Frezard F (2004) Site-specific microinjection of liposomes into the brain for local infusion of a short-lived peptide. *J. Control Release* 95:301-307.

Simões e Silva AC, Baracho NCV, Passaglio KT, Santos RAS (1997) Renal actions of angiotensin-(1–7*). Braz. J. Med. Biol. Res.* 30:503–513

Stragier B, Hristova I, Sarre S, Ebinger G, Michotte Y (2005) In vivo characterization of the angiotensin-(1–7)-induced dopamine and c-aminobutyric acid release in the striatum of the rat. *Eur. J. Neurosci.* 22: 658-664.

Strawn WB, Ferrario CM, Tallant EA (1999) Angiotensin-(1-7) reduces smooth muscle growth after vascular injury. *Hypertension* 30: 535-541.

Su Z, Zimpelmann J, Burns KD (2006) Angiotensin-(1–7) inhibits angiotensin II-stimulated phosphorylation of MAP kinases in proximal tubular cells. *Kidney Inter.* 69:2212–2218.

Tallant EA, Diz DI, Ferrario CM (1999) Antiproliferative actions of angiotensin-(1-7) in vascular smooth muscloe. *Hypertension* 34: 950-957.

Tallant EA, Clark MA (2003) Molecular mechanisms of inhibition of vascular growth by angiotensin-(1-7). *Hypertension* 42:574-579.

Tallant EA, Ferrario CM, Gallagher PE (2005) Angiotensin-(1-7) inhibits growth of cardiac myocytes through activation of the mas receptor. *Am. J. Physiol. Heart Circ. Physiol.* 289:1560-1566.

Tikellis C, Johnston CI, Forbes JM, Burns WC, Burrell LM, Risvanis J, Cooper ME (2003) Characterization of renal angiotensin-converting enzyme 2 in diabetic nephropathy. *Hypertension* 41:392–397.

Trask AJ, Ferrario CM (2007) Angiotensin-(1-7): Pharmacology and New Perspectives in Cardiovascular Treatments. Cardiovasc *Drug Rev.* 25:162–174.

Ueda S, Masumori Maemoto S, Ashino K, Nagahara T, Gotoh E, Umemura S, Ishii M (2000) Angiotensin-(1-7) attenuates vasoconstriction evoked by angiotensin II but not by noradrenaline in man. *Hypertension* 35:998–1001.

Vallon V, Richter K, Heyne N, Oswald H (1997) Effect of intratubular application of angiotensin-(1–7) on nephron function. *Kidney Blood Press. Res.* 20:233–239

van der Wouden EA, Henning RH, Deelman LE, Roks AJ, Boomsma F, de Zeeuw D (2005) Does angiotensin (1-7) contribute to the anti-proteinuric effect of ACE-inhibitors. *J. Renin. Angiotensin. Aldosterone. Syst.* 6:96-101.

Van der Wouden EA, Ochodnicky P, van Dokkum RPE, Roks AJM, Deelman LE, de Zeeuw D, Henning RH (2006) The role of angiotensin(1-7) in renal vasculature of the rat. *J. Hypertens* 24:1971-1978.

Vickers C, Hales P, Kaushik V, Dick L, Gavin J, Tang J, Godbout K, Parsons T, Baronas E, Hsieh F, et al. (2002). Hydrolysis of biological peptides by human angiotensin-converting enzyme-related carboxypeptidase. *J. Biol. Chem.* 277:14838-14843.

Walther T, Balschun D, Voigt JP, Fink H, Zuschratter W, Birchmeier C, Ganten D, Bader M (1998) Sustained long term potentiation and anxiety in mice lacking Mas protooncogene. *J. Biol. Chem.* 273: 11867-11873.

Wang J, Peng YJ, Zhu DN (2005) Amino acids modulate the hypotensive effect of angiotensin-(1–7) at the caudal ventrolateral medulla in rats. *Regul. Pept.* 129:1–7.

Wessel N, Malberg H, Heringer-Walther S, Schultheiss, Walther T (2007) The angiotensin-(1-7) receptor agonist AVE0991 dominates the circadian rhythm and baroreflex in spontaneously hypertensive rats. *J. Cardiovasc. Pharmacol.* 49; 67-73.

Widdop RE, Sampey DB, Jarrott B (1999) Cardiovascular effects of angiotensin-(1-7) in conscious spontaneously hypertensive rats. *Hypertension* 34:964-968.

Wiemer G, Dobrucki LW, Louka FR, Malinski T, Heitsch H (2002) AVE 0991, a nonpeptide mimic of the effects of angiotensin-(1–7) on the endothelium. *Hypertension* 40:847–852.

Wysocki J, Ye M, Soler MJ, Gurley SB, Xiao HD, Bernstein KE, Coffman TM, Chen S, Battle D (2006) ACE and ACE2 activity in diabetic mice. Diabetes 55:2132–2139.

Yamada K, Iyer SN, Chappell MC, Ganten D, Ferrario CM (1998)Converting enzyme determines plasma clearance of angiotensin-(1-7). *Hypertension* 32:496-502.

Yamamoto K, Chappell MC, Brosnihan KB, Ferrario CM (1992) In vivo metabolism of angiotensin I by neutral endopeptidase (EC 3.4.24.11) in spontaneously hypertensive rats. *Hypertension* 19:692-696.

Yuan L, Silver X, Torres R, Raizada MK (2006) ACE2 gene transfer attenuates hypertension-linked pathophysiological changes in the SHR. *Physiol Genomics* 27: 12–19.

Zisman LS, Keller RS, Weaver B, Lin Q, Speth R, Bristow MR, Canver CC (2003) Increased angiotensin-(1-7)-forming activity in failing human heart ventricles: evidence for upregulation of the angiotensin-converting enzyme Homologue ACE2. *Circulation* 108:1707-1712.

In: Angiotensin Research Progress
Editors: H. Miura and Y. Sasaki
ISBN: 978-1-60456-537-9

Chapter VII

The Non-Hemodynamic Effect of Angiotensin II in Renal Fibrogenesis

Bi-Cheng Lui*[*] *and Jian-Dong Zhan
Institute of Nephrology, Southeast University, Nanjing, 210009, China

Abstract

Renal fibrosis is a common pathway of progressive renal disease and parallels the reduction in organ function that necessitates dialysis or transplantation. In the last decades, accumulating experimental and clinical evidence convincingly suggests that the renin-angiotensin system (RAS) is a key player in the progression of chronic kidney disease. Angiotensin II (ANG II), the main peptide of the RAS, was classically viewed as a vasoactive factor participating in the local and systemic hemodynamic regulation. With the progression of researches, it has emerged as a multifunctional cytokine exhibiting many non-hemodynamic effects such as growth, profibrotic, matrix metabolism and even proinflammation. These new understandings of ANG II have important clinical implications. It predicted and explained why blockade of the RAS would significantly slow the progression of fibrotic disease. However, it further suggested that higher doses and/or a combination of ANG II blockade with another agent or agents might truly halt the progressive fibrosis.

This review will focus on the non-hemodynamic aspect of ANG II and its contribution in renal fibrogenesis.

Introduction

The critical role of the circulating RAS in the regulation of arterial pressure and sodium homeostasis has been recognized for many years[1]. ANG II is the most powerful biologically active product of the RAS, although there are other bioactive angiotensin peptides, including

[*] Professor Bi-Cheng Liu, MD, PhD. Email: liubc64@yahoo.com.cn

ANG III, ANG IV, and ANG 1-7. ANG II directly constricts vascular smooth muscle cells, enhances myocardial contractility, stimulates aldosterone production, stimulates release of catecholamines from the adrenal medulla and sympathetic nerve endings, increases sympathetic nervous system activity, and stimulates thirst and salt appetite. ANG II also regulates sodium transport by epithelial cells in intestine and kidney[2]. There has also been a growing appreciation of the organ-specific roles exerted by ANG II acting as a paracrine factor. In addition to its physiological roles, locally produced ANG II induces inflammation, cell growth, mitogenesis, apoptosis, migration, and differentiation, regulates the gene expression of bioactive substances, and activates multiple intracellular signaling pathways, all of which might contribute to tissue injury[3]. Clinical and preclinical studies on the pharmacological effects of ACEIs and ARBs support the notion that ANG II exerts a central role in the pathogenesis of renal injury via activation of AT1 receptors when inappropriately activated[4].

Biological Characters of Angiotensin II in the Kidney

From the classical textbook, we can clearly find that ANG II is generated from angiotensinogen--the substrate of RAS system by two steps. One is from angiotensinogen to angiotensin I with the help of renin, another is from angiotensin I to angiotensin II catalyzed by Angiotensin Converting Enzyme (ACE). Besides this concept, other ANG II-generating enzymes include the serine protease chymase, which is supposed to mediate over 80% of ANG II formation in the heart and less than 60% in the vessels[5]. ACE inhibitors do not reduce chymase activity. Upregulation of chymase, mainly in the tubules, is observed in renal biopsies of patients with diabetic nephropathy[6].

The origin of ANG II in the kidney had not been made clear until the definition of local RAS was introduced. In the lumen of the proximal tubule, ANG II levels are in the nanomolar range, or approximately 1,000-fold greater than levels in plasma[7]. A local RAAS including all its components could have been shown in the proximal tubular cells of the kidney. Proximal tubular cells actively produce ANG II and also secrete angiotensinogen into the urine. Intraluminal angiotensinogen may be converted in the distal tubules to AngII, and recent observations suggest that it leads to induction of sodium channels independent of aldosterone[8,9]. It appears that the glomerulus also contains a local RAS. Components of the RAS have been detected in cultured rat and human mesangial cells, and glomerular endothelial cells express ACE[10]. De novo synthesis of ANG II in isolated rat glomeruli has been shown by Atiyeh et al[11]. In addition, ANG II concentrations in glomerular ultrafiltrate are significantly greater than those in plasma, suggesting that glomerular ANG II synthesis occurs in vivo[12]. Renal injury can activate the local RAAS in both directly and indirectly manners[13].

The multiple effects of ANG II are mediated by different receptors. The two major ANG II receptors, AT1 and AT2, are differentially expressed within the kidney. Both are characterized by the configuration of a seven-transmembrane receptor but share only approximately 30% homology on the protein level[14].

ANG II is metabolized by peptidases, such as aminopeptidase A, into ANG III and further into ANG IV[15]. ANG III interacts with AT1 and AT2 receptors, albeit with a lower affinity, and exhibits principally similar effects as ANG II[16]. ANG IV binds to a specific receptor called AT4, which is widely expressed in the kidney, including endothelial cells and proximal as well as convoluted tubules. The AT4 receptor, by protein purification and peptide sequencing, has been identified to be insulin-regulated aminopeptidase[17].

Angiotensin II and Its Direct Effect on Kidney Cells

Resident kidney cells play a diverse role in the whole process of fibrosis. It was defined as a "victim", "accessary" and potential "saviour". Therefore, the direct influence of ANG II on resident kidney cells emerged into our sights as the first part of current discussion.

Research in the past several decades has provided ample evidence that ANG II modulates cell growth[18]. Depending on the cell type, subtype of ANG II receptors, and the presence of other growth factors and cytokines, ANG II may either stimulate growth (proliferation, hypertrophy) or act as a growth suppressor (apoptosis, phenotype change). These different cellular responses play a vital role in renal fibrosis.

In vitro and in vivo studies have provided clear evidence that ANG II exerts growth stimulatory actions. Depending on the target cell and presence of other growth factors, ANG II may induce either proliferation or hypertrophy. That various cyclin-CDK complexes must be activated and CDKIs should be suppressed in proliferation by ANG II is not surprising. In the kidney, ANG II stimulates proliferation of cultured mesangial cells, tubular epithelial cells and interstitial fibroblasts[19-21]. Mervaala et al investigated the growth effects in double-transgenic rats harboring human renin and human angiotensinogen genes. These animals have high local tissue ANG II levels, and organ injury is partly independent of hypertension. There was a significant increase in Ki-67–positive cells in the heart and kidney in double-transgenic rats; however, the exact cell type remains unclear[22].

Hypertrophy is defined as an increase in cell size, RNA content, and protein synthesis, without concomitant changes in cellular DNA content. Geisterfer et al reported in 1988 that ANG II induces hypertrophy, but not proliferation in confluently cultured rat aortic smooth muscle cells[23]. In the kidney, it was well defined that mesangial cell and proximal tubular epithelial cell could be induced to hypertrophy by ANG II too[24;25]. The hypertrophic action of ANG II is mediated through high-affinity AT_1 receptor and some growth factor such as TGF-beta1 and CTGF[26;27]. Cells undergoing ANG II-induced hypertrophy re-enter the cell cycle from G_0 but do not progress into the S phase. ANG II-mediated CDKI inductions such as $p27^{Kip1}$ play a pivotal role in this cell cycle arrest[28].

In some systems, ANG II mediates apoptosis. Although accumulating data suggest that apoptosis requires active cell cycle-dependent processes, few data exist concerning the association of ANG II-induced apoptosis and regulation of cell cycle events. Shankland's group showed a marked increase in glomerular cell apoptosis in nephritic $p27^{Kip1-/-}$ mice compared with nephritic $p27^{Kip1+/+}$ controls. Apoptosis was also increased in tubulointerstitial cells after obstruction in $p27^{Kip1-/-}$ mice compared with control $p27^{Kip1+/+}$ mice. Interestingly, inhibiting cyclin A-CDK2 activation reduced apoptosis in $p27^{Kip1-/-}$ cells. In apoptotic $p27^{Kip1-/-}$ mesangial cells, cyclin A-CDK2 was also activated without a proceeding increase in cyclin E-CDK2 activity, suggesting that uncoupling of CDK2 activity from the scheduled sequence of cell cycle protein expression may lead to an inappropriate and premature initiation of G_1/S phase transition, causing the cell to respond by apoptosis rather than proliferation[29].

ANG II induces apoptosis under certain conditions. Renal tubular apoptosis is a prominent feature of diabetic Ren-2 rats that overexpressed renin and ANG II[30]. It has been suggested that activation of AT_2 receptors is responsible for proapoptotic effect of ANG II, but this issue is controversial because AT_1 receptor antagonists also attenuated apoptosis in certain models[31;32]. Overexpression of AT_2-receptors in fibroblasts had antiproliferative effects without causing apoptosis. This effect was associated with an inhibition of CDK4 and cyclin E expression. CKI expression was not influenced in the AT_2-receptors expressing fibroblasts[33].

Finally, ANG II could exhibit a key cellular event that calls phenotypical change. These phenotypical changes could initiate in various cellular types[34-36]. One important issue is to obtain alpha-SMA expression in mesangial cells, tubular epithelial cells and interstitial fibroblasts.

In some experimental models and human membranous nephropathy, the expression of mesangial α-SMA (also called mesangioblast) is associated with the mesangial expansion and glomerular sclerosis, and blocking of ANG II could also ameliorate these changes[37]. In some aspects, the appearance of the mesangioblasts within glomeruli undergoing repair appears to reenact the embryologic steps of glomerulogenesis. The mesangioblast phenotype is also associated with changes in the ECM synthetic profile of mesangial cells. Changes in mesangial phenotype and the subsequent changes in glomerular ECM profile may also affect mesangial cell survival. It is likely to note that ANG II can also induce α-SMA expression in tubular epithelial cells and interstitial fibroblasts via direct and indirect pathways[38]. These activated tubular epithelial cells (this phenomenon is also called epithelial to mesenchymal transition, EMT) and interstitial fibroblasts will contribute as an alternative resource of the myofibroblast pool which is considered as a determinant in renal fibrosis[39].

Another important issue is the effect of ANG II on podocyte. As reported, ANG II could induce the apoptosis and detachment of podocyte[40]. It should be noticed that the interpretation of results from *in vitro* studies is largely limited, due to the properties of podocytes are largely influenced by the cell culture conditions used. By employing an isolated glomeruli, it was indicated that ANG II could depolarize podocytes, most probably by activating a Ca^{2+}-dependent Cl^- conductance, via an ANG II type 1 (AT1) receptor. Besides these direct evidences, blocking ANG II is proved to be a promising therapy to protect function of podocyte[41].

These differences in ANG II-mediated effects could be exclusively explained by activation of different ANG II receptors or different signal transduction systems. For example, depending on the cell type, hypertrophy and apoptosis could be mediated by either AT_1- or AT_2-receptor activation[42]. Improved understanding of ANG II-mediated cell cycle regulation is necessary to develop improved therapeutic strategies.

Angiotensin II and Extracellular Matrix

Excessive deposition of extracellar matrix (ECM) has been accepted as a significant feature of renal fibrosis. The amount of ECM deposited in the kidney depends on the balance between synthesis and degradation of ECM[43]. ANG II can directly influence the final deposition via mentioned two pathways.

Angiotensin II and the Synthesis of ECM Components

Nearly one decade ago, studies suggested that ANG II up-regulates collagen expression in cultured mesangial cells[44]. After that, other molecules involved in matrix expansion including fibronectin, proteoglycans and osteopontin were identified to be induced by ANG II, as well as other kidney cells such as fibroblast and tubular epithelia were documented to be potential target of ANG II[45]. Besides these molecules, it is important to note that novel proteins include splice variants of fibronectin and novel chains of laminin and type IV collagen were synthesized under the stimulation of ANG II. In addition to providing structural support, these matrix proteins may also interact with cellular receptors especially of the integrin family. As a consequence, the nature of the interstitial matrix may significantly influence the behavior of the neighboring cells. The interstitial scar can be viewed as a dynamic scaffold that changes as the renal disease progresses. Initially the scaffold is relatively unstable and susceptible to degradation. With time, cross-linking bridges add stability and confer relative resistance to protease activity. Changes in matrix composition such as the ratio of collagen I to collagen III may contribute to the stabilizing process[46].

Angiotensin II and the Degradation of ECM

Two major ECM protease systems, matrix metalloproteinases (MMP)/tissue inhibitors of matrix metalloproteinases (TIMP) system and plasminogen activator (PA)/plasmin/PA inhibitors (PAI) system, are interrelated and involved in matrix degradation. The physiology of each system is complex, and the activities are tightly regulated at many levels, including gene expression, activation, and inhibition by specific inhibitors.

MMPs are classified into six groups based on substrate and sequence homology in which collagenases, gelatinases are the most important two. Collagenases [MMP-1, -8 (neutrophil collagenase), -13, and -18 (*Xenopus laevis* collagenase)] cleave collagens I, II, and III at a specific site, generating two fragments. Gelatinases [MMP-2 (gelatinase A) and -9 (gelatinase B)] cleave the denatured collagens (gelatins) and laminin, as well as some chemokines. Tissue inhibitors of metalloproteinases (TIMPs) are endogenous, specific inhibitors that bind and inhibit MMPs[47]. ANG II is reported to be a mediator of high glucose induced alteration of MMPs/TIMPs which results in a decreased degradation ability[48]. In STZ treated rats, we found that blockade of ANG II could restore the ratio of MMP2/TIMP2 and subsequently attenuate the increasing thickness of glomerular basement[49]. Recently, Ozawa et al employed a chronic infusion system to present a direct *in vivo* evidence of influence of ANG II on degradation system. They found that after 12 days of ANG II infusion, mRNA levels of MMP-9 and tPA were significantly increased by 117 and 42%, respectively, compared with the sham-operated rats. Following cessation of ANG II infusion, mRNA levels of MMP-9 and tPA also returned to the normal range within 3 days[50].

Plasmin is generated from plasminogen by the enzymatic activity of tissue-type PA (tPA) and urokinase-type PA (uPA). PA activity is inhibited by PAI[51]. A strong positive correlation between plasmin activity and ECM degradation in cultured mesangial cells and the ability of plasmin inhibitors to inhibit ECM degradation suggest the importance of plasmin in ECM degradation by mesangial cells[52]. Plasmin not only degrades several matrix proteins such as fibronectin, laminin, proteoglycan, and type IV collagen but also activates pro-MMP. PAI-1 seems to play an important role in ECM degradation in mesangial cells, because a larger molar excess of PAI-1 over tPA and uPA is detected in mesangial cell culture supernatant and because anti-PAI-1 antibody increases ECM degradation[53;54]. Tremendous data from both *in vitro* and *in vivo* indicate that ANG II can induce PAI-1 expression in mRNA and protein levels[55-57]. Molecular studies indicate that angiotensin-dependent, TGF-beta1 independent induction of PAI-1 gene expression involves the AT1 receptor and a glucocorticoid response element in the PAI-1 promoter[58]. AT1 receptor blockade but not nonspecific antihypertensive treatment reduces renal PAI-1 expression that is induced by ANG II infusion[59;60].

Interaction between ANG II and Other Growth Factors

Numerous studies have demonstrated that renal fibrosis is orchestrated by many growth factors including ANG II. Thereby, an important issue is the relationship between ANG II and other growth factors.

Angiotensin II and TGF-Beta1

TGF-beta1 mediates tissue fibrosis and is overexpressed in fibrotic renal disease, and anti-TGF-beta1 strategies have been shown to prevent fibrosis in experimental renal fibrotic disease[61].

Chronic renal disease is almost always associated with activation of the RAS system and elevated ANG II levels, which in turn will induce expression of TGF-beta1. In cultured renal mesangial and tubular epithelial cells, ANG II induces TGF-beta1 mRNA and protein production and increases the conversion of inactive TGF-beta1 to the biologically active form, and infusion of ANG II *in vivo* induces glomerular TGF-beta1 mRNA expression[62;63] . By employing an elegant transgenic model, Fern et al showed that TGF-beta1 mRNA levels linearly increased with number of angiotensinogen genes[64]. These data show that ANG II is the main force that drives TGF-beta1 production in tissue injury.

A number of experimental studies strongly suggested that the benefit of reducing ANG II is due to reduce the TGF-beta1 levels[65;66]. It was evidenced by part of CSGCT(Collaborative Study Group Captopril Trial) that lower TGF-beta1 levels after ANG

II inhibition is beneficial to humans[67] Similar results were obtained by Laviades et al, who analyzed TGF-beta1 levels and biochemical markers of collagen metabolism in hypertensive patients treated by losartan[68]. These data indicated that TGF-beta1 as a mediator of profibrotic actions of ANG II. This deduction was strengthened by the evidence obtained by Wolf G et al. that anti-TGF-beta1 antibody abolished the release of collagen induced by ANG II[69]. Besides these, ANG II also was demonstrated to facilitate the action of TGF-β1 by increasing the expression of its receptor and activation of Smad without directly stimulation of TGF-β1[70].

In contrast, TGF-β1 can also regulate the ANG II production. The secretion of renin, which shows elevated tissue levels in some forms of fibrotic renal disease, is enhanced by TGF-β1 *in vitro* and *in vivo*[71]. Colocalized overexpression of TGF-β1 and angiotensinogen has been shown in a model of chronic renal disease induced by renal ablation[72]. Expression of ACE is increased in a model of chronic kidney disease induced by combination of ablation and hyperaldosteronism and again colocalizes with TGF-β1 expression[73]. Further evidence comes from an *in vitro* study showing that TGF-β1 directly induces expression of ACE protein and activity in cultured cardiac fibroblast[74]. If confirmed in vivo, these data suggest that TGF-β1 could enhance ANG II production by local tissue RAS and thus form a positive feedback loop with ANG II.

Angiotensin II and CTGF

Connective tissue growth factor (CTGF) is a novel profibrotic factor that is upregulated in different human kidney diseases and contributes to renal fibrosis. We observed that Angiotensin II could increase CTGF expression in mRNA and protein in tubular epithelial cell line, which was also confirmed by other groups from different cell lines[75-77]. It was reported that ANG II-mediated CTGF expression requires Erk 1,2, and Rho signaling and a calcineurin-associated pathway[77]. We have also demonstrated that the blockade of CTGF, by an antisense CTGF oligonucleotide and anti-CTGF antibody diminished AngII-induced collagen release and hypertrophy[78]. By infusing Ang II, Ruiz-Ortega et al. found that CTGF expression was elevated in aorta and kidney accompanying with excessive ECM synthesis[79]. In models of renal injury, ACE inhibitors and AT_1 antagonists diminished CTGF upregulation and fibrosis[80]. In diabetic patients, authors also analyzed that urinary CTGF could be reduced by the administration of ARB or ACEI[81]. All of these data clearly indicate that CTGF acts as a downstream mediator of ANG II induced fibrosis.

Angiotensin II and Other Factors

In vivo observations have shown that ANG II infusion increases VEGF expression in the renal tissue by regulating its transcription[82]. In podocytes, ANG II stimulates the expression of the VEGF[83]. Vascular endothelial growth factor, in turn, leads to an upregulation of the TGF-β1 receptor type II and increased Smad2 phosphorylation. The increase of VEGF induced by ANG II was also detected in renal epithelial cells.

Signal transduction pathways of PDGF and epidermal growth factor are activated by ANG II without increasing the binding of respective ligands to the receptors, and this activation should account for the mechanism of ANG II induced mitosis[84].

Angiotensin II and Inflammation

Progressive renal disease is histologically characterized by an interstitial infiltration of mononuclear cells[85]. As a source of several profibrotic molecules, monocytes and macrophages in particular are thought to directly contribute to the fibrogenic process. While this interstitial infiltrate may originate in small part from in situ proliferation of resident interstitial macrophages, most of these cells have migrated from the circulation through peritubular capillary endothelium into the interstitial space. The stimulus for this efflux is thought to be chemoattractant molecules that are either secreted across the basolateral membrane of tubules or that pass between tubular cells moving from the tubular lumina directly into the interstitium. Chemoattractant cytokines (chemokines) such as monocyte chemoattractant protein-1 (MCP-1) are thought to play an important role.

ANG II contributes to the recruitment of infiltrating cells into the kidney[86]. ANG II causes the adhesion of circulating cells to endothelial and mesangial cells, and the migration of inflammatory cells into the kidney. This process is mediated by upregulation of adhesion molecules, cytokines and chemokines. Emerging evidences suggest that transcription factors like NF-kB and novel defined toll like receptor (TLR) are also involved in ANG II induced inflammation.

Systemic ANG II infusion into normal rats caused mononuclear cell infiltration in glomeruli and interstitium, while pharmacological blockade of RAS reduced the inflammatory cells infiltration in several models of renal injury. Double-transgenic rats of renin and angiotensinogen genes presented elevated ICAM-1 and VCAM-1, as well as their ligands, such as $\alpha5\beta2$ and $\alpha4\beta1$ integrin in leukocytes[87]. In humans, the blockade of the RAS diminished the expression of adhesion molecules. In hypertensive transgenic rats only AT_1 blockade, but not the vasodilator hydralazine, diminished the number of activated monocytes adhered to the endothelium, as well as endothelial cell injury[88]. ANG II also upregulated the expression of chemokines that are small proteins with chemoattractant properties. ANG II increased monocyte chemoattractant protein type 1 (MCP-1), the main chemokine for monocyte/macrophage, and interleukin-8 and interferon gamma inducible protein-10, which are potent chemoattractants and activators of neutrophils.

I*n vivo*, ANG II activates the renal NF-κB pathway that was partially diminished by AT_1 or AT_2 antagonists alone, and was abolished by combination of both receptor antagonists and ACE inhibition[89]. In mesangial cells, NF-κB activation was mediated by both AT_1 and AT_2 receptors; in tubuloepithelial cells, this was mainly by AT_1, while in endothelial cells it was via AT_2. In the UUO model, investigators have found that blockade of renal NF-κB activation by treatment with two different NF-κB inhibitors, pyrrolidine dithiocarbamate (PDTC) and parthenolide, diminished the inflammatory cell infiltration and downregulated gene expression of several proinflammatory factors[90]. In spontaneously hypertensive rats, NF-κB inhibition attenuated renal interstitial inflammation and hypertension[91]. These data suggest that in some experimental renal diseases, the blockade of ANG II generation by an ACE inhibitor or by combined blockade of both AT_1 and AT_2 receptors, as well as by the inhibition of the NF-κB pathway, is necessary to completely stop the inflammatory process.

TLRs belong to an expanding group of innate pattern recognition receptors that can recognize microbes in all body and cellular compartments[92]. Innate immune activation via TLRs serves as an attractive research hypothesis in order to address the molecular mechanisms of infection-associated renal pathology and noninfectious inflammatory responses in the kidney. TLRs recognize molecular structures of pathogens that infect the kidney, and TLR signaling initiates local inflammation. Alternatively, TLRs trigger renal inflammation upon binding circulating components of pathogens (e.g. those embedded in immune complexes). This mechanism may contribute to the onset and progression of glomerulonephritis and possibly to other types of chronic kidney disease[93]. A new discovery by Wolf et al indicated that ANG II upregulates on mesangial cells Toll-like 4 receptors that bind LPS. This ANG II-mediated Toll-like 4 receptor upregulation resulted in enhanced NF-κB activation[94].

On the other hand, inflammatory cells express all the components of the RAS, and can produce ANG II. One of the early responses to inflammation is an upregulation of angiotensinogen levels. During the differentiation process of monocytes to macrophages, there is an activation of the RAS and re-expression of the ANG II type 2 receptor (AT_2). Moreover, monocytes/macrophages present at inflammatory sites have high ACE activity. These data show that ANG II can activate inflammatory cells, which in turn can activate the RAS and increase local ANG II generation, and therefore contribute to the progression and perpetuation of inflammation.

Recently, an elegant designed experiment showed deletion of AT1a receptor in macrophages resulted in increasing renal fibrosis[95]. Other studies, however, have suggested an opposing view because AT1a receptor-deficient mice with UUO have less fibrosis than wild-type mice[96]. Further studies obviously need to be performed to clarify this important issue of ANG II and imflammation.

Angiotensin II and Proteinuria

Proteinuria is a marker of renal injury, reflecting loss of normal permeaselectivity. Further, proteinuria itself has been proposed to contribute to progressive renal injury[97]. ANG II can not only augment proteinuria but also enhance the profibrotic signal of proteinuria.

Numerous studies in experimental animals and humans have documented that inhibition of the renin-angiotensin system by ACEIs or ARBs reduces urinary protein excretion rate and retards the development of renal injury[98]. By using isolated perfused kidney (IPK) technique, Clavant and his colleagues found that ANG II could induce clinical levels of albuminuria[99]. One possible mechanism is that ANG II preferentially raises efferent glomerular arteriole resistance resulting in enhancement of capillary filtration pressure[100]. During the past ten years, podocyte has been characterized as an important controller of secretion of proteinuria. ANG II causes podocyte apoptosis, actin rearrangement, increased

VEGF synthesis, increased TGF-β1 levels, reduction of ZO-1 and nephrin levels leading to proteinuria[101;102].

Several lines of evidence suggested a role of glomerular ultrafiltrate of plasma proteins or protein-associated factors in chronic tubulointerstitial damage. In vitro studies using proximal tubular cells as a model to assess effects of apical exposure to plasma proteins proved highly valuable to approaching direct causal relationships[103]. In the proximal tubule, albumin and other ultrafiltered proteins are reabsorbed by endocytosis involving megalin and cubulin[104]. ANG II stimulates albumin endocytosis in proximal tubule cells via AT_2 receptor–mediated protein kinase B activation[105]. Tojo A et al found ANG II infusion increased megalin expression and albumin reabsorption in the proximal tubule, while ANG II blockade restored albumin reabsorption via amelioration of megalin expression in the proximal tubules of early stage diabetic rats[106;107].

Another important issue is albumin could be also responsible for the activation of local RAS. Wolf G et al indicated that an increase in tubular albumin reabsorption increase ANG II and in turn upregulate TGF-β1 receptor expression[108]. Accordingly, a group in Chile found that components of the intrarenal RAS are selectively activated in the human diabetic kidney with different levels of proteinuria[109]. Putting these data together, it is easily to believe there lies a loop between ANG II and proteinuria. Further studies need to be performed on better understanding this loop.

Angiotensin II and Oxidative Stress

ROS are composed of a series of oxygen intermediates, including the free radical superoxide anion ($\cdot O_2^-$), the nonradical hydrogen peroxide (H_2O_2), the highly reactive hydroxyl free radical ($\cdot OH$), peroxynitrite ($ONOO^-$), and singlet oxygen ($^1O_2$), in which one of the electrons is raised to an orbital of higher energy with an inversion of spin[110]. ROS modify the activity of tyrosine kinases and phosphatases, activate MAPK, and stimulate transcription factors[111]. Moreover, adhesion molecule expression is induced and matrix metalloproteinase activity is influenced. ROS can also directly change the activity of ion channels and alter the generation of NO by uncoupling of endothelial NOS (eNOS). All of these effects contribute to renal injury by inflammation and foster the development of renal fibrosis.

More than a decade ago, it was concluded from indirect evidence that ANG II may induce ROS formation[112]. Acute ANG II infusion experiments into naive rats in the presence or absence of various free radical scavengers were performed, and these scavengers partly inhibited vascular hyperpermeability and cellular damage[113;114]. Subsequent studies demonstrated that treatment of cultured vascular smooth muscle cells with ANG II for 4 to 6 hours increased $\cdot O_2^-$.[115]. These effects were mediated by the AT_1 receptor and were induced by an activation of membrane-bound NAD(P)H oxidase because the flavoprotein inhibitor diphenylene iodinium and p22phox antisense oligonucleotides attenuated this response[116;117]. AT_1 receptor–transduced ROS formation, depending on NAD(P)H oxidase, has also been observed in several renal cells in culture[118]. Similar observations have been made in the kidney when the endogenous renin-angiotensin system was stimulated using the two kidney–one clip (2K-1C) model[119]. Using Losartan, Manucha W et al reported a significant downregulation of ROS with recovery of the SOD activity in obstructed kidneys[120]. Interestingly, Welch et al showed that Tempol, a SOD mimetic, reverses effects of prolonged ANG II treatment such as renal vasoconstriction, renal cortical hypoxia, activation of NADPH oxidase, increased expression of p22(phox), and reduced expression of EC-SOD[121].

ROS is considered as an important mediator of ANG II's actions. ANG II stimulates phosphorylation and activation of extracellular signal–regulated kinase 1,2 (ERK1,2) in renal tubular cells depending on NAD(P)H oxidase–mediated $\cdot O_2^-$ formation[122]. ANG II-mediated phosphorylation of the EGF receptor is sensitive to inhibition by antioxidants, presumably by interfering with the proteolytic cleavage of heparin-binding EGF that transactivates the EGF receptor[123]. Src tyrosine kinases are also activated by ANG II-stimulated ROS formation[124]. Various tyrosine phosphatases are susceptible to oxidation and inactivation by ROS because these enzymes contain a conserved cysteine that is oxidized to sulfenic acid in the presence of ROS. In addition, transcription factors such as NF-κB and AP-1 are induced by oxidative stress[125]. Substantial experimental evidence indicates that ANG II-induced ROS are potential regulators of adhesion molecule expression. Induction of intercellular adhesion molecule-1 and vascular cellular adhesion molecule-1 by ANG II involves NAD(P)H oxidase activation. ANG II-mediated ROS formation is important for renal growth processes that are part of an adaptive process of surviving nephrons during chronic renal injury[126].

Several studies have described the oxidative stress in the development of various human renal diseases. Women with preeclampsia have an agonistic antibody in the plasma that binds to AT1 receptors that activate NAD(P)H oxidase by inducing p22phox, p47phox, and p67phox[127]. Wolf G and others have found signs of oxidative stress associated with the apoptosis of renal cells in biopsies from patients with different types of renal allograft rejection[128]. Despite there was few studies supported that anti-oxidant therapy would benefit human kidney diseases, several studies showed that angiotensin-converting enzyme inhibitor or AT1 receptor antagonist treatment reduces oxidative stress in patients with renal diseases[129;130].

In summary, the concept that ANG II mediates oxidative stress in the kidney has been validated in experimental models, however, the exact role is still incompletely understood in human renal diseases.

Angiotensin II Receptors and Signal Transduction Pathways Involved in Renal Fibrosis

ANG II exerts its effects by binding to membrane AT_1 or AT_2 receptors[131]. AT_1 regulates blood pressure, cell proliferation, and the production of cytokines and extracellular matrix proteins[132]. AT_2 also regulates blood pressure control, renal natriuresis, cell growth inhibition, and renal inflammatory cell infiltration[133]. Although the rodent contains subtypes of the AT_1 receptor (AT_{1A}, AT_{1B}) encoded by separate genes, a single gene encodes one class of AT_1 receptors in humans. Within the kidney, AT_1 receptor distribution is widespread[134]. AT_1 receptors are present in the vasculature, on plasma membranes of smooth muscle and endothelial cells, including afferent and efferent arterioles. In the glomerulus, AT_1 receptors are present in abundance on mesangial cells, endothelial cells, and podocytes. Tubular epithelial cells along the entire nephron also express AT_1 receptors, with a predominance in the proximal tubule along apical and basolateral membranes[135].

The AT_2 receptor is expressed in relative abundance in fetal tissues, with a rapid decline in mRNA levels after birth[136]. Until recently, expression of AT_2 receptors in the adult was found to confine in certain organs, such as the heart, brain, or adrenal gland[137;138]. However, both functional and immunohistochemical studies indicate the presence of AT_2 receptors in adult kidney, accounting for up to 10% of total ANG II receptors. The cellular localization of AT_2 receptors in the adult kidney remains incompletely understood. Although Ozono et al described glomerular, tubular, and interstitial cell localization[139]. Immunohistochemical studies by Miyata et al showed AT_2 -receptor expression in all nephron segments except for the medullary thick ascending limb and glomerulus[140]. The reason for this discrepancy is unclear but may relate to variable antibody sensitivity or specificity. Within glomeruli, AT_2 -receptor expression has been found in mesangial cells and podocytes. In rat glomerular endothelial cells in culture, AT_2 receptors mediate production of the chemokine, RANTES, suggesting a role in the influx of monocytes and macrophages in glomerular inflammation[141]. In rabbit afferent arterioles, activation of AT_2 receptors causes vasodilatation, which counteracts the vasoconstricting effect of ANG II mediated by AT_1 receptors[142].

ANG II, via AT_1, activates various nuclear transcription factors, including the activator protein-1 (AP-1), STAT family of transcription factors, and cAMP-response element-binding protein (CREB)[125;143-145]. ANG II also increases calcium release and activates protein kinase C (PKC), protein tyrosine kinases (PTK), mitogen-activating protein kinases (MAPK), extracellular signal-regulated kinase (ERK), c-Jun amino terminal kinase (JNK), p38 MAP kinase (p38 MAPK), and, as mentioned before, their downstream effector AP-1[146;149]. MAPK are involved in proliferation, differentiation, fibrosis, and inflammation. In this sense, Mezzano et al have observed an elevation of renal AP-1 activity in ANG II-infused rats, which caused the tubular damage and attenuated by the AT_1 antagonist losartan[150], suggesting that ANG II may regulate the fibrotic process via AT_1/AP-1 pathway. Different intracellular mechanisms regulate the AT_2-mediated cellular responses. AT_2 is linked to inhibition of MAPK, activation of protein-phosphotyrosine phosphatase (PPase), and ceramide production[151;152]. The temporal and spatial patterns of signaling pathway activation are the most likely determinants of a particular functional response. Multiple studies show that the activation of different pathways by ANG II is time dependent. For example, activation of the G protein-dependent pathway and generation of IP_3 occurs in seconds, while MAP kinase and JAK/STAT activation occurs in minutes to hours after initial activation of AT_1R[153;154].

ANG II also cross-talks with several tyrosine kinases via AT_1Rs, including receptor tyrosine kinases[155]. Among these, transactivation of EGFR is a major mechanism by which ANG II influences growth-related signaling pathways. ANG II-mediated EGFR activation occurs in the cholesterol-rich domains of caveolae in the Src-dependent and redox-sensitive manner[156;157]. Studies show that this activation is also dependent on intact microtubules and occurs via calcium-dependent and -independent pathways[158]. These pathways lead to activation of a disintegrin and metalloproteinase (ADAMs), causing release of heparin-binding EGF (HB-EGF)[159;160]. Recently, in COS-7 cells, ANG II-activated ADAM-17 has been identified as the metalloproteinase that promotes HB-EGF shedding[161;162]. HB-EGF induces conformational changes in EGFRs, allowing them to dimerize and autophosphorylate on tyrosine. Once activated, EGFRs serve as a docking site for Grb2/Shc/Sos complexes, inducing two major transduction pathways: the PI3K/PDK1/Akt cascade, which leads to cellular metabolism, growth, survival, and remodeling, and the Ras/Raf/ERK pathway which leads to cell growth, hypertrophy, and inflammation[163]. Kagiyama et al. reported that EGFR activation is required for ANG II-mediated hypertension and left ventricular hypertrophy, both of which are attenuated when rats are treated with an intravenous infusion of antisense oligodeoxynucleotide to EGFR[164]. These studies and many others point to EGFR as an important factor in growth and hypertrophy caused by ANG II.

Other Important Roles Related to Angiotensin II

Besides of the above mentioned effects of ANG II, we would like to briefly introduce the other important issues related to ANG II, that is the novel effect of intracellular ANG II and the active fragments of ANG II.

Evidence is emerging that binding of ANG II to its cell surface AT1-receptors also activates endocytotic (or internalisation) processes that promote trafficking of both the effector and the receptor into intracellular compartments[165]. The potential role of intracellular ANG II could be traced back to an early study in which radiolabelled ANG II was found in the nuclei of vascular smooth muscle and cardiac cells following systemic administration[166]. Re defined the concept of intracrine ANG II as the effector peptide that is either synthesized within a cell via actions of renin, angiotensinogen and ACE or internalised from extracellular ANG II[167]. De Mello and colleagues have consistently shown that intracellular dialysis of renin, angiotensinogen or ANG II into hamster cardiomyocytes significantly altered cell-to-cell communications, supporting an important role for intracellular ANG II in regulating cardiac function[168]. As for the kidney, there are also quite a few lines of evidence. Li et al recently reported that extracellular ANG II accumulates in proximal tubule cells via AT_1 receptor-mediated endocytosis and that microinjection of ANG II into single proximal tubule cells stimulates intracellular calcium mobilization by acting on intracellular AT_1 receptors[169]. Interestingly, Chen et al demonstrated that ANG II induced internalization of an AT1a-receptor-green fluorescent protein complex, which is then translocated to the nucleus in transfected CHO cells[170]. It is likely that internalized Ang II activates nuclear transcription factor NF-κB by binding to cytoplasmic and nuclear AT1-receptors, thereby increasing expression of the Na^+/H^+ exchanger NHE-3, which promotes sodium transport in proximal tubules, and pro-inflammatory cytokines, chemokines and growth factors including angiotensinogen, MCP-1, IL-1β and TGF-β1[171]. NF-κB activation by intracellular ANG II may be particularly relevant because activation of NF-κB by ANG II is one of the major mechanisms by which

ANG II acts as an inflammatory cytokine to induce tubulointerstitial fibrosis. In a newly published article, Zhou et al showed that ANG II induced growth of proximal tubule cells and activated NF-κB, and these responses were blocked by inhibition of AT1-receptor-mediated endocytosis of extracellular ANG II with losartan or colchicine[172]. Taken together, these studies suggest that internalised or intracellular ANG II may play an important role in renal cellular growth and fibrotic responses by activating NF-κB signaling. This information may aid future development of drugs to prevent and treat ANG II-induced target organ injury in cardiovascular and renal diseases by blocking intracellular and/or nuclear actions of ANG II.

Although ANG II has long been considered to represent the end product of the RAS, there is accumulating evidence that this system encompasses additional effector peptides with diverse functions[173]. Shorter ANG II fragments such as Ang-(1-7) (deletion of the C terminalamino acid), angiotensin III (ANG III) (deletion of the N terminal amino acid) and angiotensin IV (ANG IV) (deletion of the two N terminal amino acids) are formed via the activity of ACE and several other peptidases and have been found to accomplish central, cardiovascular and renal actions.

ANG III plays an important role in brain and cardiovascular physiologic processes[174]. Some actions originally attributed to ANGII, such as vasopressin release, are attributable to ANG III[175]. Data suggested that ANG III could be involved in renal pathophysiologic processes. In renal cells, ANG III increases angiotensinogen, transforming growth factor-beta, fibronectin, and monocyte chemoattractant protein-1 (MCP-1) gene expression; therefore, ANG III could contribute to inflammation and fibrosis in the kidney[176;177]. ANG II and ANG III elicit cellular responses through their binding to specific receptors, namely AT_1 and AT_2 receptors. ANG III binds mainly to AT_2 receptors and exhibits lower affinity for AT_1 receptors, compared with that of ANG II[178]. By using selective nonpeptide antagonists for AT_1 and AT_2 receptors and cells from AT_1 receptor-knockout mice, Lorenzo et al demonstrated that in mesangial cells ANG III activates NF-kB mainly via AT_2 receptors.

Angiotensin-(1-7) [Ang-(1-7)] is a bioactive component of the renin-angiotensin system that is endogenously formed in the circulation and various tissues by several enzymatic pathways from either ANG I or ANG II. ANG II and Ang-(1-7) exhibit counter-regulatory effects in the systemic circulation, as well as in tissues important in cardiovascular regulation[179]. It is well documented that ANG II promotes cell proliferation and vasoconstriction, whereas Ang-(1-7) has antiproliferative actions on cardiac myocytes and vascular smooth muscle, and is a vasodilator[180]. The kidneys, which are closely linked to cardiovascular diseases, are also a main target for Ang-(1–7) actions. Ang-(1–7) is present in the kidney in concentrations comparable to ANG II and produces complex renal effects. In water-loaded rats and mice, Ang-(1–7) produces antidiuresis, probably through a mechanism involving cross-talk between Mas and AT_1, AT_2 and AVP V_2 receptors[181]. In contrast, in isolated preparation and anesthetized animals Ang-(1–7) is reported to produce natriuresis – an action which is in keeping with its *in vitro* effects on proximal tubular cells[182]. Collectively, these studies show that Ang-(1-7) exerts effects opposite to those of ANG II on the kidney, and that an attenuation of Ang-(1-7) in the kidney may be associated with human hypertension and injury.

ANG IV has sparked great interest because of its wide range of physiological effects such as memory acquisition and natriuretic effect[183]. Beyond these effects, Gesualdo et al found that ANG IV stimulates plasminogen activator inhibitor-1 expression in proximal tubular epithelial cells which may be implicated in the pathogenesis of renal interstitial fibrosis in several forms of chronic glomerulonephritides like ANG II[184]. At the molecular level, ANG IV interacts with AT1- and AT2-receptors, but only with low affinity (EC50 within the micromolar range)[185]. Yet, most of its physiological effects are already observed at nanomolar concentrations and, most importantly, they cann't be blocked by AT1- and AT2-receptor antagonists, losartan and PD 123177. This has led to the concept of a novel angiotensin receptor subtype: the 'AT4-receptor'. By using a [125] I-labelled ANG IV, Handa R identified the expression and properties of the angiotensin AT_4 receptor in epithelial HK-2 cells (an immortalized cell line derived from adult human proximal tubules)[186], and this receptor was shown to have similar pharmacologic and coupling properties as that described in more distal segments of the nephron and to be linked to the activation of several intracellular signaling and ion regulating systems[187;188].

Summary

There is now ample evidence to support the central role of RAS in the progression of chronic kidney diseases. Although hemodynamic effects of ANG II certainly contribute to renal damage, researches from the last decades clearly suggest this vasoactive agent is a multifarious factor engaged in many non-hemodynamic processes such as cell growth, apoptosis, metabolism of matrix, inflammation, proteinuria, oxidative stress, and inflammation. It should be emphasized that the non-hemodynaic effect is one of the important pathologic effect of ANG II, intervention focusing on this by larger dosage use of RAS blockers or combining use ACEIs and ARBs might be a key strategies in targeting the reverse of renal fibrosis. Detailed experiments have suggested that there are a complicated signal transduction networks in modulating the former processes including receptors, various kinases and transcription factors. Some new concepts such as intracellular ANG II and active fragments of ANG II are emerging to broaden the former field. Clarification of the mechanisms underlying the non-hemodynamic actions of ANG II will open up exciting new research avenues, and will offer the prospect of new diagnostic and therapeutic opportunities.

Acknowledgement

The authors would like to appreciate for the financial support from National Natural Scientific Fund (N0. 30471732), the key project of Jiangsu Provincial Natural Scientific Fund (BK2007709) and Jiangsu Provincial Metical Talent Program to Professor Bi-Cheng Liu.

References

[1] Ibarra-Rubio ME, Pedraza-Chaverri J: The Current Concept Of The Renin-Angiotensin System. *Rev.Invest Clin.* 45:165-177, 1993

[2] Kobori H, Nangaku M, Navar LG, Nishiyama A: The Intrarenal Renin-Angiotensin System: From Physiology To The Pathobiology Of Hypertension And Kidney Disease. Pharmacol. *Rev.* 59:251-287, 2007

[3] Kim S, Iwao H: Molecular And Cellular Mechanisms Of Angiotensin II-Mediated Cardiovascular And Renal Diseases. *Pharmacol.Rev.* 52:11-34, 2000

[4] Komers R, Anderson S: Optimal Strategies For Preventing Progression Of Renal Disease: Should Angiotensin Converting Enzyme Inhibitors And Angiotensin Receptor Blockers Be Used Together? *Curr.Hypertens.Rep.* 2:465-472, 2000

[5] Petrie MC, Padmanabhan N, Mcdonald JE, Hillier C, Connell JM, Mcmurray JJ: Angiotensin Converting Enzyme (ACE) And Non-ACE Dependent Angiotensin II Generation In Resistance Arteries From Patients With Heart Failure And Coronary Heart Disease. *J.Am.Coll.Cardiol.* 37:1056-1061, 2001

[6] Lansang MC, Stevanovic R, Price DA, Laffel LM, Hollenberg NK: ACE And Non-ACE Pathways In The Renal Vascular Response To RAS Interruption In Type 1 Diabetes Mellitus. *Kidney Int.* 67:1033-1037, 2005

[7] Van Kats JP, Schalekamp MA, Verdouw PD, Duncker DJ, Danser AH: Intrarenal Angiotensin II: Interstitial And Cellular Levels And Site Of Production. *Kidney Int.* 60:2311-2317, 2001

[8] Beutler KT, Masilamani S, Turban S, Nielsen J, Brooks HL, Ageloff S, Fenton RA, Packer RK, Knepper MA: Long-Term Regulation Of Enac Expression In Kidney By Angiotensin II. *Hypertension* 41:1143-1150, 2003

[9] Brooks HL, Allred AJ, Beutler KT, Coffman TM, Knepper MA: Targeted Proteomic Profiling Of Renal Na(+) Transporter And Channel Abundances In Angiotensin II Type 1a Receptor Knockout Mice. *Hypertension* 39:470-473, 2002

[10] Douglas JG, Hopfer U: Novel Aspect Of Angiotensin Receptors And Signal Transduction In The Kidney. *Annu.Rev.Physiol* 56:649-669, 1994

[11] Atiyeh BA, Arant BS, Jr., Henrich WL, Seikaly MG: In Vitro Production Of Angiotensin II By Isolated Glomeruli. *Am.J.Physiol* 268:F266-F272, 1995

[12] Ingelfinger JR, Dzau VJ: Molecular Biology Of Renal Injury: Emphasis On The Role Of The Renin-Angiotensin System. *J.Am.Soc.Nephrol.* 2:S9-20, 1991

[13] Navar LG, Harrison-Bernard LM, Imig JD, Wang CT, Cervenka L, Mitchell KD: Intrarenal Angiotensin II Generation And Renal Effects Of AT1 Receptor Blockade. *J.Am.Soc.Nephrol.* 10 Suppl 12:S266-S272, 1999

[14] Ardaillou R, Chansel D, Chatziantoniou C, Dussaule JC: Biology And Functions Of Renal Receptors For Angiotensin II And Its Active Fragments. *Adv.Nephrol.Necker Hosp.* 28:225-257, 1998

[15] Wolf G, Thaiss F, Scherberich JE, Schoeppe W, Stahl RA: Glomerular Angiotensinase A In The Rat: Increase Of Enzyme Activity Following Renal Ablation. *Kidney Int.* 38:862-868, 1990

[16] Reaux-Le GA, Iturrioz X, Fassot C, Claperon C, Roques BP, Llorens-Cortes C: Role Of Angiotensin III In Hypertension. *Curr.Hypertens.Rep.* 7:128-134, 2005

[17] Wright JW, Krebs LT, Stobb JW, Harding JW: The Angiotensin IV System: Functional Implications. *Front Neuroendocrinol.* 16:23-52, 1995

[18] Wolf G, Neilson EG: Angiotensin II As A Renal Growth Factor. *J.Am.Soc.Nephrol.* 3:1531-1540, 1993

[19] Wolf G, Haberstroh U, Neilson EG: Angiotensin II Stimulates The Proliferation And Biosynthesis Of Type I Collagen In Cultured Murine Mesangial Cells. *Am.J.Pathol.* 140:95-107, 1992

[20] Wolf G, Ziyadeh FN, Helmchen U, Zahner G, Schroeder R, Stahl RA: ANG II Is A Mitogen For A Murine Cell Line Isolated From Medullary Thick Ascending Limb Of Henle's Loop. *Am.J.Physiol* 268:F940-F947, 1995

[21] Schuttert JB, Liu MH, Gliem N, Fiedler GM, Zopf S, Mayer C, Muller GA, Grunewald RW: Human Renal Fibroblasts Derived From Normal And Fibrotic Kidneys Show Differences In Increase Of Extracellular Matrix Synthesis And Cell Proliferation Upon Angiotensin II Exposure. *Pflugers Arch.* 446:387-393, 2003

[22] Mervaala EM, Muller DN, Park JK, Schmidt F, Lohn M, Breu V, Dragun D, Ganten D, Haller H, Luft FC: Monocyte Infiltration And Adhesion Molecules In A Rat Model Of High Human Renin Hypertension. *Hypertension* 33:389-395, 1999

[23] Geisterfer AA, Peach MJ, Owens GK: Angiotensin II Induces Hypertrophy, Not Hyperplasia, Of Cultured Rat Aortic Smooth Muscle Cells. Circ.*Res.* 62:749-756, 1988

[24] Wolf G, Neilson EG: Angiotensin II Induces Cellular Hypertrophy In Cultured Murine Proximal Tubular Cells. *Am.J.Physiol* 259:F768-F777, 1990

[25] Orth SR, Weinreich T, Bonisch S, Weih M, Ritz E: Angiotensin II Induces Hypertrophy And Hyperplasia In Adult Human Mesangial Cells. *Exp.Nephrol.* 3:23-33, 1995

[26] Liu BC, Sun J, Liu H, Chen L, Chen Q, Fan LM: [Role Of Connective Tissue Growth Factor In Angiotensin II Induced Human Proximal Tubular Cell Hypertrophy]. Zhonghua Yi.Xue.Za Zhi. 84:1470-1474, 2004

[27] Wolf G: Link Between Angiotensin II And TGF-Beta In The Kidney. Miner.Electrolyte Metab 24:174-180, 1998

[28] Wolf G, Jablonski K, Schroeder R, Reinking R, Shankland SJ, Stahl RA: Angiotensin II-Induced Hypertrophy Of Proximal Tubular Cells Requires p27Kip1. *Kidney Int.* 64:71-81, 2003

[29] Ophascharoensuk V, Fero ML, Hughes J, Roberts JM, Shankland SJ: The Cyclin-Dependent Kinase Inhibitor p27Kip1 Safeguards Against Inflammatory *Injury. Nat.Med.* 4:575-580, 1998

[30] Kelly DJ, Cox AJ, Tolcos M, Cooper ME, Wilkinson-Berka JL, Gilbert RE: Attenuation Of Tubular Apoptosis By Blockade Of The Renin-Angiotensin System In Diabetic Ren-2 Rats. *Kidney Int.* 61:31-39, 2002

[31] Yamada T, Horiuchi M, Dzau VJ: Angiotensin II Type 2 Receptor Mediates Programmed Cell Death. *Proc.Natl.Acad.Sci.U.S.A* 93:156-160, 1996

[32] Dimmeler S, Rippmann V, Weiland U, Haendeler J, Zeiher AM: Angiotensin II Induces Apoptosis Of Human Endothelial Cells. Protective Effect Of Nitric Oxide. *Circ.Res.* 81:970-976, 1997

[33] Gingras B, Rodier G, Giasson E, Coulombe P, Chassagne C, Meloche S: Expression Of Angiotensin Type II Receptor Downregulates Cdk4 Synthesis And Inhibits Cell-Cycle Progression. *Oncogene* 22:2633-2642, 2003

[34] Long X, Xiong SD, Xiong WN, Xu YJ, Zhang ZX, Cao Y, Chen J, Xu XH: Losartan Inhibited Angiotensin Induced Human Lung Fibroblast Transformation And Collagen Synthesis. Zhonghua Jie.He.He.Hu Xi.Za Zhi. 30:273-278, 2007

[35] Kurikawa N, Suga M, Kuroda S, Yamada K, Ishikawa H: An Angiotensin II Type 1 Receptor Antagonist, Olmesartan Medoxomil, Improves Experimental Liver Fibrosis By Suppression Of Proliferation And Collagen Synthesis In Activated Hepatic Stellate Cells. *Br.J.Pharmacol.* 139:1085-1094, 2003

[36] Kobayashi S, Moriya H, Nakabayashi I, Nishiyama J, Fukuda T: Angiotensin II And IGF-I May Interact To Regulate Tubulointerstitial Cell Kinetics And Phenotypic Changes In Hypertensive Rats. *Hypertens.Res.* 25:257-269, 2002

[37] Mezzano SA, Aros CA, Droguett A, Burgos ME, Ardiles LG, Flores CA, Carpio D, Vio CP, Ruiz-Ortega M, Egido J: Renal Angiotensin II Up-Regulation And Myofibroblast Activation In Human Membranous Nephropathy. *Kidney Int.Suppls*39-S45, 2003

[38] Chen L, Liu BC, Zhang XL, Zhang JD, Liu H, Li MX: Influence Of Connective Tissue Growth Factor Antisense Oligonucleotide On Angiotensin II-Induced Epithelial Mesenchymal Transition In HK2 Cells. *Acta Pharmacol.Sin.* 27:1029-1036, 2006

[39] Wu K, Zhou T, Sun G, Wang W, Zhang Y, Zhang Y, Hao L, Chen N: Valsartan Inhibited The Accumulation Of Dendritic Cells In Rat Fibrotic Renal Tissue. *Cell Mol.Immunol.* 3:213-220, 2006

[40] Gloy J, Henger A, Fischer KG, Nitschke R, Bleich M, Mundel P, Schollmeyer P, Greger R, Pavenstadt H: Angiotensin II Modulates Cellular Functions Of Podocytes. *Kidney Int.Suppl* 67:S168-S170, 1998

[41] Nitschke R, Henger A, Ricken S, Gloy J, Muller V, Greger R, Pavenstadt H: Angiotensin II Increases The Intracellular Calcium Activity In Podocytes Of The Intact Glomerulus. *Kidney Int.* 57:41-49, 2000

[42] Sequeira Lopez ML, Gomez RA: The Role Of Angiotensin II In Kidney Embryogenesis And Kidney Abnormalities. *Curr.Opin.Nephrol.Hypertens.* 13:117-122, 2004

[43] Ha H, Lee HB: Reactive Oxygen Species And Matrix Remodeling In Diabetic Kidney. *J.Am.Soc.Nephrol.* 14:S246-S249, 2003

[44] Wolf G, Haberstroh U, Neilson EG: Angiotensin II Stimulates The Proliferation And Biosynthesis Of Type I Collagen In Cultured Murine Mesangial Cells. *Am.J.Pathol.* 140:95-107, 1992

[45] Wolf G, Killen PD, Neilson EG: Intracellular Signaling Of Transcription And Secretion Of Type IV Collagen After Angiotensin II-Induced Cellular Hypertrophy In Cultured Proximal Tubular Cells. *Cell Regul.* 2:219-227, 1991

[46] Eddy AA: Molecular Basis Of Renal Fibrosis. *Pediatr.Nephrol.* 15:290-301, 2000

[47] Bode W, Maskos K: Structural Basis Of The Matrix Metalloproteinases And Their Physiological Inhibitors, The Tissue Inhibitors Of Metalloproteinases. *Biol.Chem.* 384:863-872, 2003

[48] Zaoui P, Cantin JF, Imardani-Bessette M, Monier F, Halimi S, Morel F, Cordonnier D: Role Of Metalloproteases And Inhibitors In The Occurrence And Progression Of Diabetic Renal Lesions. *Diabetes Metab* 26 Suppl 4:25-29, 2000

[49] Liu BC, Chen Q, Luo DD, Sun J, Phillips AO, Ruan XZ, Liu NF: Mechanisms Of Irbesartan In Prevention Of Renal Lesion In Streptozotocin-Induced Diabetic Rats. *Acta Pharmacol.Sin.* 24:67-73, 2003

[50] Zhang Y, Wang C, Mizukami H, Itoh H, Kusama M, Ozawa K, Jinbu Y: Increased Expression And Activation Of Matrix Metalloproteinase-2 (MMP-2) In O-1N: Hamster Oral Squamous Cell Carcinoma With High Potential Lymph Node Metastasis. *J.Exp.Clin.Cancer Res.* 25:417-423, 2006

[51] Eddy AA, Fogo AB: Plasminogen Activator Inhibitor-1 In Chronic Kidney Disease: Evidence And Mechanisms Of Action. *J.Am.Soc.Nephrol.* 17:2999-3012, 2006

[52] Rerolle JP, Hertig A, Nguyen G, Sraer JD, Rondeau EP: Plasminogen Activator Inhibitor Type 1 Is A Potential Target In Renal Fibrogenesis. *Kidney Int.* 58:1841-1850, 2000

[53] Huang Y, Border WA, Lawrence DA, Noble NA: Noninhibitory PAI-1 Enhances Plasmin-Mediated Matrix Degradation Both In Vitro And In Experimental Nephritis. *Kidney Int.* 70:515-522, 2006

[54] Huang Y, Noble NA: PAI-1 As A Target In Kidney Disease. *Curr.Drug Targets.* 8:1007-1015, 2007

[55] Haraguchi M, Border WA, Huang Y, Noble NA: T-PA Promotes Glomerular Plasmin Generation And Matrix Degradation In Experimental Glomerulonephritis. *Kidney Int.* 59:2146-2155, 2001

[56] Rerolle JP, Hertig A, Nguyen G, Sraer JD, Rondeau EP: Plasminogen Activator Inhibitor Type 1 Is A Potential Target In Renal Fibrogenesis. *Kidney Int.* 58:1841-1850, 2000

[57] Oikawa T, Freeman M, Lo W, Vaughan DE, Fogo A: Modulation Of Plasminogen Activator Inhibitor-1 In Vivo: A New Mechanism For The Anti-Fibrotic Effect Of Renin-Angiotensin Inhibition. *Kidney Int.* 51:164-172, 1997

[58] Fintha A, Sebe A, Masszi A, Terebessy T, Huszar T, Rosivall L, Mucsi I: Angiotensin II Activates Plasminogen Activator Inhibitor-I Promoter In Renal Tubular Epithelial Cells Via The AT1 Receptor. *Acta Physiol Hung.* 94:19-30, 2007

[59] Haraguchi M, Border WA, Huang Y, Noble NA: T-PA Promotes Glomerular Plasmin Generation And Matrix Degradation In Experimental Glomerulonephritis. *Kidney Int.* 59:2146-2155, 2001

[60] Abrahamsen CT, Pullen MA, Schnackenberg CG, Grygielko ET, Edwards RM, Laping NJ, Brooks DP: Effects Of Angiotensins II And IV On Blood Pressure, Renal Function, And PAI-1 Expression In The Heart And Kidney Of The Rat. *Pharmacology* 66:26-30, 2002

[61] Border WA, Noble NA: Targeting TGF-Beta For Treatment Of Disease. *Nat.Med.* 1:1000-1001, 1995

[62] Kagami S, Border WA, Miller DE, Noble NA: Angiotensin II Stimulates Extracellular Matrix Protein Synthesis Through Induction Of Transforming Growth Factor-Beta Expression In Rat Glomerular Mesangial Cells. *J.Clin.Invest* 93:2431-2437, 1994

[63] Wolf G, Ziyadeh FN, Stahl RA: Angiotensin II Stimulates Expression Of Transforming Growth Factor Beta Receptor Type II In Cultured Mouse Proximal Tubular Cells. *J.Mol.Med.* 77:556-564, 1999

[64] Fern RJ, Yesko CM, Thornhill BA, Kim HS, Smithies O, Chevalier RL: Reduced Angiotensinogen Expression Attenuates Renal Interstitial Fibrosis In Obstructive Nephropathy In Mice. *J.Clin.Invest* 103:39-46, 1999

[65] Peters H, Noble NA, Border WA: Transforming Growth Factor-Beta In Human Glomerular Injury. *Curr.Opin.Nephrol.Hypertens.* 6:389-393, 1997

[66] Border WA, Noble NA: TGF-Beta In Kidney Fibrosis: A Target For Gene Therapy. *Kidney Int.* 51:1388-1396, 1997

[67] Strippoli GF, Craig M, Deeks JJ, Schena FP, Craig JC: Effects Of Angiotensin Converting Enzyme Inhibitors And Angiotensin II Receptor Antagonists On Mortality And Renal Outcomes In Diabetic Nephropathy: Systematic Review. *BMJ* 329:828, 2004

[68] Laviades C, Varo N, Diez J: Transforming Growth Factor Beta In Hypertensives With Cardiorenal Damage. *Hypertension* 36:517-522, 2000

[69] Wolf G, Zahner G, Schroeder R, Stahl RA: Transforming Growth Factor Beta Mediates The Angiotensin-II-Induced Stimulation Of Collagen Type IV Synthesis In Cultured Murine Proximal Tubular Cells. *Nephrol.Dial.Transplant.* 11:263-269, 1996

[70] Rodriguez-Vita J, Sanchez-Lopez E, Esteban V, Ruperez M, Egido J, Ruiz-Ortega M: Angiotensin II Activates The Smad Pathway In Vascular Smooth Muscle Cells By A Transforming Growth Factor-Beta-Independent Mechanism. *Circulation* 111:2509-2517, 2005

[71] Krag S, Nielsen AH, Wogensen L: High Plasma Concentrations Of Prorenin In A Transgenic Animal Model Of Nephropathy With Overexpression Of Transforming Growth Factor-Beta1 In The Kidneys. *Clin.Exp.Pharmacol.Physiol* 27:724-726, 2000

[72] Miyazaki Y, Tsuchida S, Fogo A, Ichikawa I: The Renal Lesions That Develop In Neonatal Mice During Angiotensin Inhibition Mimic Obstructive Nephropathy. *Kidney Int.* 55:1683-1695, 1999

[73] Gaedeke J, Peters H, Noble NA, Border WA: Angiotensin II, TGF-Beta And Renal Fibrosis. *Contrib.Nephrol.*153-160, 2001

[74] Petrov VV, Fagard RH, Lijnen PJ: Transforming Growth Factor-Beta(1) Induces Angiotensin-Converting Enzyme Synthesis In Rat Cardiac Fibroblasts During Their Differentiation To Myofibroblasts. *J.Renin.Angiotensin.Aldosterone.Syst.* 1:342-352, 2000

[75] Liu BC, Sun J, Chen Q, Ma KL, Ruan XZ, Phillips AO: Role Of Connective Tissue Growth Factor In Mediating Hypertrophy Of Human Proximal Tubular Cells Induced By Angiotensin II. *Am.J.Nephrol.* 23:429-437, 2003

[76] Ruperez M, Ruiz-Ortega M, Esteban V, Lorenzo O, Mezzano S, Plaza JJ, Egido J: Angiotensin II Increases Connective Tissue Growth Factor In The Kidney. *Am.J.Pathol.* 163:1937-1947, 2003

[77] Iwanciw D, Rehm M, Porst M, Goppelt-Struebe M: Induction Of Connective Tissue Growth Factor By Angiotensin II: Integration Of Signaling Pathways. Arterioscler.*Thromb.Vasc.Biol.* 23:1782-1787, 2003

[78] Liu BC, Chen L, Sun J, Huang HQ, Ma KL, Liu H, Liu DG, Zhang XL: Connective Tissue Growth Factor-Mediated Angiotensin II-Induced Hypertrophy Of Proximal Tubular Cells. *Nephron Exp.Nephrol.* 103:E16-E26, 2006

[79] Ruperez M, Lorenzo O, Blanco-Colio LM, Esteban V, Egido J, Ruiz-Ortega M: Connective Tissue Growth Factor Is A Mediator Of Angiotensin II-Induced Fibrosis. *Circulation* 108:1499-1505, 2003

[80] Liu N, Shimizu S, Ito-Ihara T, Takagi K, Kita T, Ono T: Angiotensin II Receptor Blockade Ameliorates Mesangioproliferative Glomerulonephritis In Rats Through Suppression Of CTGF And PAI-1, Independently Of The Coagulation System. *Nephron Exp.Nephrol.* 105:E65-E74, 2007

[81] Ruiz-Ortega M, Ruperez M, Esteban V, Rodriguez-Vita J, Sanchez-Lopez E, Carvajal G, Egido J: Angiotensin II: A Key Factor In The Inflammatory And Fibrotic Response In Kidney Diseases. *Nephrol.Dial.Transplant.* 21:16-20, 2006

[82] Rizkalla B, Forbes JM, Cooper ME, Cao Z: Increased Renal Vascular Endothelial Growth Factor And Angiopoietins By Angiotensin II Infusion Is Mediated By Both AT1 And AT2 Receptors. *J.Am.Soc.Nephrol.* 14:3061-3071, 2003

[83] Chen S, Lee JS, Iglesias-De La Cruz MC, Wang A, Izquierdo-Lahuerta A, Gandhi NK, Danesh FR, Wolf G, Ziyadeh FN: Angiotensin II Stimulates Alpha3(IV) Collagen Production In Mouse Podocytes Via TGF-Beta And VEGF Signalling: Implications For Diabetic Glomerulopathy. *Nephrol.Dial.Transplant.* 20:1320-1328, 2005

[84] Clark MA, Gonzalez N: Angiotensin II Stimulates Rat Astrocyte Mitogen-Activated Protein Kinase Activity And Growth Through EGF And PDGF Receptor Transactivation. *Regul.Pept.* 144:115-122, 2007

[85] Noronha IL, Fujihara CK, Zatz R: The Inflammatory Component In Progressive Renal Disease--Are Interventions Possible? *Nephrol.Dial.Transplant.* 17:363-368, 2002

[86] Ozawa Y, Kobori H, Suzaki Y, Navar LG: Sustained Renal Interstitial Macrophage Infiltration Following Chronic Angiotensin II Infusions. *Am.J.Physiol Renal Physiol* 292:F330-F339, 2007

[87] Theuer J, Dechend R, Muller DN, Park JK, Fiebeler A, Barta P, Ganten D, Haller H, Dietz R, Luft FC: Angiotensin II Induced Inflammation In The Kidney And In The Heart Of Double Transgenic Rats. BMC.Cardiovasc.*Disord.* 2:3, 2002

[88] Strawn WB, Gallagher PE, Tallant EA, Ganten D, Ferrario CM: Angiotensin II AT1-Receptor Blockade Inhibits Monocyte Activation And Adherence In Transgenic (Mren2)27 Rats. *J.Cardiovasc.Pharmacol.* 33:341-351, 1999

[89] Ruiz-Ortega M, Lorenzo O, Ruperez M, Blanco J, Egido J: Systemic Infusion Of Angiotensin II Into Normal Rats Activates Nuclear Factor-Kappab And AP-1 In The Kidney: Role Of AT(1) And AT(2) Receptors. *Am.J.Pathol.* 158:1743-1756, 2001

[90] Esteban V, Lorenzo O, Ruperez M, Suzuki Y, Mezzano S, Blanco J, Kretzler M, Sugaya T, Egido J, Ruiz-Ortega M: Angiotensin II, Via AT1 And AT2 Receptors And NF-Kappab Pathway, Regulates The Inflammatory Response In Unilateral Ureteral Obstruction. *J.Am.Soc.Nephrol.* 15:1514-1529, 2004

[91] Sanz-Rosa D, Cediel E, De Las HN, Miana M, Balfagon G, Lahera V, Cachofeiro V: Participation Of Aldosterone In The Vascular Inflammatory Response Of Spontaneously Hypertensive Rats: Role Of The Nfkappab/Ikappab System. *J.Hypertens.* 23:1167-1172, 2005

[92] Tse K, Horner AA: Update On Toll-Like Receptor-Directed Therapies For Human Disease. *Ann.Rheum.Dis*. 66 Suppl 3:Iii77-Iii80, 2007

[93] Anders HJ, Schlondorff D: Toll-Like Receptors: Emerging Concepts In Kidney Disease. *Curr.Opin.Nephrol.Hypertens*. 16:177-183, 2007

[94] Wolf G, Bohlender J, Bondeva T, Roger T, Thaiss F, Wenzel UO: Angiotensin II Upregulates Toll-Like Receptor 4 On Mesangial Cells. *J.Am.Soc.Nephrol.* 17:1585-1593, 2006

[95] Nishida M, Fujinaka H, Matsusaka T, Price J, Kon V, Fogo AB, Davidson JM, Linton MF, Fazio S, Homma T, Yoshida H, Ichikawa I: Absence Of Angiotensin II Type 1 Receptor In Bone Marrow-Derived Cells Is Detrimental In The Evolution Of Renal Fibrosis. *J.Clin.Invest* 110:1859-1868, 2002

[96] Satoh M, Kashihara N, Yamasaki Y, Maruyama K, Okamoto K, Maeshima Y, Sugiyama H, Sugaya T, Murakami K, Makino H: Renal Interstitial Fibrosis Is Reduced In Angiotensin II Type 1a Receptor-Deficient Mice. *J.Am.Soc.Nephrol.* 12:317-325, 2001

[97] Wolf G, Ziyadeh FN: Cellular And Molecular Mechanisms Of Proteinuria In Diabetic Nephropathy. *Nephron Physiol* 106:26-31, 2007

[98] Wolf G, Ritz E: Combination Therapy With ACE Inhibitors And Angiotensin II Receptor Blockers To Halt Progression Of Chronic Renal Disease: Pathophysiology And Indications. *Kidney Int.* 67:799-812, 2005

[99] Clavant SP, Forbes JM, Thallas V, Osicka TM, Jerums G, Comper WD: Reversible Angiotensin II-Mediated Albuminuria In Rat Kidneys Is Dynamically Associated With Cytoskeletal Organization. *Nephron Physiol* 93:51-60, 2003

[100] Denton KM, Anderson WP, Sinniah R: Effects Of Angiotensin II On Regional Afferent And Efferent Arteriole Dimensions And The Glomerular Pole. *Am.J.Physiol Regul.Integr.Comp Physiol* 279:R629-R638, 2000

[101] Chen S, Lee JS, Iglesias-De La Cruz MC, Wang A, Izquierdo-Lahuerta A, Gandhi NK, Danesh FR, Wolf G, Ziyadeh FN: Angiotensin II Stimulates Alpha3(IV) Collagen Production In Mouse Podocytes Via TGF-Beta And VEGF Signalling: Implications For Diabetic Glomerulopathy. *Nephrol.Dial.Transplant.* 20:1320-1328, 2005

[102] Langham RG, Kelly DJ, Cox AJ, Gow RM, Holthofer H, Gilbert RE: Angiotensin II-Induced Proteinuria And Expression Of The Podocyte Slit Pore Membrane Protein, Nephrin. *Nephrol.Dial.Transplant.* 19:262-263, 2004

[103] Russo LM, Bakris GL, Comper WD: Renal Handling Of Albumin: A Critical Review Of Basic Concepts And Perspective. *Am.J.Kidney Dis.* 39:899-919, 2002

[104] Christensen EI, Birn H: Megalin And Cubilin: Multifunctional Endocytic Receptors. *Nat.Rev.Mol.Cell Biol.* 3:256-266, 2002

[105] Caruso-Neves C, Kwon SH, Guggino WB: Albumin Endocytosis In Proximal Tubule Cells Is Modulated By Angiotensin II Through An AT2 Receptor-Mediated Protein Kinase B Activation. *Proc.Natl.Acad.Sci.U.S.A* 102:17513-17518, 2005

[106] Tojo A, Onozato ML, Kurihara H, Sakai T, Goto A, Fujita T: Angiotensin II Blockade Restores Albumin Reabsorption In The Proximal Tubules Of Diabetic Rats. Hypertens.*Res*. 26:413-419, 2003

[107] Tojo A, Onozato ML, Ha H, Kurihara H, Sakai T, Goto A, Fujita T, Endou H: Reduced Albumin Reabsorption In The Proximal Tubule Of Early-Stage Diabetic Rats. Histochem.*Cell Biol.* 116:269-276, 2001

[108] Wolf G, Schroeder R, Ziyadeh FN, Stahl RA: Albumin Up-Regulates The Type II Transforming Growth Factor-Beta Receptor In Cultured Proximal Tubular Cells. *Kidney Int.* 66:1849-1858, 2004

[109] Mezzano S, Droguett A, Burgos ME, Ardiles LG, Flores CA, Aros CA, Caorsi I, Vio CP, Ruiz-Ortega M, Egido J: Renin-Angiotensin System Activation And Interstitial Inflammation In Human Diabetic Nephropathy. *Kidney Int.Suppls*64-S70, 2003

[110] Fridovich I: Superoxide Anion Radical (O2-.), Superoxide Dismutases, And Related Matters. *J.Biol.Chem.* 272:18515-18517, 1997

[111] Shah SV, Baliga R, Rajapurkar M, Fonseca VA: Oxidants In Chronic Kidney Disease. *J.Am.Soc.Nephrol.* 18:16-28, 2007

[112] Wilson SK: Role Of Oxygen-Derived Free Radicals In Acute Angiotensin II--Induced Hypertensive Vascular Disease In The Rat. *Circ.Res.* 66:722-734, 1990

[113] Wang D, Chen Y, Chabrashvili T, Aslam S, Borrego Conde LJ, Umans JG, Wilcox CS: Role Of Oxidative Stress In Endothelial Dysfunction And Enhanced Responses To Angiotensin II Of Afferent Arterioles From Rabbits Infused With Angiotensin II. *J.Am.Soc.Nephrol.* 14:2783-2789, 2003

[114] Haugen EN, Croatt AJ, Nath KA: Angiotensin II Induces Renal Oxidant Stress In Vivo And Heme Oxygenase-1 In Vivo And In Vitro. *Kidney Int.* 58:144-152, 2000

[115] Griendling KK, Minieri CA, Ollerenshaw JD, Alexander RW: Angiotensin II Stimulates NADH And NADPH Oxidase Activity In Cultured Vascular Smooth Muscle Cells. *Circ.Res.* 74:1141-1148, 1994

[116] Lassegue B, Sorescu D, Szocs K, Yin Q, Akers M, Zhang Y, Grant SL, Lambeth JD, Griendling KK: Novel Gp91(Phox) Homologues In Vascular Smooth Muscle Cells : Nox1 Mediates Angiotensin II-Induced Superoxide Formation And Redox-Sensitive Signaling Pathways. *Circ.Res.* 88:888-894, 2001

[117] Griendling KK, Sorescu D, Ushio-Fukai M: NAD(P)H Oxidase: Role In Cardiovascular Biology And Disease. *Circ.Res.* 86:494-501, 2000

[118] Ushio-Fukai M, Alexander RW, Akers M, Yin Q, Fujio Y, Walsh K, Griendling KK: Reactive Oxygen Species Mediate The Activation Of Akt/Protein Kinase B By Angiotensin II In Vascular Smooth Muscle Cells. *J.Biol.Chem.* 274:22699-22704, 1999

[119] Heitzer T, Wenzel U, Hink U, Krollner D, Skatchkov M, Stahl RA, Macharzina R, Brasen JH, Meinertz T, Munzel T: Increased NAD(P)H Oxidase-Mediated Superoxide Production In Renovascular Hypertension: Evidence For An Involvement Of Protein Kinase C. *Kidney Int.* 55:252-260, 1999

[120] Manucha W, Oliveros L, Carrizo L, Seltzer A, Valles P: Losartan Modulation On NOS Isoforms And COX-2 Expression In Early Renal Fibrogenesis In Unilateral Obstruction. *Kidney Int.* 65:2091-2107, 2004

[121] Welch WJ, Blau J, Xie H, Chabrashvili T, Wilcox CS: Angiotensin-Induced Defects In Renal Oxygenation: Role Of Oxidative Stress. *Am.J.Physiol Heart Circ.Physiol* 288:H22-H28, 2005

[122] Hannken T, Schroeder R, Zahner G, Stahl RA, Wolf G: Reactive Oxygen Species Stimulate P44/42 Mitogen-Activated Protein Kinase And Induce P27(Kip1): Role In Angiotensin II-Mediated Hypertrophy Of Proximal Tubular Cells. *J.Am.Soc.Nephrol.* 11:1387-1397, 2000

[123] Thannickal VJ, Hassoun PM, White AC, Fanburg BL: Enhanced Rate Of H2O2 Release From Bovine Pulmonary Artery Endothelial Cells Induced By TGF-Beta 1. *Am.J.Physiol* 265:L622-L626, 1993

[124] Denu JM, Tanner KG: Specific And Reversible Inactivation Of Protein Tyrosine Phosphatases By Hydrogen Peroxide: Evidence For A Sulfenic Acid Intermediate And Implications For Redox Regulation. *Biochemistry* 37:5633-5642, 1998

[125] Bondeva T, Roger T, Wolf G: Differential Regulation Of Toll-Like Receptor 4 Gene Expression In Renal Cells By Angiotensin II: Dependency On AP1 And PU.1 Transcriptional Sites. *Am.J.Nephrol.* 27:308-314, 2007

[126] Wolf G: Role Of Reactive Oxygen Species In Angiotensin II-Mediated Renal Growth, Differentiation, And Apoptosis. *Antioxid.Redox.Signal.* 7:1337-1345, 2005

[127] Dechend R, Viedt C, Muller DN, Ugele B, Brandes RP, Wallukat G, Park JK, Janke J, Barta P, Theuer J, Fiebeler A, Homuth V, Dietz R, Haller H, Kreuzer J, Luft FC: AT1 Receptor Agonistic Antibodies From Preeclamptic Patients Stimulate NADPH Oxidase. *Circulation* 107:1632-1639, 2003

[128] Minz M, Heer M, Arora S, Sharma A, Khullar M: Oxidative Status In Stable Renal Transplantation. *Transplant.Proc.* 38:2020-2021, 2006

[129] Fliser D, Wagner KK, Loos A, Tsikas D, Haller H: Chronic Angiotensin II Receptor Blockade Reduces (Intra)Renal Vascular Resistance In Patients With Type 2 Diabetes. *J.Am.Soc.Nephrol.* 16:1135-1140, 2005

[130] Aslam S, Santha T, Leone A, Wilcox C: Effects Of Amlodipine And Valsartan On Oxidative Stress And Plasma Methylarginines In End-Stage Renal Disease Patients On Hemodialysis. *Kidney Int.* 70:2109-2115, 2006

[131] Inagami T: Recent Progress In Molecular And Cell Biological Studies Of Angiotensin Receptors. *Curr.Opin.Nephrol.Hypertens.* 4:47-54, 1995

[132] Griendling KK, Alexander RW: The Angiotensin (AT1) Receptor. Semin.*Nephrol.* 13:558-566, 1993

[133] Carey RM, Wang ZQ, Siragy HM: Role Of The Angiotensin Type 2 Receptor In The Regulation Of Blood Pressure And Renal Function. *Hypertension* 35:155-163, 2000

[134] Gasc JM, Shanmugam S, Sibony M, Corvol P: Tissue-Specific Expression Of Type 1 Angiotensin II Receptor Subtypes. An In Situ Hybridization Study. *Hypertension* 24:531-537, 1994

[135] Burns KD: Angiotensin II And Its Receptors In The Diabetic Kidney. *Am.J.Kidney Dis.* 36:449-467, 2000

[136] Shanmugam S, Monnot C, Corvol P, Gasc JM: Distribution Of Type 1 Angiotensin II Receptor Subtype Messenger Rnas In The Rat Fetus. *Hypertension* 23:137-141, 1994

[137] Shanmugam S, Corvol P, Gasc JM: Angiotensin II Type 2 Receptor Mrna Expression In The Developing Cardiopulmonary System Of The Rat. *Hypertension* 28:91-97, 1996

[138] Wang ZQ, Moore AF, Ozono R, Siragy HM, Carey RM: Immunolocalization Of Subtype 2 Angiotensin II (AT2) Receptor Protein In Rat Heart. *Hypertension* 32:78-83, 1998

[139] Ozono R, Wang ZQ, Moore AF, Inagami T, Siragy HM, Carey RM: Expression Of The Subtype 2 Angiotensin (AT2) Receptor Protein In Rat Kidney. *Hypertension* 30:1238-1246, 1997

[140] Miyata N, Park F, Li XF, Cowley AW, Jr.: Distribution Of Angiotensin AT1 And AT2 Receptor Subtypes In The Rat Kidney. *Am.J.Physiol* 277:F437-F446, 1999

[141] Wolf G, Schroeder R, Stahl RA: Angiotensin II Induces Hypoxia-Inducible Factor-1 Alpha In PC 12 Cells Through A Posttranscriptional Mechanism: Role Of AT2 Receptors. *Am.J.Nephrol.* 24:415-421, 2004

[142] Endo Y, Arima S, Yaoita H, Tsunoda K, Omata K, Ito S: Vasodilation Mediated By Angiotensin II Type 2 Receptor Is Impaired In Afferent Arterioles Of Young Spontaneously Hypertensive Rats. *J.Vasc.Res.* 35:421-427, 1998

[143] Chen X, Wang J, Zhou F, Wang X, Feng Z: STAT Proteins Mediate Angiotensin II-Induced Production Of TIMP-1 In Human Proximal Tubular Epithelial Cells. *Kidney Int.* 64:459-467, 2003

[144] Wolf G, Wenzel U, Burns KD, Harris RC, Stahl RA, Thaiss F: Angiotensin II Activates Nuclear Transcription Factor-Kappab Through AT1 And AT2 Receptors. *Kidney Int.* 61:1986-1995, 2002

[145] Nahman NS, Jr., Rothe KL, Falkenhain ME, Frazer KM, Dacio LE, Madia JD, Leonhart KL, Kronenberger JC, Stauch DA: Angiotensin II Induction Of Fibronectin Biosynthesis In Cultured Human Mesangial Cells: Association With CREB Transcription Factor Activation. *J.Lab Clin.Med.* 127:599-611, 1996

[146] Feng Z, Wei C, Chen X, Wang J, Cheng H, Zhang X, Hong Q, Shi S, Fu B, Wei R: Essential Role Of Ca2+ Release Channels In Angiotensin II-Induced Ca2+ Oscillations And Mesangial Cell Contraction. *Kidney Int.* 70:130-138, 2006

[147] Venema RC, Venema VJ, Eaton DC, Marrero MB: Angiotensin II-Induced Tyrosine Phosphorylation Of Signal Transducers And Activators Of Transcription 1 Is Regulated By Janus-Activated Kinase 2 And Fyn Kinases And Mitogen-Activated Protein Kinase Phosphatase 1. *J.Biol.Chem.* 273:30795-30800, 1998

[148] Venema RC, Ju H, Venema VJ, Schieffer B, Harp JB, Ling BN, Eaton DC, Marrero MB: Angiotensin II-Induced Association Of Phospholipase Cgamma1 With The G-Protein-Coupled AT1 Receptor. *J.Biol.Chem.* 273:7703-7708, 1998

[149] Marrero MB, Schieffer B, Li B, Sun J, Harp JB, Ling BN: Role Of Janus Kinase/Signal Transducer And Activator Of Transcription And Mitogen-Activated Protein Kinase Cascades In Angiotensin II- And Platelet-Derived Growth Factor-Induced Vascular Smooth Muscle Cell Proliferation. *J.Biol.Chem.* 272:24684-24690, 1997

[150] Ruiz-Ortega M, Ruperez M, Lorenzo O, Esteban V, Blanco J, Mezzano S, Egido J: Angiotensin II Regulates The Synthesis Of Proinflammatory Cytokines And Chemokines In The Kidney. *Kidney Int.Suppl*12-22, 2002

[151] Esteban V, Ruperez M, Vita JR, Lopez ES, Mezzano S, Plaza JJ, Egido J, Ruiz-Ortega M: Effect Of Simultaneous Blockade Of AT1 And AT2 Receptors On The Nfkappab Pathway And Renal Inflammatory Response. *Kidney Int.Suppls*33-S38, 2003

[152] Ruiz-Ortega M, Esteban V, Suzuki Y, Ruperez M, Mezzano S, Ardiles L, Justo P, Ortiz A, Egido J: Renal Expression Of Angiotensin Type 2 (AT2) Receptors During Kidney Damage. *Kidney Int.Suppls*21-S26, 2003

[153] Marrero MB, Fulton D, Stepp D, Stern DM: Angiotensin II-Induced Insulin Resistance And Protein Tyrosine Phosphatases. Arterioscler.Thromb.Vasc.Biol. 24:2009-2013, 2004

[154] Schmitz U, Ishida T, Ishida M, Surapisitchat J, Hasham MI, Pelech S, Berk BC: Angiotensin II Stimulates P21-Activated Kinase In Vascular Smooth Muscle Cells: Role In Activation Of JNK. *Circ.Res.* 82:1272-1278, 1998

[155] Yin G, Yan C, Berk BC: Angiotensin II Signaling Pathways Mediated By Tyrosine Kinases. *Int.J.Biochem.Cell Biol.* 35:780-783, 2003

[156] Ushio-Fukai M, Hilenski L, Santanam N, Becker PL, Ma Y, Griendling KK, Alexander RW: Cholesterol Depletion Inhibits Epidermal Growth Factor Receptor Transactivation By Angiotensin II In Vascular Smooth Muscle Cells: Role Of Cholesterol-Rich Microdomains And Focal Adhesions In Angiotensin II Signaling. *J.Biol.Chem.* 276:48269-48275, 2001

[157] Ushio-Fukai M, Griendling KK, Becker PL, Hilenski L, Halleran S, Alexander RW: Epidermal Growth Factor Receptor Transactivation By Angiotensin II Requires Reactive Oxygen Species In Vascular Smooth Muscle Cells. Arterioscler.Thromb.*Vasc.Biol.* 21:489-495, 2001

[158] Eguchi S, Numaguchi K, Iwasaki H, Matsumoto T, Yamakawa T, Utsunomiya H, Motley ED, Kawakatsu H, Owada KM, Hirata Y, Marumo F, Inagami T: Calcium-Dependent Epidermal Growth Factor Receptor Transactivation Mediates The Angiotensin II-Induced Mitogen-Activated Protein Kinase Activation In Vascular Smooth Muscle Cells. *J.Biol.Chem.* 273:8890-8896, 1998

[159] Andreev J, Galisteo ML, Kranenburg O, Logan SK, Chiu ES, Okigaki M, Cary LA, Moolenaar WH, Schlessinger J: Src And Pyk2 Mediate G-Protein-Coupled Receptor Activation Of Epidermal Growth Factor Receptor (EGFR) But Are Not Required For Coupling To The Mitogen-Activated Protein (MAP) Kinase Signaling Cascade. *J.Biol.Chem.* 276:20130-20135, 2001

[160] Blobel CP: Adams: Key Components In EGFR Signalling And Development. *Nat.Rev.Mol.Cell Biol.* 6:32-43, 2005

[161] Mifune M, Ohtsu H, Suzuki H, Nakashima H, Brailoiu E, Dun NJ, Frank GD, Inagami T, Higashiyama S, Thomas WG, Eckhart AD, Dempsey PJ, Eguchi S: G Protein Coupling And Second Messenger Generation Are Indispensable For Metalloprotease-Dependent, Heparin-Binding Epidermal Growth Factor Shedding Through Angiotensin II Type-1 Receptor. *J.Biol.Chem.* 280:26592-26599, 2005

[162] Ohtsu H, Dempsey PJ, Frank GD, Brailoiu E, Higuchi S, Suzuki H, Nakashima H, Eguchi K, Eguchi S: ADAM17 Mediates Epidermal Growth Factor Receptor Transactivation And Vascular Smooth Muscle Cell Hypertrophy Induced By Angiotensin II. Arterioscler.*Thromb.Vasc.Biol.* 26:E133-E137, 2006

[163] Prenzel N, Zwick E, Daub H, Leserer M, Abraham R, Wallasch C, Ullrich A: EGF Receptor Transactivation By G-Protein-Coupled Receptors Requires Metalloproteinase Cleavage Of Prohb-EGF. *Nature* 402:884-888, 1999

[164] Kagiyama S, Eguchi S, Frank GD, Inagami T, Zhang YC, Phillips MI: Angiotensin II-Induced Cardiac Hypertrophy And Hypertension Are Attenuated By Epidermal Growth Factor Receptor Antisense. *Circulation* 106:909-912, 2002

[165] Zhuo JL, Li XC: Novel Roles Of Intracrine Angiotensin II And Signalling Mechanisms In Kidney Cells. *J.Renin.Angiotensin.Aldosterone.Syst.* 8:23-33, 2007

[166] Robertson AL, Jr., Khairallah PA: Angiotensin II: Rapid Localization In Nuclei Of Smooth And Cardiac Muscle. *Science* 172:1138-1139, 1971

[167] Re RN: On The Biological Actions Of Intracellular Angiotensin. *Hypertension* 35:1189-1190, 2000

[168] De Mello WC: Intracellular Angiotensin II Regulates The Inward Calcium Current In Cardiac Myocytes. *Hypertension* 32:976-982, 1998

[169] Li XC, Zhuo JL: In Vivo Regulation Of AT1a Receptor-Mediated Uptake Of [125I]-Val5-Angiotensin II In The Kidney And Adrenals Of AT1a Receptor-Deficient Mice. *Am.J.Physiol Renal Physiol* 2007 Nov 28; [Epub Ahead Of Print]

[170] Chen R, Mukhin YV, Garnovskaya MN, Thielen TE, Iijima Y, Huang C, Raymond JR, Ullian ME, Paul RV: A Functional Angiotensin II Receptor-GFP Fusion Protein: Evidence For Agonist-Dependent Nuclear Translocation. *Am.J.Physiol Renal Physiol* 279:F440-F448, 2000

[171] Mehta PK, Griendling KK: Angiotensin II Cell Signaling: Physiological And Pathological Effects In The Cardiovascular System. *Am.J.Physiol Cell Physiol* 292:C82-C97, 2007

[172] Zhuo JL, Li XC: Novel Roles Of Intracrine Angiotensin II And Signalling Mechanisms In Kidney Cells. *J.Renin.Angiotensin.Aldosterone.Syst.* 8:23-33, 2007

[173] Ardaillou R, Chansel D: Synthesis And Effects Of Active Fragments Of Angiotensin II. *Kidney Int.* 52:1458-1468, 1997

[174] Reaux-Le GA, Iturrioz X, Fassot C, Claperon C, Roques BP, Llorens-Cortes C: Role Of Angiotensin III In Hypertension. *Curr.Hypertens.Rep.* 7:128-134, 2005

[175] Reaux A, Fournie-Zaluski MC, Llorens-Cortes C: Angiotensin III: A Central Regulator Of Vasopressin Release And Blood Pressure. Trends *Endocrinol.Metab* 12:157-162, 2001

[176] Ruiz-Ortega M, Lorenzo O, Egido J: Angiotensin III Increases MCP-1 And Activates NF-Kappab And AP-1 In Cultured Mesangial And Mononuclear Cells. *Kidney Int.* 57:2285-2298, 2000

[177] Ruiz-Ortega M, Lorenzo O, Egido J: Angiotensin III Up-Regulates Genes Involved In Kidney Damage In Mesangial Cells And Renal Interstitial Fibroblasts. *Kidney Int.Suppl* 68:S41-S45, 1998

[178] Lorenzo O, Ruiz-Ortega M, Suzuki Y, Ruperez M, Esteban V, Sugaya T, Egido J: Angiotensin III Activates Nuclear Transcription Factor-Kappab In Cultured Mesangial Cells Mainly Via AT(2) Receptors: Studies With AT(1) Receptor-Knockout Mice. *J.Am.Soc.Nephrol.* 13:1162-1171, 2002

[179] Santos RA, Ferreira AJ: Angiotensin-(1-7) And The Renin-Angiotensin System. *Curr.Opin.Nephrol.Hypertens.* 16:122-128, 2007

[180] Trask AJ, Ferrario CM: Angiotensin-(1-7): Pharmacology And New Perspectives In Cardiovascular Treatments. Cardiovasc.*Drug Rev.* 25:162-174, 2007

[181] Baracho NC, Simoes-E-Silva AC, Khosla MC, Santos RA: Effect Of Selective Angiotensin Antagonists On The Antidiuresis Produced By Angiotensin-(1-7) In Water-Loaded Rats. *Braz.J.Med.Biol.Res.* 31:1221-1227, 1998

[182] Caruso-Neves C, Lara LS, Rangel LB, Grossi AL, Lopes AG: Angiotensin-(1-7) Modulates The Ouabain-Insensitive Na+-Atpase Activity From Basolateral Membrane Of The Proximal Tubule. *Biochim.Biophys.Acta* 1467:189-197, 2000

[183] Hamilton TA, Handa RK, Harding JW, Wright JW: A Role For The Angiotensin IV/AT4 System In Mediating Natriuresis *In The Rat. Peptides* 22:935-944, 2001

[184] Gesualdo L, Grandaliano G, Schena FP: Angiotensin IV And Renal Diseases. Contrib.*Nephrol.*63-71, 2001

[185] Dulin N, Madhun ZT, Chang CH, Berti-Mattera L, Dickens D, Douglas JG: Angiotensin IV Receptors And Signaling In Opossum Kidney Cells. *Am.J.Physiol* 269:F644-F652, 1995

[186] Handa RK, Krebs LT, Harding JW, Handa SE: Angiotensin IV AT4-Receptor System In The Rat Kidney. *Am.J.Physiol* 274:F290-F299, 1998

[187] Chai SY, Fernando R, Peck G, Ye SY, Mendelsohn FA, Jenkins TA, Albiston AL: The Angiotensin IV/AT4 Receptor. *Cell Mol.Life Sci.* 61:2728-2737, 2004

[188] Krishnan R, Hanesworth JM, Wright JW, Harding JW: Structure-Binding Studies Of The Adrenal AT4 Receptor: Analysis Of Position Two- And Three-Modified Angiotensin IV Analogs. *Peptides* 20:915-920, 1999

In: Angiotensin Research Progress
Editors: H. Miura and Y. Sasaki
ISBN: 978-1-60456-537-9

Chapter VIII

The Role of Vitamin D in Modulation of the Pancreatic Renin-Angiotensin System

***Po Sing Leung*[1] *and Barbara J. Boucher*[2]**
[1]Department of Physiology, Faculty of Medicine,
The Chinese University of Hong Kong, Hong Kong, China
[2]Queen Mary School of Medicine and Dentistry,
University of London, London, U.K.

Introduction

We have identified the existence of a local renin-angiotensin system (RAS) in the pancreas which is upregulated in animal models of type 2 diabetes mellitus (T2DM) (Lau et al., 2004; Chu et al., 2006). We found that inhibition of this local pancreatic RAS improves islet beta-cell function and glucose tolerance and reduces pro-inflammatory factor activity; these findings suggest a mechanism for the protective effects of RAS antagonism, including those for T2DM, in humans. Upregulation of the pancreatic islet RAS leads to impaired islet beta-cell function and glucose intolerance, and increases pro-inflammatory factor gene expression and protein production in experimental animal models of T2DM (see review by Leung and Gasparo, 2006; Leung, 2007a and 2007b).

Vitamin D has several known physiological functions beyond calcium homeostasis. Relevant to the present discussion, insulin secretion is reduced in vitamin D insufficiency (hypovitaminosis D) and can be restored by replacement of vitamin D both experimentally and in humans without established diabetes. Furthermore vitamin D supplementation has been reported to improve glycemia in human T2DM (see discussion below). Meanwhile vitamin D deficiency can amplify insulin resistance and thus contribute to beta-cell dysfunction. The association between hypovitaminosis D and glucose intolerance suggests that hypovitaminosis D may contribute to the pathogenesis of T2DM (see review by Boucher 1998; Mathieu et al., 2005).

Despite the clear association described above, the mechanistic explanation of the effects of vitamin D on beta cells is currently limited to enhancement of availability of the ionized calcium required by some insulin-releasing proteases. More than two decades ago, Resnick and Laragh (1985) found abnormal variations in circulating levels of activated vitamin D and parathyroid hormone in hypertensive patients, as defined by renin responses. On the other hand, vitamin D was recently shown to negatively regulate the RAS via suppression of renin gene activity, presumably by binding of activated vitamin D complexes with its receptor (VDR) to vitamin D receptor response elements (VDREs) in the promoter region of the renin gene (Li et al., 2002; Li et al., 2004). Both vitamin D supplementation and exposure to vitamin D synthesis-inducing wavelength ultraviolet B radiation can reduce blood pressure in humans. Some clinical studies suggest that RAS blockade can reduce the risk of developing T2DM in hypertensive patients. Whether this risk reduction varies with vitamin D repletion status is not known, but if so, it may explain the differences among reported studies.

In view of these findings, we have postulated that vitamin D modulates pancreatic islet RAS expression and activity in the same way that it modulates the general RAS, and that this modulation contributes to the effects of vitamin D on islet cell function. This hypothesis is being tested by *in vitro* and *in vivo* experiments investigating the effects of activated vitamin D on islet RAS activity, on RAS-induced inhibition of islet beta-cell function, and on circulating RAS activity. Consistent with our hypothesis, preliminary *in vitro* works have shown that activated vitamin D reduces the abnormally high RAS gene expression seen in mouse islets incubated in high glucose concentration conditions. A discussion of the mechanism(s) by which vitamin D-dependent RAS may modulate pancreatic islet function, and thus T2DM risk, follows.

T2DM accounts for more than 90% of diabetes cases world-wide. It imposes huge demands on healthcare resources in Western countries and is also becoming a burden in developing countries. Hypovitaminosis D is prevalent in Western countries due to lifestyle and sun avoidance for reduction of skin cancer. T2DM and hypovitaminosis D are now being reported in developing countries, especially in places where people work long hours indoors or women are veiled. Elucidation of the mechanisms involved in vitamin D modulation of islet cell function, local RASs and blood pressure, could contribute to the development of novel strategies for cost-effective prevention and management of diabetes. In this article, we provide a critical appraisal of progress in research examining the roles of vitamin D and pancreatic islet RAS in diabetes, with particular emphasis on putative roles of vitamin D in the regulation of pancreatic islet RAS, in relation to T2DM.

Vitamin D and Its Effects

Vitamin D has long been known to be involved in calcium homeostasis, by promoting calcium absorption and ensuring calcium availability for bone mineralization. More recently, many 'non-classical' effects of vitamin D have been identified in tissues that express the hydroxylase activating vitamin D and also the vitamin D receptor; indeed VDREs have been identified in the promoter regions of more than 200 genes. Such tissues include colon, pancreas, breast, heart, blood vessel, lung, brain, circulating mononuclear cells and the placenta, presumably to promote vitamin D availability to the fetus (see review by Norman, 2006). Thus, not surprisingly, vitamin D has multiple physiological effects, including (1) promotion of myocardial contractility, (2) promotion of cell differentiation, (3) suppression of cell proliferation (in normal and cancerous cells), (4) immune modulation with reductions in autoimmune responses, (5) increases in protective cellular responses to infection, such as secretion of cathelicidin which can destroy tuberculosis and other micro-organisms when circulating 25-(OH) vitamin D concentrations exceed 50 nmol/l (see review by Holick, 2007).

Progress in Research Examining Vitamin D in Diabetes

Although higher maternal and early life vitamin D intake have been associated with reduced incidence of type 1 DM (T1DM), a prospective randomized controlled trial (RCT) has yet to be carried out to confirm these findings. The low circulating concentrations of vitamin D in newly diagnosed T1DM patients may reflect increased risk of T1DM in hypovitaminosis D or effects of T1DM on vitamin D metabolism (Pozzilli et al., 2005; Hyppönen et al., 2001; Stene et al. 2000; EURODIAB, 1999; Schneider et al., 1977). On the other hand, inverse associations between vitamin D status and T2DM risk have been reported (Scragg et al., 1990 and 2004; Hintzpeter et al., 2007); similar associations have been made with all features of the metabolic syndrome, Syndrome 'X' (Boucher, 1998, Chiu et al., 2004; Hypponen et al., 2007). RCTs have suggested that higher intakes of vitamin D and calcium may prevent worsening of impaired glucose tolerance and of insulin resistance (Pittas et al., 2006 and 2007a). Furthermore, a recent meta-analysis suggested that a high intake of, or supplementation with, both calcium and vitamin D may reduce the risk of developing T2DM (Pittas et al., 2007b). Large scale RCTs of vitamin D supplementation with a wider range of doses in healthy people are needed to determine how effective vitamin D repletion may be in reducing the burden of T2DM in populations with a high prevalence of hypovitaminosis D. In this regard, a most recent report has shown that vitamin D is inversely associated with metabolic syndrome while this inverse relationship with IGF-1 was found only among those without hypovitaminosis D. These results suggest that metabolic syndrome prevalence is the lowest when both vitamin D and IGF-1 are high (Hypponen et al., 2007).

The recent identification of receptors for $1\alpha,25(OH)_2D_3$ (i.e. VDR) and of the 1α-OHase vitamin D activating enzyme in immune cells, vasculature and pancreatic islet beta cells has led to the recognition of vitamin D's many physiological roles beyond its direct role in calcium and bone homeostasis (Holick, 2003 and 2007; Mathieu et al., 1994; Hewison et al., 2007). Binding of $1\alpha,25(OH)_2D_3$ to VDRs in the cell nucleus or membrane caveolae generates relatively slow genomic responses or rapid non-genomic responses, respectively (Norman, 2006; Lips 2006). Insulin secretion by pancreatic islet beta cells, a non-genomic caveolae-associated VDR effect (Kajikawa et al., 1999), is reduced in hypovitaminosis D and restored, often to normal, by adequate replacement of vitamin D in isolated islets, in experimental animals and in non-diabetic human subjects (Mathieu et al., 2005; Bourlon et al., 1999; Boucher et al., 1995). Glucose-stimulated islet impairment of insulin secretion in both early and sustained vitamin D deficiency can be improved, or restored, by adequate dietary supplementation with vitamin D, both *in vitro* and *in vivo* (Kadowski and Norman, 1984; Cade and Norman, 1986).

In keeping with these findings, clinical studies have shown that an increased prevalence of T2DM is closely associated with vitamin D-deficiency in humans (Scragg 1981; Scragg et al., 1990and2004; Forman et al., 2007). In addition, hypovitaminosis D is associated with the defining abnormalities of the metabolic syndrome and with increased risks of both T2DM and cardiovascular disease (Boucher 1998; Chiu et al., 2004; Mathieu et al., 2005; Hypponen et al, 2007). There are additional supportive data showing an association between reduced exposure to ultraviolet light and increased risk of T2DM attributed to abnormal immune function (Dixon et al., 2007; Staples et al., 2003); the proposed pathways are normally involved in the stimulation of vitamin D synthesis and rapid non-genomic photoprotective effects in skin. Nevertheless, the mechanisms by which vitamin D contributes to maintenance of normal islet cell physiology remain obscure; proposed pathways include increasing intracellular calcium ion concentration, which is necessary for activation of some proteases that release insulin from proinsulin (Davidson et al., 1988; Sergeev and Rhoten, 1995). Furthermore, circulating 25-hydroxyvitamin D_3 can be hydroxylated to the active form, $1\alpha,25(OH)_2D_3$, in VDR-expressing islets. Thus, vitamin D repletion can exert a direct action on insulin secretory function and other protective effects through both genomic and non-genomic effects (Clark et al., 1987; Ishida and Norman, 1988; Billaudel et al., 1989; Lee et al., 1994; Boucher et al., 1995; Bland et al., 2004; Lips, 2006).

Clinical and epidemiological studies have suggested that hypovitaminosis D is associated with increases in blood pressure in adult humans and in obese juveniles (Scragg et al., 2007; Chiu et al., 2004; Smotkin-Tangorra et al., 2007). The ability of vitamin D to reduce renal RAS activity through suppression of renin secretion targeted to the islet beta cells (Li, 2003; Li et al., 2004) may account for the reductions in blood pressure reported with vitamin D supplementation in humans, especially those with hypovitaminosis D (Lind et al., 1988; Lind et al. 1989). Accordingly, RAS activity is increased several-fold in VDR-null mice (Li et al., 2002 and 2004; Kong and Li, 2003; Li, 2007). Converging data have prompted us to propose that vitamin D is a previously unrecognized negative endocrine mediator of local RAS activity, in the pancreas in this instance, as well as playing a critical role in the regulation of electrolytes, fluid and blood pressure homeostasis by suppression of circulating RAS activity that accounts for reductions in blood pressure with vitamin D supplementation.

Circulating and Local RAS

The circulating RAS plays a pivotal endocrine role in the regulation of circulatory and fluid homeostasis. This, the 'classical' RAS, consists of the liver-derived precursor angiotensinogen and two critical enzymes for the system, namely kidney renin and membrane-bound pulmonary angiotensin-converting enzyme (ACE). The sequential actions of these two enzymes generate plasma angiotensin I (1-10) and angiotensin II (1-8), respectively, the latter being a physiologically active RAS element. In addition, alternative enzymes to renin and ACE (such as chymase, cathepsin G, ACE2, endopeptidases and aminopeptidases) produce a number of bioactive peptides including angiotensin III (2-8), angiotensin IV (3-8) and angiotensin (1-7). Angiotensin II and these bioactive peptides mediate their specific functions via cellular transmembrane receptors in their target tissues and organs (Leung, 2004). Most, if not all, of the actions of angiotensin II are mediated by the two major receptors, angiotensin II receptor type 1 (AT_1 receptor) and type II (AT_2 receptor).

Local RASs, distinct from the circulating RAS, have been identified in diverse cells and tissues, such as adipose tissue, placenta, bone, brain, muscle, and more recently in the pancreas. Various autocrine, paracrine and intracrine roles, including regulation of cell proliferation, apoptosis, oxidative stress and both inflammatory and fibrotic responses have been attributed to local RASs (Lavoie and Sigmund, 2003; Paul, 2006). Newly discovered functions of locally generated angiotensin II in tissues and organs are of particular physiological interest; amongst them the putative novel roles for the RAS in the heart (Xiang et al., 2005) and in the pancreas (Leung, 2007a).

Existence of Functional Local RAS in Pancreas

It is now known that the pancreas possesses a local cellular RAS (in dog: Chappell et al., 1991and1992; in rat: Ghiani et al., 1995; Leung et al., 1997and1999; in mouse: Leung 1997and1998; and in human: Tahmasebi et al., 1999; Lam and Leung, 2002) that produces RAS products locally for regulating pancreatic exocrine and endocrine functions (Leung and Carlsson, 2001; Leung and Chappell, 2003). All major RAS components have been detected throughout the endocrine and exocrine portions of the pancreas. Indeed, angiotensinogen, angiotensin II, AT_1 and AT_2 receptors, and renin have each been localized to distinct pancreatic cell types, such as endothelial, ductal and acinar (Leung et al., 1997and1998), and islet cells (Tahmasebi et al., 1999, Lam and Leung, 2002). Consistent with these findings, RAS components have been demonstrated to be expressed in isolated islets from both the mouse and from humans (Lau et al., 2004; Lupi et al., 2006; Ramracheya et al., 2006), in isolated rat acinar cells (Tsang et al., 2004a) and rat pancreatic stellate cells (Ko et al., 2006), as well as in mouse MIN6 and rat pancreatic AR42J cells (Ramracheya et al., 2006; Chappell et al., 1995; Cheung et al., 1999). Of particular interest in this context is the regulation of the pancreatic RAS components in response to various physiological and pathophysiological conditions, including hypoxia, pancreatitis, pancreatic tumors, islet transplantation, hyperglycemia and T2DM (Leung, 2007a).

Recent advances in this research suggest that pancreatic microcirculation, acinar enzyme secretion, and ductal anion secretion are regulated by a local RAS, probably via AT_1 receptors (Carlsson et al., 1998; Tsang et al., 2004a; Fink et al., 2002). In the exocrine pancreas, AT_1 receptors and AT_2 receptors are predominantly localized to the pancreatic ducts, blood vessels and acinar cells (Leung et al., 1997; Leung et al., 1999). We recently found that experimental pancreatitis upregulates the expression of RAS components, especially the AT_1 receptor and ACE, in the pancreas (Leung et al., 2000; Ip et al., 2003a). Prophylactic administration of saralasin, a nonspecific AT_1 and AT_2 receptor antagonist, reduces the severity of acute pancreatitis (Tsang et al., 2003b), suggesting that upregulation of the pancreatic RAS may contribute to oxidative stress, which leads to pancreatic inflammation and injury (Ip et al., 2003b). Moreover, activation of the pancreatic RAS during the development of pancreatitis has been associated with pro-inflammatory AT_1 receptor-mediated NADPH oxidase activity (Tsang et al., 2004b). NADPH oxidase activity induced by pancreatitis can be alleviated by prophylactic or therapeutic treatment with the AT_1 receptor antagonist losartan, but not the AT_2 receptor antagonist PD123319 (Tsang et al., 2004b). Thus the protective role of AT_1 receptor antagonism in pancreatitis may be mediated by inhibition of NADPH oxidase-dependent generation of reactive oxygen species. Interestingly, further research has also shown that AT_1 receptor antagonism can ameliorate acute pancreatitis-induced pulmonary damage (Chan and Leung, 2006); this protective mechanism could also be involved in AT_1 receptor-mediated NADPH oxidase-dependent NFκB activation and thus it may ultimately promote pro-inflammatory responses in pancreatitis (Chan and Leung, 2007). Despite these observations, a functional islet RAS has yet to be confirmed in mammals in general and the potential roles of a local pancreatic RAS have not been fully established.

Novel Roles for a Pancreatic Islet RAS

Evidence suggests that a local islet RAS regulates pancreatic islet blood flow, oxygen tension, islet (pro)insulin biosynthesis and oxidative stress; these functions affect glucose-stimulated insulin secretory responses and other aspects of islet cell behavior (Leung, 2007a). Angiotensin II dose-dependently inhibits whole pancreatic blood flow, including islet blood flow, and also insulin secretion. Meanwhile selective RAS inhibition by AT_1 receptor antagonism or ACE inhibition enhances islet blood flow and glucose tolerance (Carlsson et al., 1998; Kampf et al., 2005; Huang et al., 2006and2007). Moreover, RAS blockade has been shown to improve function in pancreatic islet grafts (Kampf et al., 2005). On the other hand, local pancreatic islet RAS activation provides angiotensin II which inhibits glucose-stimulated insulin secretion (Lau et al., 2004).

Our laboratory demonstrated that exposing isolated islets to angiotensin II induces a dose-dependent inhibition of glucose-stimulated insulin release (figure 1A) and inhibits (pro)insulin biosynthesis (figure 1B). The inhibitory action of the highest dose of angiotensin II used (100 nmol/l) was blocked by pre-incubation of the pancreatic islets with the specific AT_1 receptor antagonist losartan. At 100 nmol/l, angiotensin II inhibited islet (pro)insulin biosynthesis under high (16.7 mmol/l), but not low (1.7 mmol/l), glucose conditions (figure 1C) whilst islet total (pro)insulin synthesis was not affected by angiotensin II under either condition (data not show).

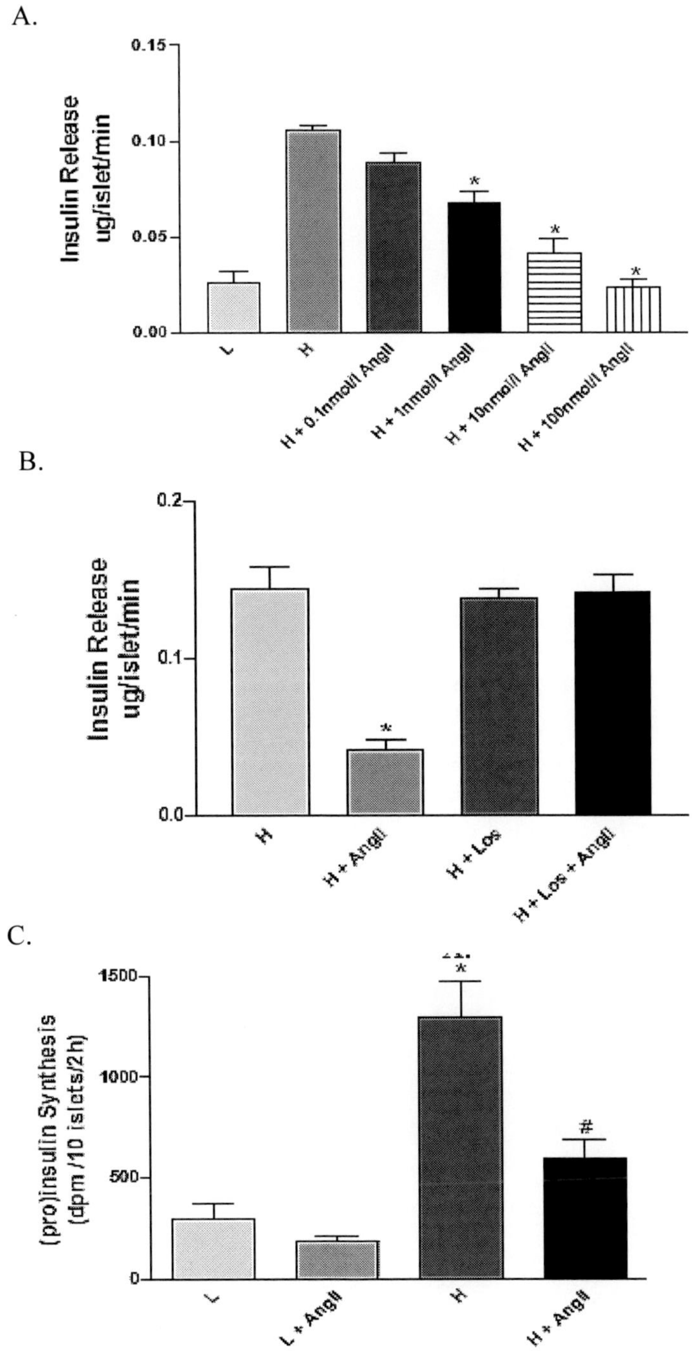

Figure 1. Effects of angiotensin II and losartan on insulin release and (pro)insulin biosynthesis in the presence of low (L; 1.7 mmol/l) or high (H; 16.7 mmol/l) glucose from isolated mouse islets. A. Dose-dependent effect angiotensin II on insulin release. B. Effect of losartan (Los, 1 μmol/l) and angiotensin II (100 nmol/l) on insulin release. C. Effect of angiotensin II (100 nmol/l) on (pro)insulin biosynthesis. All data are expressed as means ± SEM; n = 5-8 each group. *$p < 0.005$ when compared to islets exposed to 16.7 mmol/l glucose only (A andB); *$p < 0.05$ when compared with islets exposed to 1.7 mmol/l glucose whereas #$p < 0.05$ when compared to islets exposed to 16.7 mmol/l glucose only (C). Data are adapted from Lau et al., 2004.

Immunohistochemical observations indicate that the inhibitory effects of angiotensin II on pancreas function are mediated by AT_1 receptors located on the surface of islet beta cells. Key RAS components, including the AT_1 receptor, have been observed on isolated human pancreatic islets (Lupi et al., 2006). In contrast to the inhibitory effect of extended angiotensin II exposure (1 h) on glucose-stimulated insulin release (Lau et al., 2004), short-term exposure (2 min) of human islets to angiotensin II induced phase 1 insulin secretion through a rapid change in calcium depolarization, probably via a phospholipase C-sensitive mechanism and calcium influx through voltage-gated calcium channels (Ramracheya et al., 2006). These opposing observations can likely be attributed, at least in part, to the different durations of angiotensin II exposure which may affect either phase 1 or both phase 1 and 2 insulin secretions. A batch-type incubation of isolated mouse islets was employed for examining the longer-term (1 h) action of angiotensin II on glucose-stimulated secretion and it was observed that the inhibitory effect of angiotensin II using that experimental protocol produced a decrease in (pro)insulin biosynthesis (Lau et al., 2004).

Although isolated mouse islets express the AT_2 receptor, the failure of PD123319 treatment to attenuate angiotensin II-mediated effects on glucose-stimulated insulin secretion suggest that the AT_2 receptor is not critical for these effects (Lau *et al.* 2004). The AT_2 receptor is expressed along the outer region of islets, co-localizes with somatostatin-secreting cells in the endocrine pancreas and is expressed by the immortalized rat pancreatic cell lines RIN-m and RIN-14B (Wong *et al.* 2004). In RIN-14B cells, angiotensin II stimulates somatostatin secretion in a dose-dependent manner and this action seems to be AT_2 receptor-dependent (Wong *et al.* 2004). These data indicate that the pancreatic islet RAS has a functional role in the regulation of pancreatic islet cell insulin and somatostatin secretion and is, therefore, physiologically involved in glucose homeostasis.

Basic Studies of Islet RAS and T2DM: Recent Developments

Several animal models for studying islet RAS involvement in T2DM have been developed recently. Long-term (24 weeks) administration of the ACE inhibitor ramipril to diabetic Otsuka Long-Evans Tokushima fatty (OLETF) rats can prevent islet destruction by fibrosis, as evidenced by the expression profile of TGF-β and its downstream signaling molecules (Ko *et al.* 2004). Likewise, chronic treatment (10 weeks) with either an ACE inhibitor (perindopril) or AT_1 receptor blocker (irbesartan) attenuates islet fibrosis and reduces islet cell apoptosis and oxidative stress in Zucker diabetic fatty rats (Tikellis *et al.* 2004). Interestingly, hyperglycemia, comparable to that in T2DM, has also recently been shown to activate the RAS in pancreatic islet and stellate cells whilst RAS blockade ameliorates the angiotensin II-induced pancreatic inflammation and fibrosis that is exacerbated by chronic exposure to high glucose levels (Lupi *et al.* 2006, Ko *et al.* 2006). This convergence of findings suggests that islet RAS activation may be involved in oxidative stress-induced islet cell apoptosis and islet fibrosis, and thus that excessive RAS activation may lead to islet dysfunction.

Despite these persuasive data, the precise mechanism(s) by which the local pancreatic RAS modulates islet function has yet to be elucidated. The hypothesis that there is a change in AT_1 receptor expression in T2DM which enables endogenous levels of angiotensin II to impair islet function was recently tested in obesity-induced db/db mice (Chu *et al.* 2006). Though the pancreatic islet AT_1 receptors had no obvious effects on normal islet function, they were upregulated in the obesity-induced db/db T2DM mice; this upregulation has negative effects on glucose-stimulated islet cell insulin secretion and (pro)insulin biosynthesis and on islet blood flow (Chu *et al.* 2006). Figure 2 shows the effects of angiotensin II and its AT_1 receptor blocker (losartan) on glucose-stimulated insulin secretion and (pro)insulin biosynthesis from isolated islets of both control and diabetic mice. These findings provide a novel explanation for the reduced incidence of T2DM that has been observed in a number of clinical trials of AT_1 receptor blockade in individuals at high risk for the disease (Chu et al., 2006; Leung and De Gasparo, 2006). AT_1 receptor antagonism (for 8 weeks) has also been shown to attenuate NADPH oxidase-induced oxidative stress in isolated islets from db/db mice; this in turn results in a down-regulation of uncoupling protein 2 (UCP2) expression, which is associated with augmented beta-cell insulin secretion and reduced loss of beta cells to apoptosis (Chu and Leung, 2007). In keeping with these findings, chronic AT_1 receptor antagonism improves islet cell function and structure in OLETF rats and db/db mice, an effect that is apparently mediated by NADPH oxidase-induced oxidative stress (Nakayama *et al.* 2005; Shao *et al.* 2006). Taking all these data together, it is plausible to propose that AT_1 receptor activation in T2DM mediates increases in the oxidative stress following the resultant reductions in UCP2 activity which, in turn, leads to pancreatic islet beta-cell dysfunction. Figure 3 summarizes this proposed mechanism of AT_1 receptor-mediated islet cell function and structure, and insulin resistance.

A.

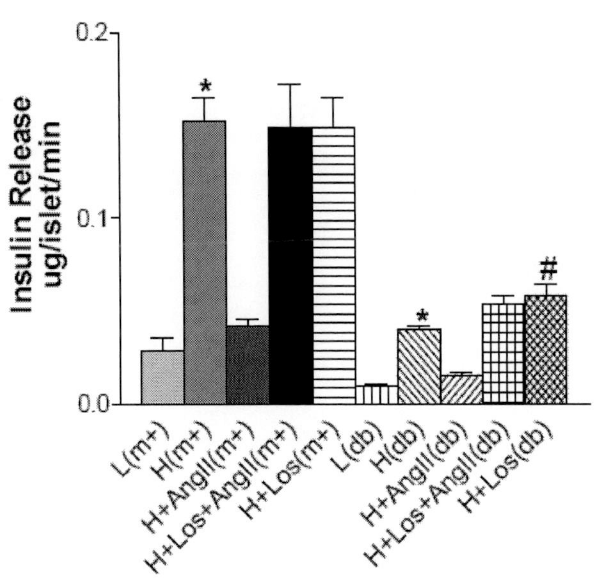

Figure 2. (Continued)

B.

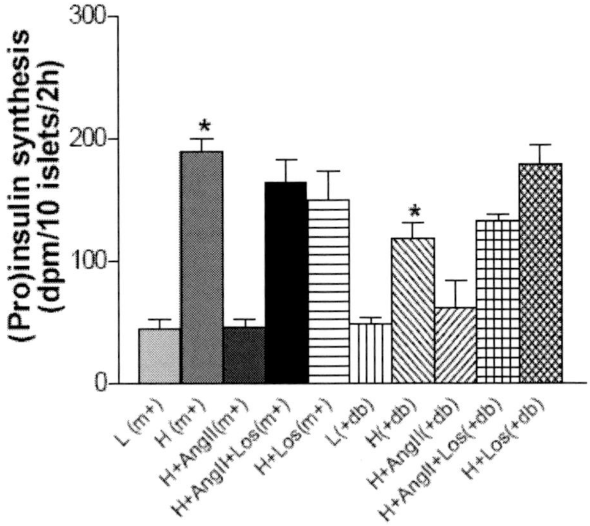

Figure 2. Effects of angiotensin II and losartan on insulin release and (pro)insulin biosynthesis in response to low (L; 1.7 mmol/l) or high (H; 16.7 mmol/l) glucose from isolated islets of control and db/db mice. All values are given as means ± S; n = 5-8 each group. *p < 0.001 vs. islets exposed to 1.7 mmol/l glucose only; †p < 0.05, *vs.* islets exposed to 16.7 mM glucose only. Data are adapted from Chu et al., 2006.

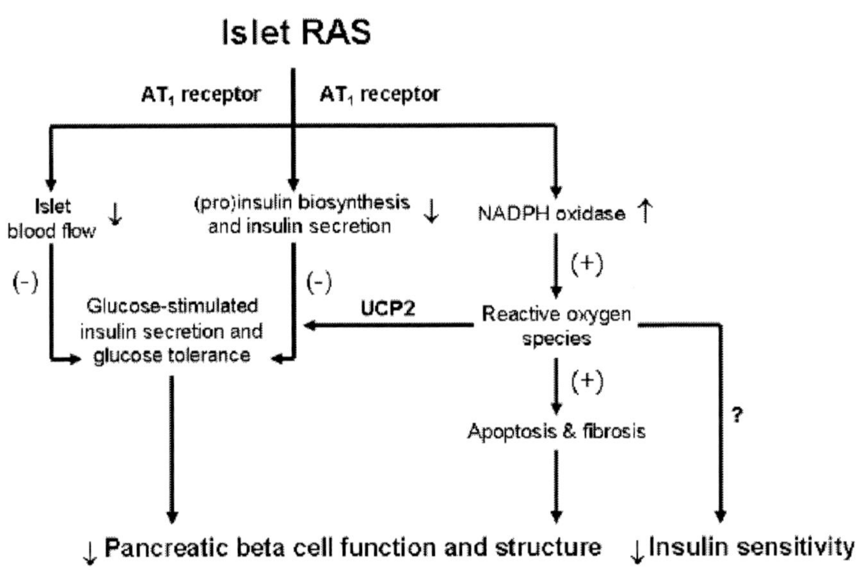

Figure 3. Proposed pathways for local pancreatic islet RAS in beta cell function and structure, and insulin resistance. According to this model, activation of AT_1 receptors leads to modulation of AT_1 receptor-mediated islet blood flow, (pro)insulin biosynthesis and NADPH oxidase-dependent oxidative stress, triggering apoptosis, fibrosis and thus beta-cell dysfunction and insulin resistance. (+) denotes stimulatory effect; (-) denotes inhibitory effect.

Advances in Clinical Studies of the RAS and T2DM

Complementary to the experimental work, the following large clinical trials have reported that RAS blockade can delay and/or prevent the onset of T2DM in "at risk" patients with hypertension and chronic heart disease (Scheen, 2004a and 2004b): Heart Outcomes Prevention Evaluation (HOPE, Yusurf et al., 2000), Losartan for Interventions for Endpoints in Hypertension (LIFE, Dahlof et al., 2002), Nateglimide And Valsartan in Impaired Glucose Tolerance Outcomes Research (NAVIGATOR, Califf, 2003), and Captopril Prevention Project (CAPP, Hansson et al., 1999). These findings were further supported by a meta-analysis of randomized clinical studies (Abuissa et al., 2005). Each of the twelve clinical studies included cardiovascular prognosis as a primary endpoint and the incidence of T2DM as a secondary endpoint, or as a post-hoc analysis. The mean duration of follow-up ranged from 1 to 6 years, and together these studies included 116,220 patients. This meta-analysis led to the conclusion that the mean risk for developing T2DM was reduced by 27% with ACE inhibitor treatments, by 23% with ARBs treatments, and by 25% overall in a pooled analysis of the data for these RAS blocker types.

Although the DREAM trial did not reveal a protective effect of ramipril (an ACE inhibitor) treatment for 3 years against T2DM, its findings did show enhanced regression of impaired glucose tolerance back to normoglycemia (DREAM Trial Investigators, 2006). This dissimilar finding may reflect its short duration and restriction solely to hypertensive subjects. Furthermore, a recent meta-analysis of hypertension treatment trials (Elliott and Meyer, 2007a and 2007b) suggests that angiotensin receptor blockers, but not ACE inhibitors, can reduce the risk of T2DM, a finding that persists after adjustment for random effects (Lam and Owen, 2007). In light of the clinical and basic research findings discussed above, we postulate that the protective effects for glycemia induced by RAS blockade involve the local pancreatic-islet RAS; the postulated protective mechanisms include improved pancreatic islet cell function and structure, mediated through enhancement of islet blood flow and (pro)insulin biosynthesis with concomitant reductions in ROS-induced apoptosis and fibrosis and also reduced peripheral tissue insulin sensitivity mediated through an increase in blood flow of skeletal muscle (Leung, 2007b; Leung and De Gasparo, 2006; Henriksen, 2007).

Potential Interactions of Vitamin D, Islet RAS Axis and T2DM Risk

The precise mechanism(s) by which hypovitaminosis D-dependent islet RAS activation may reduce islet beta-cell function have yet to be resolved. As a first step in testing our hypothesis, we are examining dose-dependent effects of activated vitamin D on changes in RAS expression in isolated islets incubated with high and normal glucose concentrations. Pancreatic islets were harvested from 8-9 week-old C57BL/6 mice by intra-ductal collagenase digestion and Histopaque gradient separation. The islets were then incubated with RPMI medium containing either 28 mM or 5.6 mM glucose for 48 h in the presence of different concentrations of calcitriol. AT_1 receptor and ACE mRNA expression were examined by real-time reverse transcriptase polymerase chain reaction (RT-PCR) after extraction of total RNA from islets using the Qiagen RNeasy mini kit (Qiagen Inc., USA). First strand cDNA was generated from 2 μg RNA using the oligo (dT) primer and used for real-time PCR by SYBR Green PCR. The β-actin gene was co-amplified as a housekeeping control gene for sample size standardization. Real-time PCR was performed using an iQ5 Multicolor Real-Time PCR Detection System (Bio-Rad Laboratories, Inc., Hercules, CA, USA). The data were calculated using the comparative C_T method (Lau et al., 2004) and are expressed as relative expression, normalized as percentage of β-actin.

Preliminary results showed that expression of both the AT_1 receptor and ACE genes were upregulated under high glucose conditions (28 mM) relative to low glucose conditions (5.6 mM), and that co-exposure to the activated adduct of vitamin D (calcitriol) may have beneficial effects on islet function by down-regulating expression of RAS components in the hyperglycemic state. These preliminary data suggest that 10^{-9} M calcitriol suppresses transcription of RAS component genes in the 5.6 mM condition more effectively than either 10^{-8} M or 10^{-13} M calciferol (figure 4, inverted-U dose response curve). The loss of efficacy at higher concentrations may be due to hyper-physiological concentrations of the active form of vitamin D inhibiting cell proliferation and inducing apoptosis as has been described in certain cancers and in human pancreatic islets (Chatterjee, 2001; Riachy et al., 2006; Trump et al., 2007). In addition, concerns have been raised about possible adverse effects of local activation in cardiovascular tissues (Richart et al., 2007); thus we may find that there are adverse effects of exposure of islets to excessive, as well as abnormally low, concentrations of calcitriol.

There are likely to be effects of ACE polymorphisms on the interactions between the RAS, vitamin D axes and blood pressure (Kulah et al., 2007), given that ACE inhibitors appear to reduce circulating calcitriol concentrations in the presence of certain ACE polymorphisms (Perez-Castrillon et al., 2006). In addition, the effects of calcitriol may interact with those of hypercalcemia (Ryan et al., 2007) as hypercalcemia may itself influence the effect of calcitriol on renin (Hall and Brands, 1992) and calcitriol increases free calcium availability in cells (Shan et al., 1993). The most effective concentration of calcitriol on islet RAS activity appears to depend on glucose level. This relationship may result from glucose conditions altering calcitriol induced effects in the cell, possibly by reducing VDR availability (Inaba et al., 1995). Additional mechanisms may also be involved; the observation of circulating 25(OH)D falls with induction of experimental hyperglycemia suggests that additional as yet undiscovered mechanisms may also be involved (Schneider et al., 1977).

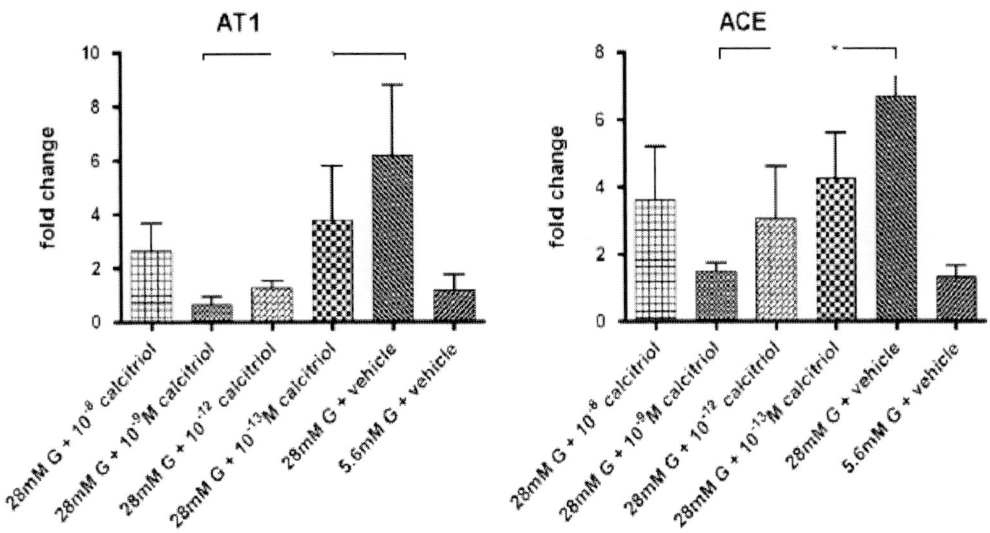

Figure 4. Real-time PCR analysis of AT_1 receptor and ACE transcription in isolated islets following treatment with different concentrations of calcitriol under high (28 mM) or low (5.6 mM) glucose [G] conditions for 4 d. n = 4 for each group. (***$p < 0.001$).

Summary

The existence of an islet RAS in the pancreas has recently been identified and found to be upregulated in T2DM. Furthermore, its inhibition significantly improves islet beta-cell function and glucose tolerance and reduces pro-inflammatory factor activity. Such findings offer a mechanistic explanation for the protective effect of RAS antagonism against T2DM in humans. Meanwhile, insulin secretion is reduced in hypovitaminosis D and can, in non-diabetic subjects, be restored by replacement of vitamin D. This effect could be mediated through regulation of the pancreatic RAS since vitamin D has recently been shown to be a negative regulator of the renin gene in the kidney, especially in hypertensive patients. Thus, we have proposed that hypovitaminosis D drives inappropriate activation of the local pancreatic islet RAS which ultimately contributes to pancreatic islet beta cell dysfunction. If so, insulin secretion may be modulated by the availability of vitamin D acting as a regulator for a local activation system in pancreatic islets via genomic and rapid non-genomic effects on insulin secretion in humans. Since blood pressure can also be reduced through this mechanism, the findings arising from this study suggest that formal assessments of the postulated benefits of adequate vitamin D supplementation on susceptibility to hypertension, diabetes and cardiometabolic syndrome should be pursued. There is currently a high worldwide prevalence of hypovitaminosis D as well as T2DM and its complications, and T2DM incidence is grossly increased in the presence of hypertension. At the same time there is evidence that improved vitamin D status is associated with reductions in subsequent all-cause mortality and that vitamin D supplementation is associated with small but meaningful reductions in all-cause mortality in prospective studies.

In conclusion, the evidence presented identifies mechanisms that provide further support for the avoidance of hypovitaminosis D as a safe and cost-effective public health measure suitable for use at the population level.

References

Abuissa H, Jones PG, Marso, SP and O'Keefe JH. Angiotensin-converting enzyme inhibitors or angiotensin receptor blockers for prevention of type 2 diabetes: a meta-analysis of randomized clinical trials. *J. Am. Coll. Cardiol.* 46:821-826, 2005.

Billaudel B, Faure A, Labriji-Mestaghanmi H and Sutter BC. Direct in vitro effect of 1,25-dihydroxyvitamin D3 on islets insulin secretion in vitamin deficient rats: influence of vitamin D3 pretreatment. *Diabetes Metab.*15:85-87, 1989.

Bland R, Maskovic D, Hills CE, Hughes SV, Chan SL, Squires PE and Hewison M. Expression of 25-hydroxyvitamin D_3-1-a-hydroxylase in pancreatic islets. *J. Steroid Biochem. Mol. Biol.* 89-90:121-125, 2004.

Boucher BJ, Mannan N, Noonan K, Hales CN and Evans SJW. Glucose intolerance and impairment of insulin secretion in relation to vitamin D deficiency in east London Asians. *Diabetologia* 38:1239-1245, 1995

Boucher BJ. Inadequate vitamin D status: does it contribute to the disorders comprising syndrome 'X'? *Br. J. Nutr.*79:315-327, 1998.

Bourlon PM, Faure-Dussert A and Billaudel B. The *de novo* synthesis of numerous proteins is decreased during vitamin D_3 deficiency and is gradually restored by 1,25-dihydroxyvitamin D_3 repletion in the islets of Langerhans of rats. *J. Endocrinol.* 162:101-109, 1999.

Cade C and Norman AW. Vitamin D3 improves impaired glucose tolerance and insulin secretion in the vitamin D-deficient rat in vivo. *Endocrinology* 119:84-90, 1986.

Califf RM. Insulin resistance: a global epidemic in need of effective therapies. *Eur Heart J.* 5(Supplement C):13-18, 2003.

Carlsson PO, Berne C, Jansson L. Angiotensin II and the endocrine pancreas: effects on islet blood flow and insulin secretion in rats. *Diabetologia* 41:127-33, 1998.

Chan YC and Leung PS. AT_1 receptor antagonism ameliorates acute pancreatitis-associated pulmonary injury. *Regul. Pept.* 134:46-53, 2006.

Chan YC and Leung PS. AT_1 receptor-dependent NFκB activation-mediated proinflammatory actions in a rat model of obstructive acute pancreatitis. *J. Pharmacol. Exp. Ther.*323:8-18, 2007.

Chatterjee M. Vitamin D and genomic stability. *Mutat Res.* 475:69-87. 2001

Chappell MC, Milsted A, Diz DI, Brosnihan KB, Ferrario CM. Evidence for an intrinsic angiotensin system in the canine pancreas. *J Hypertens* 9: 751-759, 1991.

Chappell MC, Diz DI, Jacobsen DW. Pharmacological characterization of angiotensin II binding sites in the canine pancreas. *Peptides* 13: 311-318, 1992.

Chappell MC, Jacobsen DW and Tallant EA. Characterization of angiotensin II receptor subtypes in pancreatic acinar AR42J cells. *Peptides* 16: 741-747, 1995.

Cheung WT, Yeung SY, Yiu AKL, Ip TM, Wan DCC, Luk SKS and Ho WKK. Characterization of a functional AT_{1A} angiotensin receptor in pancreatoma AR4-2J cells. *Peptides* 20: 829-836, 1999.

Chiu KC, Chu A, Go VL and Saad MF. Hypovitaminosis D is associated with insulin resistance and beta cell dysfunction. *Am. J. Clin. Nutr.* 79:820-825, 2004.

Chu KY, Lau T, Carlsson PO and Leung PS. Angiotensin II type 1 receptor blockade improves beta-cell function and glucose tolerance in a mouse model of type 2 diabetes. *Diabetes* 55:367-374, 2006.

Chu KY and Leung PS. Angiotensin II type 1 receptor antagonism mediates uncoupling protein 2-driven oxidative stress and ameliorates pancreatic islet β-cell function in young type 2 diabetic mice. *Antioxid. Redox Signal.* 9:869-878, 2007.

Clark SA, Stumpf WE, Sar M and DeLucca HF. 1,25-dihydroxyvitamin D3 target cells in immature pancreatic islets. *Am. J. Physiol.* 253:E99-E105, 1987.

Dahlöf B, Devereux RB, Kjeldsen SE, et al. Cardiovascular morbidity and mortality in the Losartan Intervention for Endpoints reduction in hypertension study (LIFE): a randomized trial against atenolol. *Lancet* 359:995-1003, 2002.

Davidson HW, Rhodes CJ and Hutton JC. Intra-organellar calcium and pH control proinsulin cleavage in the pancreatic beta cell via two distinct site-specific endopeptidases. *Nature* 333:93-96, 1988.

Dixon KM, deo SS, Norman AW, Bishop JE, Halliday GM, Reeve VE, Mason RS. In vivo relevance for photoprotection by teh vitamin D rapid response pathway. *J Steroid Biochem Mol Biol.* 103:451-456. 2007.

DREAM Trial Investigators. Effect of ramipril on the incidence of diabetes. *N. Engl. J. Med.* 355:1551-1562, 2006.

Elliott WJ and Meyer PM. Incident diabetes in clinical trials of antihypertensive drugs: a network meta-analysis. *Lancet* 369:201-207, 2007a (Correction ibid: 1518).

Elliott WJ and Meyer PM. Incident diabetes in clinical trials of antihypertensive drugs: a network meta-analysis. Correction. *Lancet* 369:1518, 2007b.

EURODIAB Substudy 2 Study Group: Vitamin D supplements in early childhood and risk for Type I (insulin-dependent) diabetes mellitus. The EURODIAB Substudy 2 Study Group. *Diabetologia.* 42:51-54, 1999.

Fink AS, Wang Y, Mendez T, Worrell RT, Eaton D, Nguyen TD, Lee SP. Angiotensin II evokes calcium-mediated signaling events in isolated dog pancreatic epithelial cells. *Pancreas* 25:290-295, 2002.

Forman JP, Giovannucci E, Holmes MD, Bishcoff-Ferrari HA, Tworoger SS, Willett WC and Curhan GC. Plasma 25-hydroxyvitamin D levels and risk of incident diabetes. *Hypertension* 49:1063-1069, 2007.

Ghiani BU, Masini MA. Angiotensin II binding sites in the rat pancreas and their modulation after sodium loading and depletion. *Comp. Biochem. Physiol.* 111A: 439-444, 1995.

Hall JE, Brands MW: The renin-angiotensin-aldosterone systems; in Seldin DW (Ed): The Kidney: *Physiology and Pathophysiology*, Ed 2. New York, Raven Press, pp1455-1503, 1992.

Hansson L, Lindholm DH, Niskanen L, et al. Effects of angiotensin-converting-enzyme inhibition compared with conventional therapy on cardiovascular morbidity and mortality in hypertension: the Captopril Prevention Project (CAPP) randomized trial. *Lancet* 353:611-616, 1999.

Henriksen EJ. Improvement of insulin sensitivity by antagonism of the renin-angiotensin system. *Am. J. Physiol.* 293:R974-R980, 2007.

Hewison M, Burke F, Evans KN, Lammas DA, Sansom DM, Liu P, Modlin RL and Adams JS. Extra-renal 25-hydroxyvitamin D3-1alpha-hydroxylase in human health and disease. *J. Steroid Biochem. Mol. Biol.* 103:316-321, 2007.

Hintzpeter B, Mensink GBM, Thierfelder W, Muller MJ, Scheldt-Nave C. Vitamin D status and health correlates among German adults. *Eur. J. Clin. Nutr.* (Epub ahead of print, May 30, 2007).

Holick MF. Vitamin D: a millennium perspective. *J. Cell Biochem.* 88:296-307, 2003.

Holick MF. Vitamin D deficiency. *New Eng J Med.* 357:266-281. 2007.

Hyppönen E, Läärä E, Reunanen A, Järvelin MR, Virtanen SM.. Intake of vitamin D and risk of type 1 diabetes: a birth-cohort study. *Lancet.* 358:1500-1503. 2001.

Hyppönen E, Boucher BJ, Berry DJ, Power C. 25-hydroxyvitamin D, insulin-like growth factor 1 and metabolic syndrome at age 45: cross-sectional study in the 1958 British birth cohort. *Diabetes* 57:298-305.

Huang Z, Jansson L, Sjoholm A. Pancreatic islet blood flow is selectively enhanced by captopril, irbesartan and pravastatin, and suppressed by palmitate. *Biochem. Biophys. Res. Commun.* 346:26-31, 2006.

Huang Z, Jansson L, Sjoholm A. Vasoactive drugs enhance pancreatic islet blood flow , augment insulin secretion and improve glucose tolerance in female rats. *Clin. Sci.* 112:69-76, 2007.

Inaba M, Terada M, Koyama H, Yoshida O, Ishimura E, Kawagishi T, Okuno Y, Nishizawa Y, Otani S, Morii H. Influence of high glucose on 1,25-dihydroxyvitamin D3-induced effect on human osteoblast-like MG-63 cells. *Bone Miner Res.* 10:1050-1056, 1995.

Ip SP, Kwan PC, Williams CH, Pang S, Hooper NM and Leung PS. Changes of angiotensin-converting enzyme activity in the pancreas of chronic hypoxia and acute pancreatitis. *Int. J. Biochem. Cell Biol.* 35:944-954, 2003a.

Ip SP, Tsang SW, Wong TP, Che CT, Leung PS. Saralasin, a nonspecific angiotensin II receptor antagonist, attenuates oxidative stress and tissue injury in cerulein-induced acute pancreatitis. *Pancreas* 26:224-229, 2003b.

Ishida H and Norman AW. Demonstration of high affinity receptors for 1,25-dihydroxyvitamin D3 in rat pancreas. *Mol. Cell. Endocrinol.* 60:109-117, 1988.

Kadowski S and Norman AW. Dietary vitamin D is essential for normal insulin secretion from the perfused rat pancreas. *J. Clin. Invest.* 73:759-766, 1984.

Kajikawa M, Ishida H, Fujimoto S, Mukai E, Nishimura M, Fujita J, Tsuura Y, Okamoto Y, Norman AW and Seino Y. An insulinotropic effect of vitamin D analog with increasing intracellular Ca^{2+} concentration in pancreatic β-cells through nongenomic signal transduction. *Endocrinology* 140:4760-4712, 1999.

Kampf C, Lau T, Olsson R, Leung PS and Carlsson PO. Angiotensin II type 1 receptor inhibition markedly improves the blood perfusion, oxygen tension and first phase of glucose-stimulated insulin secretion in revascularised syngeneic mouse islet grafts. *Diabetologia* 48:1159-1167, 2005.

Ko SH, Kwon HS, Kim SR, Moon SD, Ahn YB, Song KH, Son HS, Cha BY, Lee KW, Son HY, Kang SK, Park CG, Lee IK and Yoon KH. Ramipril treatment suppresses islet fibrosis in Otsuka Long-Evans Tokushima fatty rats. *Biochem Biophys Res Commum* 316: 114-122, 2004.

Ko SH, Hong OK, Kim JW, Ahn YB, Song KH, Cha BY, Son HY, Kim MJ, Jeong IK and Yoon KH. High glucose increases extracellular matrix production in pancreatic stellate cells by activating the renin-angiotensin system. *J. Cell. Biochem.* 98:343-55, 2006.

Kong J and Li YC. Effect of angiotensin II type 1 receptor antagonist and angiotensin-converting inhibitor on vitamin D receptor-null mice. *Am. J. Physiol.* 285:R255-R261, 2003.

Kulah E, Dursun A, Aktunc E, Acikgoz S, Aydin M, Can M, Dursun A. Effects of angiotensin-converting enzyme gene polymorphisms and serum vitamin D levels on ambulatory blood pressure measurement and left ventricular mass in Turkish hypertensive population. *Blood Press Monit.* 12:207-213. 2007

Lam KY and Leung PS. Regulation and expression of renin-angiotensin system in human pancreas and pancreatic endocrine tumours. *Eur J Endocrinol* 146: 567-572, 2002.

Lam SK and Owen A. Incident diabetes in clinical trials of antihypertensive drugs. *Lancet* 369:1513-1514, 2007.

Lau T, Carlsson PO and Leung PS. Evidence for a local angiotensin-generating system and dose-dependent inhibition of glucose-stimulated insulin release by angiotensin II in isolated pancreatic islets. *Diabetologia* 47: 240-248, 2004.

Lavoie JL and Sigmund CD. Minireview: Overview of the renin-angiotensin system: an endocrine and paracrine system. *Endocrinology* 144:2179-2183, 2003.

Lee S, Clark SA, Gill RK and Christakos S. 1,25-dihydroxyvitamin D3 and pancreatic β-cell function: vitamin D receptors, gene expression, and insulin secretion. *Endocrinology* 134:1602-1610, 1994.

Leung PS, Chan HC, Fu LXM, Wong PYD. Localization of AT1 and AT2 receptors in the pancreas of rodents. *J. Endocrinol.* 153:269-274, 1997.

Leung PS, Chan HC, Wong PYD. Immunohistochemical localization of angiotensin II in the mouse pancreas. *Histochemical J* 30: 21-25, 1998.

Leung PS, Chan WP, Wong TP, Sernia C. Expression and localization of the renin angiotensin system in the rat pancreas *J. Endocrinol.* 160:13-19, 1999.

Leung PS, Chan WP, Nobiling R. Regulated expression of pancreatic renin-angiotensin system in pancreatitis. *Mol. Cell. Endocrinol.* 166:121-128, 2000.

Leung PS and Carlsson PO. Tissue renin-angiotensin system: its expression, localization, regulation and potential role in the pancreas. *J. Mol. Endocrinol.*26:155-164, 2001.

Leung PS and Chappell MC. A local pancreatic renin-angiotensin system: endocrine and exocrine roles. *Int. J. Biochem. Cell. Biol.* 35:838-846, 2003.

Leung PS. The peptide hormone angiotensin II: its new functions in tissues and organs. *Curr. Protein Pept. Sci.* 5:267-273, 2004.

Leung PS and De Gasparo M. Involvement of the pancreatic renin-angiotensin system in insulin resistance and the metabolic syndrome. *J. Cardiometabol. Syndrome* 1:197-203, 2006.

Leung PS. The physiology of a local renin-angiotensin system in the pancreas. *J. Physiol.* 580: 31-37, 2007a.

Leung PS. Mechanisms of protective effects induced by blockade of the renin-angiotensin system: novel role of the pancreatic islet angiotensin-generating system in type 2 diabetes. *Diabet. Med.* 24:110-116, 2007b.

Li YC. Vitamin D regulation of the renin-angiotensin system. *J Cell Biochem.* 88:327-331. 2003

Li YC, Inhibition of renin: an updated review of the development of renin inhibitors. *Curr Opin Investig drugs.* 8:750-757. 2007

Li YC, Kong J, Wei M, Chen ZF, Liu SQ and Cao LP. 1,25-dihydroxyvitamin D3 is a negative endocrine regulator of the renin angiotensin system. *J. Clin. Invest.* 110:229-238, 2002.

Li YC, Qiao G, Uskokovic M, Xiang W, Zheng W and Kong J. Vitamin D: a negative endocrine regulator of the renin-angiotensin system and blood pressure. *J. Steroid Biochem. Mol. Biol.* 89-90:387-392, 2004.

Lind L, Lithell H, Skafors E, Wide L and Ljunghall S. Reduction in blood pressure by treatment with alphacidol: A double-blind, placebo-controlled study in subjects with impaired glucose tolerance. *Acta Med. Scand.* 223:211-217, 1988.

Lind L, Wengle B, Wide L and Ljunghall S. Reduction of blood pressure during long-term treatment with active vitamin D (alphacalcidol) is dependent on plasma renin activity and calcium status. A double-blind, placebo-controlled study. *Am. J. Hypertens.* 2:20-25, 1989.

Lips P. Vitamin D physiology. *Prog. Biophys. Mol. Biol.* 92:4-8, 2006.

Lupi R, Guerra SD, Bugliani M, Boggi U, Mosca F, Torri S, Prato SD, Marchetti P. The direct effects of the angiotensin-converting enzyme inhibitors, zofenoprilat and enalaprilat, on isolated human pancreatic islets. *Eur. J. Endocrinol.*154:355-61, 2006.

Mathieu C, Waer M, Laureys J, Rutgeerts O and Bouillon R. Prevention of autoimmune diabetes in NOD mice by 1,25-dihydroxyvitamin D3. *Diabetologia* 37:552-558, 1994.

Mathieu C, Gysemans C, Giulietti A and Bouillon R. Vitamin D and diabetes. *Diabetologia* 48:1247-1257, 2005.

Nakayama M, Inoguchi T, Sonta T, Maeda Y, Sasaki S, Sawada F, Tsubouchi H, Sonoda N, Kobayashi K, Sumimoto H and Nawata H. Increased expression of NAD(P)H oxidase in islets of animal models of Type 2 diabetes and its improvement by an AT1 receptor antagonist. *Biochem Biophys Res Commun* 332: 927-933, 2005.

Norman AW. Vitamin D receptor: new assignments for an already busy receptor. *Endocrinology* 147:5542-5548, 2006.

Paul M, Mehr AP and Kreutz R. Physiology of local renin-angiotensin systems. *Physiol Rev* 86: 747-803, 2006.

Perez-Castrillon JL, Justo I, Sanz A, De Luis D, Duenas A. Effect of angiotensin converting enzyme inhibitors on 1,25-(OH)2D levels of hypertensive patients. Relationships with ACE polymorphisms. Horm Metab Res. 38;812-816, 2006.

Pittas AG, Harris SS, Stark PC, Dawson-Hughes B. The effects of calcium and vitamin D supplementation on blood glucose and markers of inflammation in nondiabetic adults. *Diabetes Care*. 30:980-986. 2007a

Pittas AG, Lau J, Hu FB, Dawson-Hughes B. The role of vitamin D and calcium in type 2 diabetes. A systematic review and meta-analysis. *J Clin Endocrinol Metab.* 92:2017-2029. 2007b.

Pittas AG, Dawson-Hughes B, Li T, Van Dam RM, Willett WC, Manson JE, Hu FB. Vitamin D and calcium intake in relation to type 2 diabetes in women. *Diabetes Care.* 29:650-656. 2006

Pozzilli P, Manfrini S, Crinò A, Picardi A, Leomanni C, Cherubini V, Valente L, Khazrai M, Visalli N; IMDIAB group. Low levels of 25-hydroxyvitamin D3 and 1,25-dihydroxyvitamin D3 in patients with newly diagnosed type 1 diabetes. *Horm Metab Res.* 37:680-683, 2005.

Ramracheya RD, Muller DS, Wu Y, Whitehouse BJ, Huang GC, Amiel SA, Karalliedde J, Viberti G, Jones PM, Persaud SJ. Direct regulation of insulin secretion by angiotensin II in human islet of Langerhans. *Diabetologia* 49:321-331, 2006.

Resnick LM, Laragh JH. Renin, calcium metabolism and the pathophysiologic basis of antihypertensive therapy. *Am J Cardiol* 56:68H-74H, 1985.

Riachy R, Vanderwalle B, Moerman E, Belaich S, Luklowiak B, Gmyr V, Muharram G, Kerr Conte J, Pattou F. 1,25-Dihydroxyvitamin D3 protects human pancreatic islets against cytokine-induced apoptosis via down-regulation of the Fas receptor. *Apoptosis* 11:151-159, 2006

Richart T, Li Y, Staessen JA. Renal versus Extrarenal activation of vitamin D in relation to atherosclerosis, arterial stiffening, and hypertension. *Am J Hypertens.* 20:1007-1015. 2007.

Ryan M. Fryer, Pamela A. Rakestraw, Masaki Nakane, Doug Dixon, Patricia N. Banfor, Kristin A. Koch, J. Ruth Wu-Wong. Differential inhibition of renin mRNA expression by paricalcitol and calcitriol in C57/BL6 Mice. *Nephron Physiol.* 106:76-81, 2007.

Scheen AJ. Renin-angiotensin system inhibition prevents type 2 diabetes mellitus: a meta-analysis of randomized clinical trials. *Diabetes Metab.* 30:487-496, 2004a.

Scheen AJ. Renin-angiotensin system inhibition prevents type 2 diabetes mellitus: overview of physiological and biochemical mechanisms. *Diabetes Metab.* 30:498-505, 2004b.

Schneider LE, Schedl HP, McCain T, Haussler MR. Experimental diabetes reduces circulating 1,25-dihydroxyvitamin D in the rat. *Science* 196:1452-1454, 1977.

Scragg R. Seasonality of cardiovascular disease mortality and the possible protective effect of ultra-violet irradiation. *Int. J. Epidemiol.* 10:337-341, 1981.

Scragg R, Jackson R, Holdaway IM, Lim T and Beaglehole R. Myocardial infarction is inversely associated with plasma 25-hydroxyvitamin D3 levels: a community-based study. *Int. J. Epidemiol.* 19:559-563, 1990.

Scragg R, Sowers M and Bell C. Serum 25-hydroxyvitamin D, diabetes and ethnicity in the Third National Health and Nutrition Examination Survey. *Diabetes Care* 27:2813-2818. 2004.

Scragg R, Sowers M and Bell C. Serum 25-hydroxyvitamin D, ethnicity and blood pressure in the Third National Health and Nutrition Examination Survey. *Am. J. Hypertens.* 20:713-719, 2007.

Sergeev IN and Rhoten WB. 1,25-Dihydroxyvitamin D_3 evokes oscillations of intracellular calcium in a pancreatic beta-cell line. *Endocrinology* 136:2852-2861, 1995.

Shan J, Resnick LM, Lewanczuk RZ, Karpinski E, Li B, Pang PK. 1,25-Dihydroxyvitamin D as a cardiovascular hormone. Effects on calcium current and cytosolic free calcium in vascular smooth muscle cells. *Am J Hypertens* 6:983-988. 1993

Shao J, Iwashita N, Ikeda F, Ogihara T, Uchida T, Shimizu T, Uchino H, Hirose T, Kawamori R and Watada H. Beneficial effects of candesartan, an angiotensin II type 1 receptor blocker, on beta-cell function and morphology in db/db mice. *Biochem Biophys Res Commun* 344: 1224-1233, 2006.

Smotkin-Tangorra M, Purushothaman R, Gupta A, Nejati G, Anhalt H, Ten S. Prevalence of vitamin D deficiency in obese children and adolescents. *J Pediatr Endocrinol Metab,* 20:817-823, 2007.

Staples JA, Ponsonby AL, Lim L and McMichael AJ. Ecological analysis of some immune-related disorders, including type 1 diabetes, in Australia: latitude, regional ultraviolet radiation, and disease prevalence. *Environmental Health Perspectives* 111, 518-523, 2003.

Stene LC, Ulriksen J, Magnus P, Joner G. Use of cod liver oil during pregnancy associated with lower risk of Type I diabetes in the offspring. *Diabetologia.* 43:1093-1098. 2000.

Tahmasebi M, Puddefoot JR, Inwang ER, Vinsion GP. The tissue renin-angiotensin system in human pancreas. *J Endocrinol* 161: 317-322, 1999.

Tikellis C, Wookey, PJ, Candido, R, Andrikopoulos S, Thomas MC and Cooper ME. Improved islet morphology after blockade of the renin-angiotensin system in the ZDF rat. *Diabetes* 53: 989-997, 2004.

Trump DL. Double-blinded randomized study of high-dose calcitriol plus docetaxel compared with placebo plus docetaxel in androgen-independent prostate cancer: A report from the ASCENT investigators, *Urol. Oncol.* 25:445-446, 2007.

Tsang SW, Ip SP, Wong TP, Che CT, Leung PS. Differential effects of saralasin and ramiprilat, the inhibitors of renin-angiotensin system, on cerulein-induced acute pancreatitis. *Regul. Pept.* 111:47-53, 2003.

Tsang SW, Cheng CHK, Leung PS. The role of pancreatic renin-angiotensin system in acinar digestive enzyme secretion and in acute pancreatitis. *Regul. Pept.* 119:213-219, 2004a.

Tsang SW, Ip SP and Leung PS. Prophylactic and therapeutic treatments with AT_1 and AT_2 receptor antagonists and their effects on changes in severity of pancreatitis. *Int. J. Biochem. Cell Biol.* 36:330-339, 2004b.

Wong PF, Lee SS and Cheung WT. Immunohistochemical colocalization of type II angiotensin receptors with somatostatin in rat pancreas. *Regul Pept* 117: 195-205, 2004.

Xiang W, Kong J, Chen S, Cao LP, Qiao G, Zheng W, Liu W, Li X, Gardner DG, Li YC. Cardiac hypertrophy in vitamin D receptor knockout mice: role of the systemic and cardiac renin-angiotensin systems. *Am J Physiol Endocrinol Metab.* 288:E125-E132. 2005

Yusuf S, Sleight P, Pogue J, et al. The Heart Outcomes Prevention Evaluation Study, effects of an angiotensin-converting enzyme inhibitor ramipril on cardiovascular events in high-risk patients. *N. Engl. J. Med.* 342:145-53, 2000.

In: Angiotensin Research Progress
Editors: H. Miura and Y. Sasaki
ISBN: 978-1-60456-537-9

Chapter IX

Differential Pharmacotherapeutic Effects of Angiotensin II Type 1 Receptor Blockers

Shin-ichiro Miura [1,2*]***, Masahiro Fujino*** [1]***,***
Sadashiva S. Karnik [2] ***and Keijiro Saku*** [1]
[1]Department of Cardiology, Fukuoka University School of Medicine, Fukuoka, Japan.
[2]Department of Molecular Cardiology, Lerner Research Institute,
Cleveland Clinic Foundation, Cleveland, OH, USA

Abstract

Angiotensin II (Ang II) type 1 (AT_1) receptor blockers (ARBs) are highly selective for the AT_1 receptor, which is a member of the G protein-coupled receptor superfamily, and block the diverse effects by Ang II. In addition to their beneficial effects against hypertension, ARBs have also been associated with a decrease in cardiovascular morbidity and mortality. Many ARBs are available for clinical use and several studies have shown that these agents are safe and effective. Over the past decade, some reports have discussed the different degrees of the beneficial effects of ARBs. Since ARBs may not have the same effects, the benefits conferred by ARBs may also be different. These differential effects may be due to differences in the molecular characteristics of ARBs, such as insurmountability, dissociation, binding affinity and inverse agonism, in addition to their chemical structures. These findings represent an exciting new area in ARB treatment. A better understanding of the differential molecular mechanisms of each ARB could be useful for the treatment of cardiovascular diseases.

* Tel: 81-92-801-1011; Fax: 81-91-865-2692; E-mail: miuras@cis.fukuoka-u.ac.jp

1. Introduction

The renin-angiotensin system octapeptide hormone angiotensin II (Ang II), which is potentially cleaved from Ang I by angiotensin converting enzyme and chymase [1], plays a central role as a major regulator of blood pressure, electrolyte balance and endocrine function related to cardiovascular disease. Ang II binds to the Ang II type 1 (AT_1) receptor which is a member of the G-protein coupled receptor superfamily (GPCR) [also referred to as seven-membrane spanning receptor] [2].

Losartan was developed as a non-peptidergic AT_1 receptor blocker (ARB), and considerable clinical experience with losartan has been accumulated. Several new ARBs have been introduced and clinically developed over the past decade. Some reports have discussed the different degrees of the beneficial effects of ARBs [3]. Since ARBs may not have the same effects, the benefits conferred by ARBs may also be different. We discuss here the basic and clinical characteristics of ARBs.

2. Basic Characteristics of ARBs

2-1. Chemical Structure of ARBs

As shown in figure 1, all ARBs have a common chemical structure, a biphenyl-tetrazole group, which interacts with Lys^{199} and secondarily with His^{256} in the AT_1 receptor [4]. In addition, Exp 3174, candesartan, valsartan and olmesartan, but not losartan, contain an α-carboxyl group, which may be important for inverse agonism. The α-carboxyl group may interact with a positively-charged or hydrogen bonding residue in the AT_1 receptor. In fact, the α-carboxyl group of candesartan has been shown to interact with Gln^{257} in the AT_1

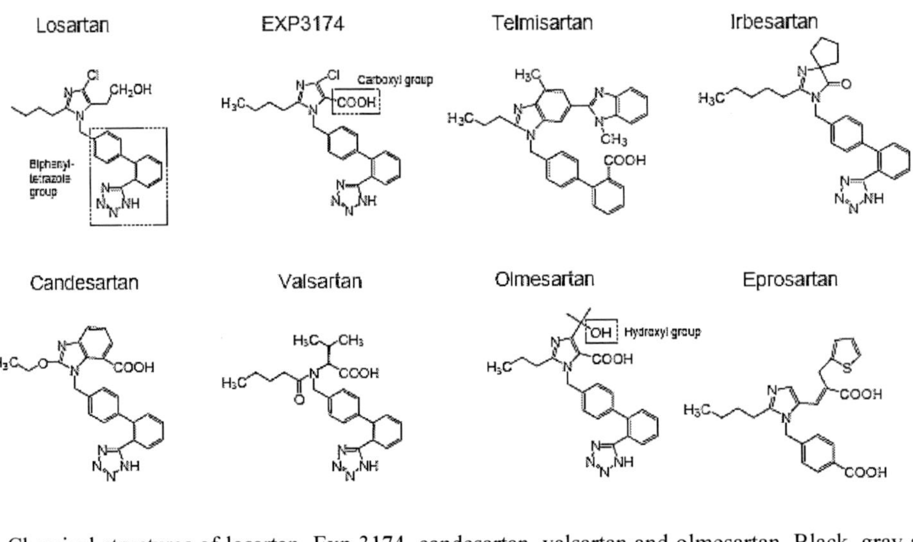

Figure 1. Chemical structures of losartan, Exp 3174, candesartan, valsartan and olmesartan. Black, gray and dotted squares indicate hydroxylate, α-carboxylate and tetrazole groups, respectively.

receptor [5]. In addition, losartan does not contain a hydroxylate group and does not show inverse agonism, while olmesartan contains a hydroxyl group in addition to an α-carboxyl group.

2-2. Kinetics, Receptor Selectivity and Surmountability of ARBs

Table 1 shows the pharmacological characteristics of ARBs. Most ARBs have been reported to be competitive antagonists. The kinetic properties of ARBs are rather similar, except for losartan. In addition, all ARBs are highly selective for the AT_1 receptor. For most ARBs, the affinity for the AT_1 receptor is >10,000 times higher than that for the AT_2 receptor.

Table 1. Pharmacological characteristics of AT_1 receptor blockers (ARBs)

	Losartan	Telmisartan	Irbesartan	Candesartan Cilexetil	Valsartan	Eprosartan	Olmesartan Medoxomil
Active metabolite	EXP3174	(-)	(-)	Candesartan	(-)	(-)	Olmesartan
Selectivity (fold)	>1,000 (Losartan) >44,000 (EXP)	>27,000	>8,500	>17,000	>30,000	>10,000	>15,000
half life (hr)	2 (Losartan) 6-9 (EXP)	20-24	11-15	9	4-6	5-9	12-18
Binding mode	surmountable(Los) /insurmountable (EXP)	insurmountable	insurmountable	insurmountable	insurmountable	insurmountable	insurmountable
Dissociation constant (nM)	10.6±3.0(Los) /4.7±1.0(EXP)	6.3±0.9	1.9±1.1	3.8±1.6	7.3±2.4	6.1±2.1	3.7±1.1

Since Ang II-stimulated AT_2 receptor signals have been shown to induce vasodilatation in animal models [6], this mechanism may also be important for counteracting the vasoconstriction induced by the stimulation of AT_1 receptor by ARBs. Although all ARBs have a long duration of action, which enables significant blood pressure control over 24 hours, telmisartan has a longer half-life in blood. In addition, there are two binding modes for AT_1 receptor: 'surmountable' and 'insurmountable' (figure 2).

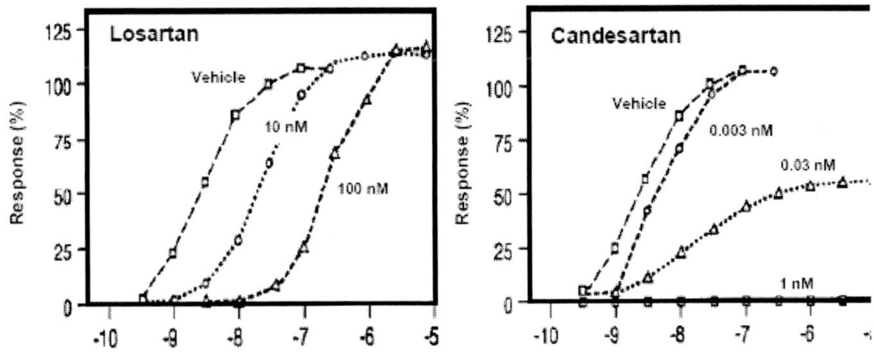

Figure 2. Response to Ang II, expressed as percentage of maximum contraction, in rabbit aortic strips preincubated with various concentrations of losartan and candesartan.

While insurmountable antagonism refers to antagonism with depression of the maximal agonist response, surmountable antagonism is associated with a shift in the agonist dose-response curve. Although losartan is a surmountable antagonist, most ARBs show insurmountable antagonism.

Le *et al.* attempted to elucidate the molecular characteristics that underlie insurmountability, dissociation and affinity of ARBs and provided a new perspective in research on the AT_1 receptor [7]. The differential characteristics of the ARBs olmesartan and telmisartan was discussed and they suggested that the higher degree of insurmountability, slower dissociation, and higher affinity of olmesartan compared to telmisartan for AT_1 receptor may help it to form a tight binding complex with AT_1 receptor.

2-3. Inverse Agonism of ARBs

Several naturally occurring and disease-causing G-protein coupled receptor (GPCR) mutants with increased constitutive activity relative to wild-type (WT) GPCRs have been identified [8]. To date, more than 60 wild-type GPCRs have been found to exhibit constitutive activity (table 2) [9]. The most important pharmacological tools for analyzing constitutive activity are inverse agonists. Inverse agonists may be useful in the treatment of cardiovascular disease and disease caused by constitutively active GPCRs.

Table 2. Constitutively active wild-type G-protein coupled receptors (GPCRs)

α1adrenergic receptor(AR), α2AR, β1AR, β2AR, 5-hydroxytryptamine (HT)1R, 5-HT2R, 5-HT4R, 5-HT7R, Dopamine(D)1R, D2R, D3R, Histamine(H)1R, H2R, H3R, H4R, Angiotensin II (Ang II) type 1 (AT1) receptor, AT2R, Muscarinic acetylcholine (M)1R, M2R, M3R, M4R, M5R, Cannabinoid (CB)1R, CB2R, Endothelial differentiation gene(edg)1R, edg2R, Opioid R, Bradykinin2R, Formyl peptide R, Chemoattractant5aR, CXCR2, CCR5, Neurokinin1R, Neuropeptide(NPY)1R, NPY2R, NPY4R, Neurotensin(NT)R1, NTR2, Vasopressin2R, Cholecystokinin 2R, Melanocortin(MC)R1, MCR3, MCR4, MCR5, Calcitonin(CT)R-1, CTR-2, etc.

Although there is no naturally occurring mutation in AT_1 receptor that can induce constitutive activity, we propose here that AT_1-WT receptor may have such activity. To confirm whether AT_1-WT receptor has constitutive activity for inositol phosphate (IP) production, we used endogenous AT_1-WT receptor expressed in vascular smooth muscle cells [10]. Valsartan significantly suppressed basal IP production in a dose-dependent manner (figure 3). When the residue at position Asn^{111} in AT_1-WT receptor was increased in size to Ile^{111} (N111I) or Tyr^{111} (N111Y) in in vitro mutagenesis studies, these receptors showed a reduction of basal activity (figure 4), which indicated that AT_1-WT receptor may be in the partially active state and may have slight constitutive activity.

Since AT_1 receptor has constitutive activity, we can use such activity to classify many ARBs. AT_1-N111G is thought to mimic the pre-activated state of the WT receptor, and by comparing the efficacy and affinity of ligands for such mutated receptors we may facilitate the distinction of partial and inverse agonists from true antagonists [11-13]. The AT_1 receptor antagonist Exp 3174 (an active metabolite of losartan) [14], but not losartan, has been shown to be an inverse agonist against constitutive activity of the AT_1-N111G mutant receptor [11, 15-17].

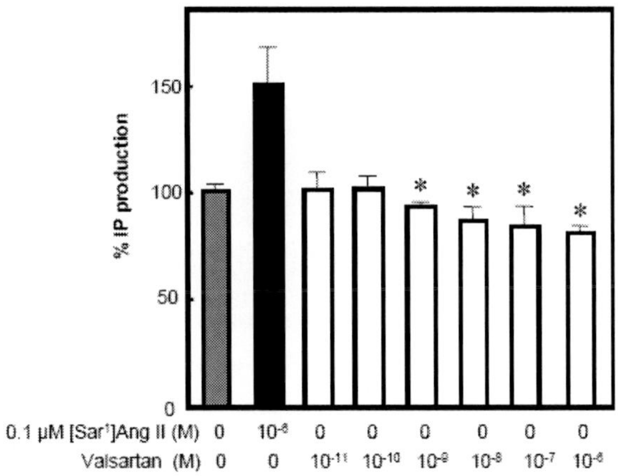

Figure 3. Percentage inositol phosphate (IP) production with or without 0.1 μM [Sar[1]]Ang II or the indicated concentration of valsartan in native vascular smooth muscle cells expressing AT_1-WT receptor. % IP production adjusted for basal IP production (570 cpm) as 100 %. *p<0.05 vs. without [Sar[1]]Ang II.

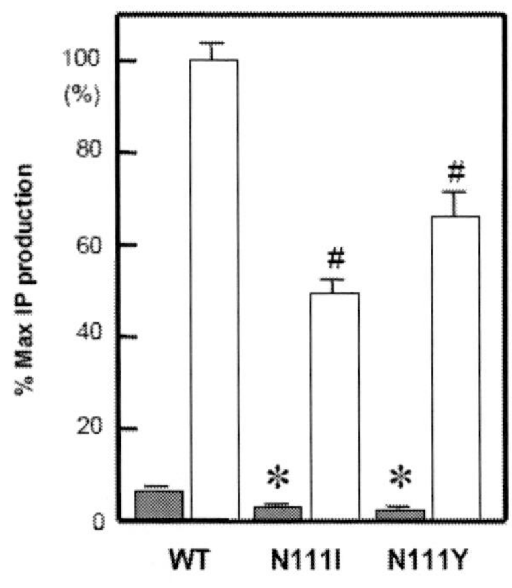

Figure 4. Percentage maximum (% Max) inositol phosphate (IP) production with (open bars) or without (gray bars) 0.1 μM [Sar[1]]Ang II in COS1 cells transiently expressing AT_1-WT, N111I, N111Y and N111G receptors. *p<0.05 versus no treatment on AT_1-WT. #p<0.05 versus [Sar[1]]Ang II on AT_1-WT.

In addition, the ARB olmesartan was shown to have stronger inverse agonist activity using the super-constitutive activity of the AT_1-F77A/N111G mutant receptor. Treatment with the ARB Exp 3174 increased phospho-extracellular signal-activated kinase (ERK) 1/2 levels in AT_1-N111G receptor-expressing cell lines, which confirms the existence of an inhibitory signal [18]. This implies that the signal produced by the constitutive activity of the AT_1-N111G receptor inhibits ERK 1/2 phosphorylation, suggesting that Exp 3174 is an inverse agonist toward ERK 1/2 activity. AT_1 receptor can activate tyrosine kinases involved in growth signaling [1]. In acute and chronic hypertrophic conditions, there is evidence of impaired angiotensinergic stimulation [19, 20].

Ang II can promote the growth of human cardiac muscle cells in vivo and is a mediator of stretch-induced myocyte hypertrophy. Ang II also promotes fibroblast proliferation, collagen deposition and a decrease in heart compliance. In fact, blockade of the renin-angiotensin system is a highly effective therapy for hypertension and heart failure [21]. Since Exp 3174 may block both cell hypertrophy and hyperplasia, inverse agonism may be an important issue for clinical use. Olmesartan and valsartan also showed stronger inverse agonism than other ARBs in the AT_1-N111G mutant receptor (figure 5) [10, 15, 22]. Moreover, olmesartan has stronger inverse agonistic actions toward IP production and extracellular-signal-regulated kinase activation independent of Ang II stimulation (we refer to this notion as "dual inverse agonism".) [23].

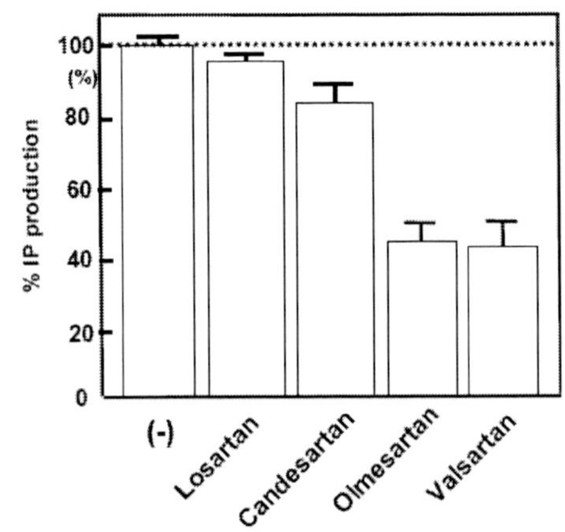

Figure 5. Percent of basal level of inositol phosphate (IP) production by various ARBs (10^{-6}M) in constitutively active AT_1-N111G receptor-transfected COS cell systems. Olmesartan or valsartan showed stronger inverse agonist activity. Basal activity of 100 indicated AT_1-N111G receptor-transfected COS cells without ARBs.

Since the AT_1 receptor has constitutive activity, we should consider the results of two very interesting studies. Losartan prevented Ang II-induced, but not mechanical stretch-induced, vascular endothelial growth factor (VEGF) protein secretion in human mesangial cells [24]. Ang II and stretching can each induce VEGF production, suggesting that both Ang II- and stretch-induced AT_1 receptor signals may exist. If losartan has a strong inverse agonist, the stretch-induced signal might be blocked. To resolve this question, Zou et al. [25] reported that mechanical stress activates AT_1 receptor independent of Ang II, and this activation can be inhibited by the inverse agonist candesartan. They examined pressure overload on the heart by constricting the transverse aorta of adult male angiotensinogen knock-out mice. Although treatment with candesartan did not reduce blood pressure, candesartan significantly attenuated the development of cardiac hypertrophy. These results suggest that mechanical stress can induce cardiac hypertrophy in vivo by activating the AT_1 receptor independent of Ang II. We also suggested a mechanism by which mechanical stress activates the AT_1 receptor without Ang II [26]. Stretching of the cell membrane may directly change the conformation of the AT_1 receptor and candesartan may prevent such changes as an inverse agonist [27]. It is clear that treatment with an antagonist could be useful, but not as effective as treatment with an inverse agonist. The constitutive activity of AT_1 receptor is relatively slight. Since hypertension induces congestive heart failure, and since it takes a couple of decades [28], it may be important to block constitutive activity by an inverse agonist (figure 6). Therefore, we should use ARBs which have stronger inverse agonism.

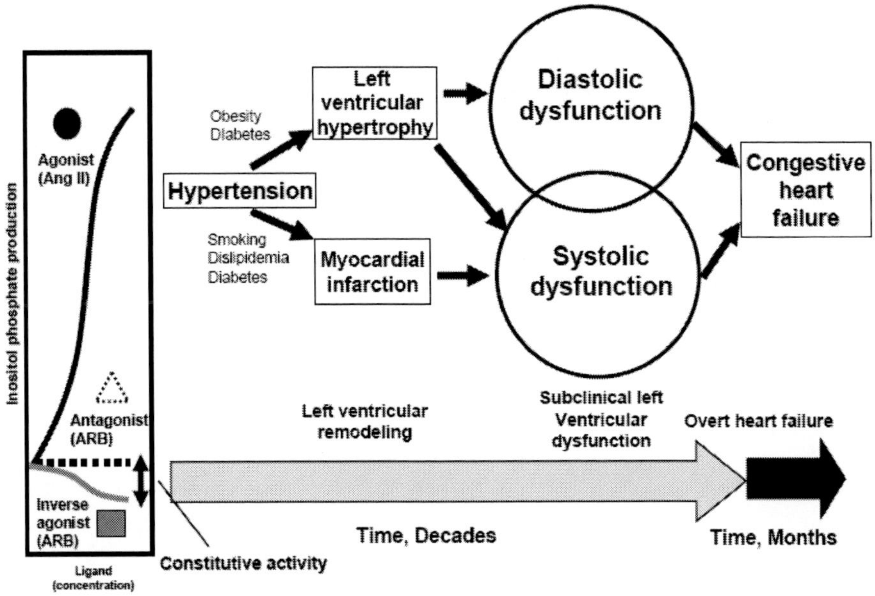

Figure 6. Progression from hypertension to congestive heart failure.

As shown in figure 1, the carboxyl group of candesartan has been shown to interact with Gln^{257} in the AT_1 receptor [5]. In addition, losartan does not contain a hydroxylate group and does not show inverse agonism. Since olmesartan contains a hydroxyl group in addition to a carboxyl group, this ARB showed stronger inverse agonism. The carboxyl group and hydroxyl group may interact with two positively-charged residues in the AT_1 receptor. We refer to this notion as the "double-chain domain" [3, 26].

3. Clinical Characteristics of ARBs

Over the past several years, the efficacies of ARBs have been compared and differences have been observed, as shown in table 3, except with regard to the lowering of blood pressure (as a pleiotropic effect). Picca et al. reported that valsartan is more effective than losartan at reducing left ventricular mass in hypertensive patients with concentric left ventricular hypertrophy [29]. Diabetes mellitus is associated with an increase in the renal production of Ang II, while the renal production of nitric oxide is reduced in streptozotocin-induced diabetic rats [30].

The decrease in renal nitric oxide (NO) is reversed more strongly by valsartan than losartan. Valsartan induces greater renal NO production than losartan in hypertensive patients with chronic renal disease [31]. Koh et al. [32] reported that treatment with candesartan but not losartan significantly lowered plasma levels of plasminogen activator inhibitor type-1 antigen and monocyte chemoattractant protein-1 in patients with mild to moderate hypertension. As for the effects of ARBs on coronary artery disease, Ang II-induced contraction of human coronary artery is mediated mainly by AT_1 receptor. Exp 3174 is the most efficacious ARB at preventing coronary artery contraction (Exp 3174 > candesartan = valsartan > losartan) [33]. Two hundred fifty patients with type B2/C coronary lesions were randomized in an open-label clinical study (Val-PREST) with respect to age, gender, lesion type and indication for percutaneous coronary intervention to receive the chronic administration of 80 mg/day valsartan or placebo (β-blocking agents and/or angiotensin converting enzyme inhibitors) [34]. Valsartan was shown to be effective at preventing coronary restenosis. In addition, our clinical trial showed that valsartan decreased the incidence of target lesion revascularization after stenting compared to losartan [35]. On the other hand, one study showed that losartan is more effective than valsartan or candesartan. Human platelet activation was significantly reduced by losartan in a dose-dependent manner. Only higher doses of valsartan reduced such activation. Candesartan failed to modify platelet activation [36]. Thus, these abilities of ARBs may not be class effects.

Table 3. Direct comparison of the efficacies of AT_1 receptor blockers (ARBs)

Effects	Efficacies
Reducing left ventricular mass	valsartan > losartan
Increasing nitric oxide production	valsartan > losartan
Lowering plasma plasminogen activator inhibitor type-1 antigen and monocyte chemoattractant protein-1	Candesartan > losartan
Preventing coronary artery contraction	Exp 3174 > candesartan = valsartan > losartan
Preventing corornary restenosis	valsartan > losartan
Inhibition of platelet activation	losartan > valsartan > candesartan

There is another important issue to address. Several clinical trials have evaluated the effects of ARBs on morbidity and mortality in patients with heart failure. The Evaluation of Losartan in the Elderly (ELITE) II trial suggested that treatment with losartan is not superior to treatment with captopril, but is significantly better tolerated [37]. The effect of valsartan on the outcome in patients with chronic and symptomatic heart failure was evaluated in a large-scale, placebo-controlled clinical study: the Valsartan in Heart Failure Trial (Val-HeFT) [38]. Although overall mortality was similar in the valsartan and placebo groups, valsartan significantly reduced the combined end-point of mortality and morbidity. These results were confirmed and extended by the Candesartan in Heart failure: Assessment of Reduction in Mortality and morbidity (CHARM) trial [39], which demonstrated that candesartan was beneficial in patients with heart failure. Although there are important differences in the design and hypothesis of these three trials that must be taken into account when comparing their results and interpreting their clinical impact, we should also consider whether these are class effects of ARBs. The inverse agonistic activity of ARBs might also be important for their efficacy in the long-term treatment of heart disease, such as heart failure including cardiac remodeling, independent of any blood pressure-lowering effect. On the other hand, in mortality trials after myocardial infarction, losartan (OPTIMAAL, optimal trial in myocardial infarction with the angiotensin II antagonist losartan) [40] and valsartan (VALANT, valsartan in acute myocardial infarction) have been shown to have similar effects [41].

So far, all ARBs have shown very slight side-effects that do not differ from those with placebo treatment, even if a high dose of ARB is given.

4. Characteristic Features of Each ARB

4-1. Losartan

Losartan was the first ARB developed clinically and acts as a surmountable antagonist. Losartan is a prodrug that is converted into its active metabolite Exp3174, which is an insurmountable antagonist and inverse agonist. The side-effect profile does not differ from that of placebo, and losartan is well tolerated and may be beneficial in the treatment of renal disease [42]. The Reduction of Endpoints in NIDDM with the Angiotensin II Antagonist Losartan (RENAAL) study showed that losartan has renoprotective effects in terms of endo-stage renal disease or death in patients with type 2 diabetes and nephropathy. The renoprotective effects were induced beyond the effects on blood pressure control.

4-2. Irbesartan

Irbesartan acts as an insurmountable antagonist and is well tolerated. It is not influenced by food. Irbesartan and its metabolites are eliminated via both the liver and kidney. Irbesartan also has renoprotective effects independent of changes in blood pressure [Irbesartan in Patients with Type 2 Diabetes and Microalbuminuria (IRMA)-2 study] [43].

4-3. Candesartan Cilexetil

Candesartan cilexetil is a prodrug that is converted into the active compound candesartan after ingestion. Candesartan is an insurmountable antagonist and contains a carboxyl group, which is important for inverse agonism after stretching [27]. Candesartan is generally well tolerated, like other ARBs, and has been shown to significantly reduce cardiovascular deaths and hospital admissions for heart failure [44].

4-4. Eprosartan

Eprosartan is a selective AT_1 receptor blocker and an insurmountable antagonist. Its antihypertensive effect and tolerability are comparable to those of other ARBs.

4-5. Valsartan

Valsartan is an insurmountable antagonist and contains a carboxyl group, which is important for inverse agonism. Food appears to reduce its oral bioavailability but does not influence its antihypertensive effect. The main outcome of cardiac disease did not differ between the treatment groups. Patients with heart failure of New York Heart Association (NYHA) class II, III, or IV were randomly assigned to receive 160 mg/day of valsartan or placebo. Treatment with valsartan resulted in significant improvements in NYHA class, ejection fraction, signs and symptoms of heart failure, and quality of life compared to treatment with placebo [45].

4-6. Telmisartan

Telmisartan is more lipophilic than the other ARBs and is an insurmountable antagonist. Telmisartan may be an insulin-sensitizer and activates peroxisome-proliferator activated receptor-γ [46] and may resemble glitazones. The half-life of telmisartan in blood is greater than those of other ARBs.

4-7. Olmesartan

Olmesartan medoxomil is a prodrug that is converted into the active compound olmesartan. Olmesartan is an insurmountable antagonist and an inverse agonist. This ARB is not metabolized by the cytochrome P450 enzyme; about 60 % is eliminated through the liver and the rest is eliminated via the kidney. Food does not influence its bioavailability.

5. Conclusions and Perspectives

The clinical evidence available to date is consistent with a class effect. Possible differences in outcome appear to be based on the limitations of basic experiments and small clinical trials. For example, the importance of inverse agonism for the clinical effects of ARBs is just beginning to be understood. These findings represent an exciting new area in ARB treatment. ARB-based therapy holds the promise of greatly reducing the incidence of cardiovascular disease.

References

[1] Miyazaki M, Takai S. Local angiotensin II-generating system in vascular tissues: the roles of chymase. *Hypertens Res*. 2001; 24: 189-193.

[2] De Gasparo M, Catt K, Inagami T, Wright JW, Unger T. International Union of Pharmacology. XXIII. The angiotensin II receptors. *Pharmacol. Rev.* 2000; 52: 415-472.

[3] Miura S, Fujino M, Saku K. Angiotensin II receptor blocker as an inverse agonist: a current perspective. *Curr. Hypertens. Rev.* 1:115-121:2005.

[4] Noda K, Saad Y, Kinoshita A, Boyle TP, Graham RM, Husain A, Karnik SS. Tetrazole and carboxylate groups of angiotensin receptor antagonists bind to the same subsite by different mechanisms. *J. Biol. Chem.* 1995; 270: 2284-2289.

[5] Takezako T, Gogonea C, Saad Y, Noda K, Karnik SS. "Network leaning" as a mechanism of insurmountable antagonism of the angiotensin II type 1 receptor by non-peptide antagonists. *J. Biol. Chem.* 2004; 279: 15248-15257.

[6] Kivlighn SD, Zingaro GJ, Gabel RA, Broten TP, Chang RS, Ondeyka DL, Mantlo NB, Gibson RE, Greenlee WJ, Siegl PK.. In vivo pharmacology of an angiotensin AT1 receptor antagonist with balanced affinity for AT2 receptors. *Eur. J. Pharmacol.* 1995; 294: 439-450.

[7] Le MT, Pugsley MK, Vauquelin G, Van Liefde I. Molecular characterization of the interactions between olmesartan medoxomil and telmisartan and the human angiotensin II AT_1 receptor. *Br. J. Pharmacol.* 2007; 151: 952-962.

[8] Bais C, Santomasso B, Coso O, Arvanitakis L, Geras-Raaka E, Gutkind JS, Asch AS, Cesarman E, Gershengorn MC, Mesri EA, Gershengorn MC. G-protein coupled receptor of Kaposi's sarcoma-associated herpesvirus is a viral oncogene and angiogenesis activator. *Nature* 1998; 391: 86-89.

[9] Seifert R, Wenzel-Seifert K. Constitutive activity of G-protein-coupled receptors: cause of disease and common property of wild-type receptors. Naunyn Schmiedebergs *Arch. Pharmacol.* 2002; 366: 381-416.

[10] Miura S, Kiya Y, Kanazawa T, Imaizumi S, Fujino M, Matsuo Y, Karnik SS, Saku K. Differential bonding interactions of inverse agonists of Angiotensin II type 1 receptor in stabilizing the inactive-state. *Mol. Endocrinol.* 2007. in press.

[11] Noda K, Feng YH, Liu XP, Saad Y, Husain A, Karnik SS. The active state of the AT1 angiotensin receptor is generated by angiotensin II induction. *Biochemistry* 1996; 35: 16435-42.

[12] Miura S, Feng YH, Husain A, Karnik SS. Role of aromaticity of agonist switches of angiotensin II in the activation of the AT1 receptor. *J. Biol. Chem.* 1999; 274: 7103-10.

[13] Miura S, Karnik SS. Constitutive activation of angiotensin II type 1 receptor alters the orientation of transmembrane Helix-2. *J. Biol. Chem.* 2002; 277: 24299-305.

[14] Suzuki J, Ohta H, Hanada K, Kawai N, Ikeda T, Nakao M, Ikemoto F, Nishikibe M. Acute effects of E-3174, a human active metabolite of losartan, on the cardiovascular system in tachycardia-induced canine heart failure. *Hypertens Res.* 2001; 24: 65-74.

[15] Miura S, Saku K, Karnik SS. Molecular analysis of the structure and function of the angiotensin II type 1 receptor. *Hypertens Res.* 2003; 26: 937-43.

[16] Miura S, Feng, YH, Husain A, Karnik SS. Mechanism of constitutive activation of the AT1 receptor: influence of the size of the agonist switch binding residue Asn(111). *Biochemistry* 1998; 37: 15791-8.

[17] Miura S, Zhang J, Boros J, Karnik SS. TM2-TM7 interaction in coupling movement of transmembrane helices to activation of the angiotensin II type-1 receptor. *J. Biol. Chem.* 2003; 278: 3720-5.

[18] Miura S, Ramchandran R, Chang CS, Karnik SS. Changes in cell cycle regulation during angiotensin II type 1 receptor-mediated cell growth. *FASEB J.* 1999; **13:** A1537 (Abstract).

[19] Dostal DE, Baker KM. The cardiac renin-angiotensin system: conceptual, or a regulator of cardiac function? *Circ. Res.* 1999; 85: 643-50.

[20] Sadoshima J, Xu Y, Slayter HS, Izumo S. Autocrine release of angiotensin II mediates stretch-induced hypertrophy of cardiac myocytes in vitro. *Cell* 1993; 75: 977-84.

[21] McKelvie RS, Yusuf S, Pericak D, Avezum A, Burns RJ, Probstfield J, Tsuyuki RT, White M, Rouleau J, Latini R, Maggioni A, Young J, Pogue J. Comparison of candesartan, enalapril and their combination in congestive heart failure: randomized evaluation of strategies for left ventricular dysfunction (RESOLVD) pilot study: the RESOLVD pilot study investigators. *Circulation* 1999; 100: 1056-1064.

[22] Le MT, Vanderheyden PM, Szaszák M, Hunyady L, Kersemans V, Vauquelin G. Peptide and nonpeptide antagonist interaction with constitutively active human AT1 receptors. *Biochem. Pharmacol.* 2003; 65: 1329-1338.

[23] Miura S, Saku K. Do angiotensin II type 1 receptor blockers all have the same beneficial effects ? *Br. J. Pharmacol.* 2007; 151: 912-913.

[24] Gruden G, Thomas S, Burt D, Zhou W, Chusney G, Gnudi L, Viberti G. Interaction of angiotensin II and mechanical stretch on vascular endothelial growth factor production by human mesangial cells. *J. Am. Soc. Nephrol.* 1999; 10: 730-737.

[25] Zou Y, Akazawa H, Qin Y, Sano M, Takano H, Minamino T, Makita N, Iwanaga K, Zhu W, Kudoh S, Toko H, Tamura K, Kihara M, Nagai T, Fukamizu A, Umemura S, Iiri T, Fujita T, Komuro I. Mechanical stress activates angiotensin II type 1 receptor without the involvement of angiotensin II. *Nat. Cell. Biol.* 2004; 6: 499-506.

[26] Miura S, Fujino M, Hanzawa H, Kiya Y, Imaizumi S, Matsuo Y, Tomita S, Uehara Y, Karnik SS, Yanagisawa H, Koike H, Komuro I, Saku K. Molecular mechanism underlying inverse agonist of angiotensin II type 1 receptor. *J. Biol. Chem.* 2006; 281: 19288-19295.

[27] Yasuda N, Miura S, Akazawa H, Tanaka T, Qin Y, Kiya Y, Imaizumi S, Fujino M, Ito K, Zou Y, Ge J, Fukuhara S, Kunimoto S, Mochizuki N, Fukuzaki K, Sato T, Nakaya H, Saku K, Komuro I. Conformational switch of angiotensin II type 1 receptor underlying mechanical stress-induced activation. *EMBO Rep.* 2008 (in press)

[28] Vasan RS, Levy D. The role of hypertension in the pathogenesis of heart failure. A clinical mechanistic overview. *Arch. Intern. Med.* 1996;156:1789-1796.

[29] Picca M, Agozzino F, Pelosi G. Effects of losartan and valsartan on left ventricular hypertrophy and function in essential hypertension. *Adv. Ther.* 2004; 21: 76-86.

[30] Awad AS, Webb RL, Carey RM, Siragy HM. Renal nitric oxide production is decreased in diabetic rats and improved by AT1 receptor blockade. *J. Hypertens.* 2004; 22: 1571-7.

[31] Matsuda H, Hayashi K, Homma K, Yoshioka K, Kanda T, Takamatsu I, Tatematsu S, Wakino S, Saruta T. Differing anti-proteinuric action of candesartan and losartan in chronic renal disease. *Hypertens. Res.* 2003; 26: 875-80.

[32] Koh KK, Han SH, Chung WJ, Ahn JY, Jin DK, Kim HS, Park GS, Kang WC, Ahn TH, Shin EK. Comparison of effects of losartan, irbesartan, and candesartan on flow-mediated brachial artery dilation and on inflammatory and thrombolytic markers in patients with systemic hypertension. *Am. J. Cardiol.* 2004; 93: 1432-5, A10.

[33] Pantev E, Stenman E, Wackenfors A, Edvinsson L, Malmsjo M. Comparison of the antagonistic effects of different angiotensin II receptor blockers in human coronary arteries. *Eur. J. Heart. Fail.* 2002; 4: 699-705.

[34] Peters S, Gotting B, Trummel M, Rust H, Brattstrom A. Valsartan for prevention of restenosis after stenting of type B2/C lesions: the VAL-PREST trial. *J. Invasive. Cardiol.* 2001; 13: 93-97.

[35] Iwata A, Miura S, Imaizumi S, Kiya Y, Nishikawa H, Zhang B, Shimomura H, Kumagai K, Matsuo K, Shirai K, Saku K.Do valsartan and losartan have the same effects in the treatment of coronary artery disease? *Circ. J.* 2007; 71: 32-38.

[36] Nunez A, Gomez J, Zalba LR, Monton M, Jimenez A, Velasco S, Lopez-Blaya A, Uriarte AC, Casado S, Lopez-Farre A. Losartan inhibits in vitro platelet activation: comparison with candesartan and valsartan. *J. Renin. Angiotensin. Aldosterone Syst.* 2000; 1: 175-9.

[37] Pitt S, Poole-Wilson PA, Segal R, Martinez FA, Dickstein K, Camm JA, Konstam MA, Riegger G, Klinger GH, Neaton J, Sharma D, Thiyagarajan B. Randomised trial of losartan versus captopril on mortality in patients with symptomatic heart failure: the losartan heart failure survival study - ELITE II. *Lancet* 2000; 355: 15827.

[38] Cohn J, Tognoni G, Glazer R, Sporman D, Hester A. Rationale and design of the Valsartan Heart Failure Trial: a large multinational trial to assess the effects of valsartan, an angiotensin-receptor blocker, on morbidity and mortality in chronic congestive heart failure. *J. Card. Fail.* 1999; 5: 15560.

[39] Pfeffer MA, Swedberg K, Granger CB, Held P, McMurray JJ, Michelson EL, Olofsson B, Ostergren J, Yusuf S, Pocock S. CHARM Investigators and Committees. Effects of candesartan on mortality and morbidity in patients with chronic heart failure: the CHARM-Overall programme. *Lancet* 2003; 362:759-66.

[40] Dickstein K, Kjekshus J; OPTIMAAL Steering Committee of the OPTIMAAL Study Group. Effects of losartan and captopril on mortality and morbidity in high-risk patients after acute myocardial infarction: the OPTIMAAL randomised trial. Optimal Trial in Myocardial Infarction with Angiotensin II Antagonist Losartan. *Lancet* 2002; 360: 752-60.

[41] Pfeffer MA, McMurray JJ, Velazquez EJ, Rouleau JL, Kober L, Maggioni AP, Solomon SD, Swedberg K, Van de Werf F, White H, Leimberger JD, Henis M, Edwards S, Zelenkofske S, Sellers MA, Califf RM; Valsartan in Acute Myocardial Infarction Trial Investigators. Valsartan, captopril, or both in myocardial infarction complicated by heart failure, left ventricular dysfunction, or both. *N. Engl. J. Med.* 2003; 349: 1893-06.

[42] Brenner BM, Cooper ME, de Zeeuw D, Keane WF, Mitch WE, Parving HH, Remuzzi G, Snapinn SM, Zhang Z, Shahinfar S; RENAAL Study Investigators. Effects of losartan on renal and cardiovascular outcomes in patients with type 2 diabetes and nephropathy. *N. Engl. J. Med.* 2001; 345: 861-819.

[43] Parving HH, Lehnert H, Bröchner-Mortensen J, Gomis R, Andersen S, Arner P; Irbesartan in Patients with Type 2 Diabetes and Microalbuminuria Study Group. The effect of irbesartan on the development of diabetic nephropathy in patients with type 2 diabetes. *N. Engl. J. Med.* 2001; 345:870-878.

[44] Pfeffer MA, Swedberg K, Granger CB, Held P, McMurray JJ, Michelson EL, Olofsson B, Ostergren J, Yusuf S, Pocock S; CHARM Investigators and Committees.Effects of candesartan on mortality and morbidity in patients with chronic heart failure: the CHARM-Overall programme. *Lancet* 2003; 362:759-766.

[45] Cohn JN, Tognoni G; Valsartan Heart Failure Trial Investigators. A randomized trial of the angiotensin-receptor blocker valsartan in chronic heart failure. *N. Engl. J. Med.* 2001; 345:1667-1675.

[46] Schupp M, Janke J, Clasen R, Unger T, Kintscher U. Angiotensin type 1 receptor blockers induce peroxisome proliferator-activated receptor-gamma activity. *Circulation* 2004;109:2054-2057.

Index

A

B

C

D

E

F

G

H

I

M

N

O

P

Q

R

S

T

U

V

W

X

Y

Z